Seaweeds of the British Isles

Seaweeds of the British Isles

A collaborative project of the British Phycological Society
and the Natural History Museum

Volume 1 Rhodophyta

Part 1 Introduction, Nemaliales, Gigartinales

Peter S Dixon & Linda M Irvine

Natural History Museum, London

First published by the Natural History Museum,
Cromwell Road, London SW7 5BD

This edition printed and published by Pelagic Publishing, 2011,
in association with the Natural History Museum, London

ISBN 978-1-907807-08-4

A catalogue record for this book is available from the British Library.

Seaweeds of the British Isles

Volume 1 Rhodophyta

To be published in three parts:

Part 1:	*Introduction* *Florideophyceae:*	 *Nemaliales* *Gigartinales*
Part 2:	*Florideophyceae:*	*Cryptonemiales* *Palmariales* *Rhodymeniales*
Part 3:	*Florideophyceae:* *Bangiophyceae:* Key to genera Glossary Index	*Ceramiales* *Porphyridiales* *Bangiales*

Contents

Preface

At the first Annual General Meeting of the British Phycological Society, held in London in January 1953, a resolution was adopted that the activities inaugurated by the informal Group, which had first met at Bangor in September 1951, should be continued by the Society. These included 'the preparation of a revised Check-list of the marine algae of the British Isles and the accumulation of information for an eventual comprehensive new British marine algal Flora', and it was agreed that this project should be entrusted to a committee which was established under the Chairmanship of Dr M. Parke. This committee arranged for various people to take responsibility for the main divisions of the marine algae, and Dr K. M. Drew Baker and Dr M. T. Martin undertook the work on the Rhodophyta.

Following the death of Dr Drew Baker in September 1957, Dr P. S. Dixon was asked to help Dr Martin with responsibility for this group. In 1967 Dr Martin resigned from active participation in the production of the 'Flora' and at about the same time Mrs L. M. Irvine, who had been assisting Dr Dixon, took on the role of joint author for the Rhodophyta.

The present volume represents the first part of the new 'Flora' and deals with the red algae. Its purpose is to present a critical appraisal of those marine Rhodophyta which have so far been found in the British Isles, to furnish detailed descriptions of their morphology and reproductive structures, to provide brief notes on their ecology and distribution and to summarize present views on their taxonomic relationships. Resolution of many of the questions concerning the organisms under consideration was simply not possible but it has been the authors' aim at least to draw attention to them if not to provide all the answers.

An undertaking of this magnitude depends to a very great extent on assistance of many kinds and from many sources. It could not have been carried through without the continued support of the successive Presidents and Councils of the British Phycological Society and especially of its Flora Committee.

In 1960 the Nature Conservancy provided a grant to the British Phycological Society for the employment of a research assistant to Dr Dixon. This position was filled successively by Dr M. A. Allen (1961–62), Dr A. Archer, now Mrs Taylor, (1963–64) and Mrs L. M. Irvine (1964–67). In 1967 Mrs Irvine was appointed to the staff of the British Museum (Natural History) and became a joint author. From 1969 to 1970, Miss W. D. Armstrong, now Mrs Fremmerlid, gave full-time assistance to Mrs Irvine, working on the extraction of distribution records from literature, whilst she was on the staff of the British Museum (Natural History). A grant to the British Phycological Society from the Natural Environment Research Council enabled a research assistant to be employed for a further period to carry on this work. Miss L. Otway, now Mrs Wooldridge, and Miss A. Webster filled this post successively. We are grateful for these grants and to the research assistants for their contribution to the work.

Grants in support of travel and research by Dr P. S. Dixon were made by the Joint Committee on Research at the University of Liverpool, the Browne Fund of

the Royal Society, the Intercampus Research and Travel Fund of the University of California and the National Science Foundation (USA) and we are very grateful for these. We also wish to put on record our thanks to:
Dr M. T. Martin, Dr M. Parke F.R.S., Mr R. Ross, Dr F. E. Round, Prof. M. de Valéra for guidance and advice;
Dr W. H. Adey (Corallinaceae), Prof. A. D. Boney (Acrochaetiaceae), Dr B. T. Gittins, Mr W. F. Farnham (Cryptonemiaceae), Dr. D. E. G. Irvine, Dr H. W. Johansen (Corallinaceae), Dr P. R. Newroth (Phyllophoraceae) for help during the production of the text;
Dr M. C. H. Blackler, Dr E. M. Burrows, Dr G. T. Boalch, Dr J. J. P. Clokie, Dr J. P. Cullinane, Dr R. C. Duerden, Dr L. V. Evans, Dr R. L. Fletcher, Dr M. D. Guiry, Dr D. J. Guiterman, Mr R. G. Hooper, Dr D. M. John, Dr W. E. Jones, Dr J. M. Kain, Dr G. W. Lawson, Dr J. R. Lewis, Dr H. McAllister, Mr O. Morton, Dr B. Moss, Dr T. A. Norton, Mrs J. M. Pope, Mr J. H. Price, Mr H. T. Powell, Dr G. Russell, Mr R. M. Smith, Dr G. R. South, Mr I. Tittley, Miss A. M. Webster, for the loan or gift of specimens, records, etc.

We are indebted to the Directors and Curators of the following institutions for working facilities, for permission to examine specimens or borrow material:
Department of Botany, The University, Aberdeen ABD
School of Plant Biology, University College of North Wales, Bangor UCNW
North Devon Athenaeum, Barnstaple BPL
Ulster Museum, Belfast BFT
Centre d'Etudes et de Recherche Scientifique, Biarritz (France)
Laboratoire de Botanique, Faculté des Sciences, Caen (France) CN
Farlow Herbarium of Cryptogamic Botany, Harvard University, Cambridge (USA) FH
Department of Botany, National Museum of Wales, Cardiff NMW
Public Library, Museum and Art Gallery, Carlisle CLE
Laboratoire de Biologie Marine, Concarneau (France) CO
Botanical Museum and Herbarium, Copenhagen (Denmark) C
Department of Botany, Trinity College, Dublin TCD
Royal Botanic Garden, Edinburgh E
Istituto Botanico, Firenze (Italy) FIAF
Department of Botany, The University, Glasgow GL
Department of Natural History, Glasgow Art Gallery & Museums, Glasgow GLAM
Department of Botany, University of Strathclyde, Glasgow
Guernsey Museum, St Peter Port, Guernsey STP
Ilfracombe Museum, Ilfracombe ILF
Royal Botanic Gardens, Kew The algae previously at K are now on permanent loan to BM (as BM-K).
Rijksherbarium, Leiden (Netherlands) L
Botanical Institute, Leningrad (USSR) LE
City of Liverpool Museum, Liverpool LIV
British Museum (Natural History), London BM
Linnean Society, London LINN
Passmore Edwards Museum, Newham, London
South London Botanical Institute, London SLBI

Department of Botany, Queen Mary College, University of London, London QMC
Botanical Museum, Lund (Sweden) LD
Manchester Museum, The University, Manchester MANCH
Marine Science Laboratories, University College of North Wales, Menai Bridge
New York Botanical Garden, New York (USA) NY
Botanisk Museum, Oslo (Norway) O
Fielding-Druce Herbaria, The University, Oxford OXF
Laboratoire de Cryptogamie, Muséum National d'Histoire Naturelle, Paris (France) PC
Marine Biological Station, Port Erin
Marine Biological Association, Plymouth MBA
Bibliothèque Municipale, Quimper (France)
Museum of Buteshire Natural History Society, Rothesay
Saffron Walden Museum, Saffron Walden
Naturhistoriska Riksmuseum, Stockholm (Sweden) S
Institut de Botanique, Faculté des Sciences, Strasbourg (France) STR
Somerset County Museum, Taunton
Museum of the Royal Norwegian Society for Science and Letters, Trondheim (Norway) TRH
Institute of Systematic Botany, University of Uppsala, Uppsala (Sweden) UPS
Museo Civico di Storia Naturale, Venezia (Italy)
Naturhistorisches Museum, Wien (Austria) W
Yorkshire Museum, York YRK

The Standard Herbarium Abbreviations are given above and are used throughout the text to indicate the location of type specimens.

Finally, to any others who helped but whose names are omitted from this list of acknowledgements we can only offer our apologies and our thanks.

A considerable number of the drawings of microscopic details are by Dr Dixon. Most of the remaining illustrations were prepared by Miss V. C. Gordon and, in the Nemaliales, by Mr Richard Miller. Users of the work will appreciate how much they have contributed.

Introduction

THE red algae are a very specialized group of plants, often omitted from general textbooks of botany. The present introduction is intended to serve as a concise survey of the group. It provides a brief but critical summary of structure and morphology, reproduction and life histories, economic utilization, systematics, ecology, occurrence and distribution necessary for a floristic treatment of the marine Rhodophyta which occur in the British Isles. References are cited only for the most critical aspects. For sources relating to more general information, the reader is referred to the treatments of Oltmanns (1922, 1923), Fritsch (1945), Drew (1951), Dixon (1973) and Stewart (1974).

CELL STRUCTURE

The cell wall

Except for a few details, cell structure is remarkably uniform throughout the red algae. The cell wall is composed of randomly arranged microfibrils in an amorphous matrix, both components being composed of polysaccharide. At the surface of the thallus, a 'cuticle' may be apparent; this often displays several layers and appears to be pectic in nature. The carbohydrate constituents of the wall may be divided into two parts, an inner layer which is only slightly extractable by hot water and an outer portion which is more readily soluble. The water-soluble components, often termed 'mucilages', may form 70 per cent of the dry weight. Some of these mucilages, such as agar and carrageenan, have been used extensively for centuries. The cell walls of certain Rhodophyta are impregnated with calcium carbonate. Field-collected specimens of the Corallinaceae always have thalli which are calcified, although in culture calcification may be reduced or even lacking.

Pit connections

The occurrence of structures referred to as 'pit connections' between certain red algal cells has been known for many years. Investigations of fine structure have shown (Ramus, 1969) that they are neither a 'pit' nor an intercellular 'connection', although the term continues to be used. Pit connections vary considerably in size and shape (Fig. 1). Their diameter ranges from 0·2 μm (some Delesseriaceae) to 30–40 μm between axial cells of *Ptilota* or *Bonnemaisonia*. The thickness is equally variable, ranging from a very flat disc to an almost spherical structure. For many years it was considered that one difference between the Bangiophyceae and Florideophyceae was the presence of pit connections in the latter and their

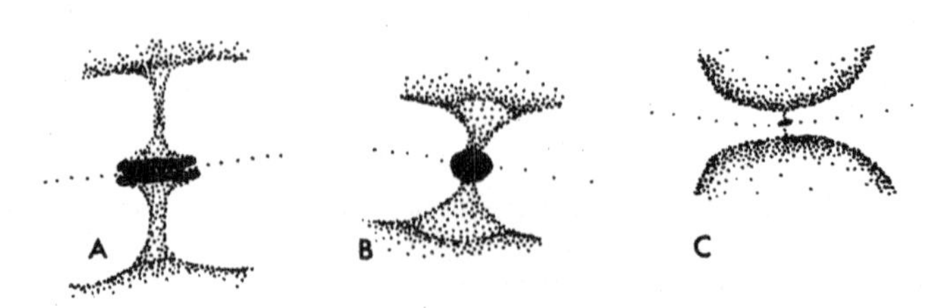

Fig. 1 Pit connections in Florideophyceae
A. *Antithamnion plumula* ×1000; B. *Champia parvula* ×1000; C. *Nitophyllum punctatum* ×1000.

absence in the former. More recently, pit connections have been reported consistently in some Bangiophyceae. Fine-structural investigations have shown that the structures in the *Conchocelis* phase of *Porphyra* are similar to the pit connections of the Florideophyceae.

Cytoplasm and vacuolation

Cell shape, size and appearance in red algae show considerable variation. In newly-formed cells the cytoplasm is dense and viscous, although during enlargement the cell becomes highly vacuolate with the cytoplasm occupying a peripheral position. In cells with a single nucleus this is usually located at the basal pole of the cell.

The cytoplasm of red algae is susceptible to environmental changes, although some species exhibit remarkable capacity for survival. Specimens of *Bangia* and *Porphyra* are dried out and remoistened, often every day for a considerable period, apparently without harm. The capacity to withstand low temperatures is said to run parallel to the capacity to withstand hypertonic solutions. The harmful effects of dilution are seen in estuaries where the number of species diminishes rapidly (see p. 46) with decrease in salinity.

The nucleus

In the Bangiophyceae, the cells are usually uninucleate. In the Florideophyceae, the apical cells are either uninucleate or multinucleate. In some genera the cells remain uninucleate whereas in others there is an increase in the number of nuclei as a consequence of nuclear division without cell division. Where the apical cells are multinucleate, mature cells contain many nuclei and the number may continue to increase with age. The largest numbers of nuclei are found in cells of some species of *Griffithsia* whose apical cells contain from 50 to 75 nuclei, with 3000 to 4000 nuclei in the oldest cells. In the Bangiophyceae, the single nucleus is central in young cells and parietal in older vacuolate cells. A similar situation occurs in those Florideophyceae where a single nucleus is present in each cell, although where each cell contains several nuclei, these always lie at the periphery, whether the cell is vacuolate or not. The average size of nuclei is of the order of 3 μm. The smallest nuclei occur in some members of the Cryptonemiales and Gigartinales where the diameter is often less than 1 μm while the largest (*c.* 15 μm) occur in some genera of the Ceramiales.

Chloroplasts, pigments and coloration

A single chloroplast with a prominent pyrenoid occurs in every cell in many Bangiophyceae (Fig. 2). Each cell in the simpler Nemaliales contains a single chloroplast, with a pyrenoid. A single parietal chloroplast without a pyrenoid

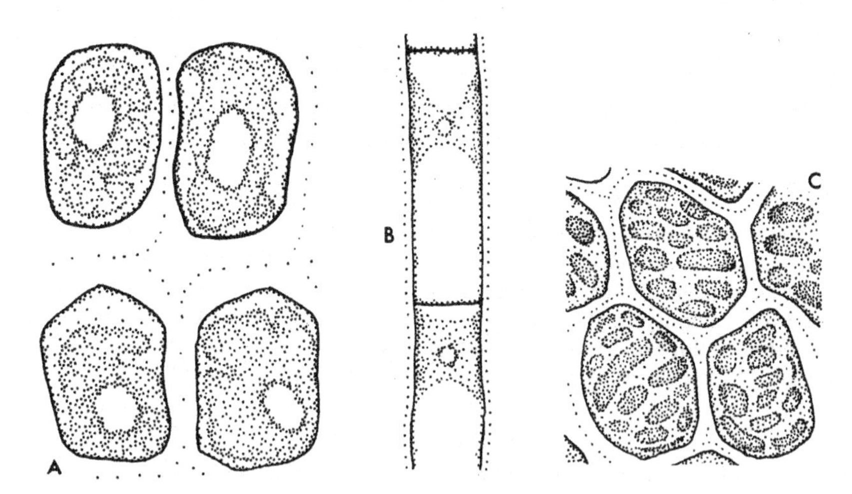

Fig. 2 Chloroplast types in Rhodophyta
A. *Porphyra* sp., a single stellate chloroplast with pyrenoid ×1600; B. *Audouinella thuretii,* a single parietal chloroplast with pyrenoid ×1000; C. *Ceramium ciliatum,* numerous discoid chloroplasts without pyrenoids ×1000.

occurs in young cells of many of the more advanced Florideophyceae, becoming increasingly lobed with age and eventually fragmenting. Many Florideophyceae always contain several discoid chloroplasts in each cell in all parts of the thallus (Fig. 2).

The pigments which occur are of three principal kinds, chlorophylls, carotenoids and phycobilins. Five chlorophylls have been identified but only one, chlorophyll *a,* is of universal occurrence in red algae. Chlorophyll *d* has been reported for some although it appears to be of erratic occurrence. The principal carotene of red algae is β-carotene although α-carotene has been reported in *Phycodrys*. Of the xanthophylls, lutein appears to occur in all cases examined. The phycobilins comprise phycoerythrin and phycocyanin, red or blue biliproteins readily soluble in water.

The Rhodophyta are usually spoken of as the 'red algae' and most are red, crimson or pink. This colour is due to the predominance of the red phycoerythrin. However, the red algae contain green chlorophyll, yellow, red or orange carotenoids and both red and blue phycobilins which can produce a whole range of colours from blue-green through green-brown, purple or black in addition to the predominant red. Some are often of characteristic colour. For example, certain species of *Porphyra* may be green in colour, while the thalli of *Furcellaria lumbricalis* are often brownish and those of *Ahnfeltia plicata* purplish black. In addition, some species are of different colours in different seasonal and ecological situations, due either to the induction of a different blend of pigments or the photodestruction of those present at an earlier stage of development. Drying out can also change the colour so that the colour of a dried herbarium specimen may differ considerably from that of the same specimen in the living condition. Certain red algae which are usually interpreted as being partially or completely parasitic contain reduced quantities of pigments or even none at all.

Reserve materials and storage products

The characteristic reserve storage material found in red algae is floridean starch, which occurs in grains, often of small size (*c.* 1 μm) although larger grains (*c.* 25 μm) have been recorded. These grains lie free in the cytoplasm and although formed in proximity to the chloroplast are never formed within it. On treatment with iodine solution, grains of floridean starch are stained yellow or brown although a blue coloration may appear after prolonged treatment.

Specialized cells

Secretory cells Secretory (or 'vesicular') cells occur in several Florideophyceae although not reported in the Bangiophyceae (Dixon, 1973). They occur extensively in members of the Rhodymeniales and widely in the Cryptonemiales and Gigartinales. In the Nemaliales, secretory cells occur only in the Bonnemaisoniaceae while in the Ceramiales they are restricted to a few Ceramiaceae. Where the thallus is diffuse the secretory cells are located in a lateral position while in more compact thalli they may be formed internally. The few secretory cells which have been examined for their fine structure or chemical composition display considerable diversity. Some examples *(Antithamnion, Bonnemaisonia)* appear to have a connection with halide metabolism, while those in cells of the Rhodymeniales and *Halymenia* have been interpreted as mucilage-producing cells. Those of *Peyssonnelia* appear to contain crystalline material suggesting a function equivalent to 'cystoliths'. In *Schizymenia dubyi,* the secretory cells have homogeneous contents and may appear as brilliant points in face view of the thallus.

Iridescent bodies and iridescence Thalli of certain red algae show a marked blue or green iridescence when viewed in reflected light, very different from the normal red coloration seen in transmitted light. The phenomenon is most common in genera of the Ceramiales and Rhodymeniales although a few cases are also known in the Nemaliales, Gigartinales and Bangiophyceae. Morphological studies have shown that iridescence is usually due to the presence of inclusions within cells (Dangeard, 1940). Where no morphological cause can be established the iridescence would appear to be due to interference phenomena in the wall or the cytoplasm. The occurrence of iridescent cells or iridescence is insufficiently consistent for this to be used as a criterion of taxonomic importance.

Hairs Hairs occur in many red algae, formed from superficial cells. The enlarging hair is cylindrical, with a large central vacuole and cytoplasm at the tip, but by maturity the cytoplasm has disappeared. Hairs are ephemeral structures. The frequency of occurrence of hairs in species for which they have been reported varies enormously from specimen to specimen and fluctuates with season and environmental conditions. They appear to be lacking in the Bangiophyceae, Gigartinaceae and Phyllophoraceae (Gigartinales), Delesseriaceae, Dasyaceae and Rhodomelaceae (Ceramiales).

MORPHOLOGY

There is much variation in the level of organization and degree of complexity of the thalli throughout the Rhodophyta. The thalli of the marine Bangiophyceae are relatively simple, while thalli of the Florideophyceae are much more complex and elaborate. The differences between these two classes are discussed on page 60.

Morphology of the Bangiophyceae

Thallus morphology in members of the Bangiophyceae ranges from unicellular to parenchymatous. Two genera of the British marine Bangiophyceae consist of unicells or cell masses resulting from the products of division failing to separate. Individual cells of *Porphyridium* are globose, with a prominent stellate chloroplast and central pyrenoid. Species of *Goniotrichum* and *Asterocytis* have small epiphytic thalli which become multiseriate with age in some species of the former. Species of *Erythrotrichia* are small epiphytes consisting of a simple uniseriate filament sometimes with secondary development of a prostrate system. In *Erythropeltis* and *Erythrocladia* the thallus consists of little more than a prostrate disc, although short upright filaments can arise from this. The thallus of *Porphyropsis* passes through a discoid stage of development from which a small saccate structure develops. This ultimately breaks to give a sheet of cells. Genera of the Bangiaceae have more complex thalli with no trace of a heterotrichous base although a stipe may form by the aggregation of rhizoids. In *Bangia*, the young thallus consists of a uniseriate row of cells which can become multiseriate. Thalli of *Porphyra* also develop initially as uniseriate filaments but this stage quickly passes and large parenchymatous sheets of cells, one or two cells thick, are produced by intercalary cell division.

Morphology of the Florideophyceae

Florideophycean thalli are obviously filamentous or are formed by the aggregation of filaments so that they are pseudoparenchymatous. The filaments are formed by the division of apical cells although intercalary cell division also occurs in members of the Delesseriaceae and Corallinaceae and there are indications that it may occur elsewhere. It has been customary to interpret thallus organization in the Florideophyceae on the basis of two categories depending on whether there is a single principal apical cell and a single axial filament or several principal apical cells and a group of axial filaments in each axis. The former is termed uniaxial and the latter multiaxial. This scheme is applicable *only* to the erect thalli. From the developmental standpoint there are two types of construction in the Florideophyceae, delimited not on the basis of axial organization but on whether the thallus represents the elaboration of the erect or the prostrate parts of a heterotrichous system, giving a thallus which is either erect or prostrate.

Prostrate thalli Prostrate thalli of simple heterotrichous construction occur in the Gigartinales and Cryptonemiales. The thalli consist of a basal layer of radiating filaments from which upright filaments arise. The upright filaments are simple or branched and arise either at right angles or at an acute angle and then curve to an upright position. The upright filaments are either held together loosely by mucilage *(Cruoria)* or compacted tightly into dense pseudoparenchyma *(Hildenbrandia)*, with all intergrades between these two extremes. The thalli are sometimes attached to the substrate by short rhizoids which arise from the underside of the prostrate filaments.

Erect thalli The erect thalli of the Florideophyceae exhibit much greater diversity of form and organization than the encrusting thalli. Erect thalli can be treated most conveniently by considering first those examples in which a heterotrichous organization is present. In some, the prostrate system is well developed; in others it forms a simple attachment disc at the base of the thallus and the heterotrichous nature of the latter can be demonstrated only by sectioning or by careful study of spore germination.

The simplest heterotrichous organization is that found in the genus *Audouinella* where the prostrate system may even be reduced in some species to a single cell formed from the spore from which the thallus has been derived. A more elaborate construction is found in most erect Florideophyceae, with a clear differentiation between principal and lateral filaments. The latter surround the former, the degree of aggregation varying from minimal (as in *Dudresnaya* or *Nemalion*) to the more compact arrangements found in cartilaginous thalli. It is necessary to distinguish between these two types of filament in order to explain the patterns of cell division and cell enlargement by which a thallus develops. It has not been appreciated that there are three levels of differentiation in florideophycean thalli and that the terminology used must be discriminatory. Of the three levels of differentiation, the first is where there is no differentiation between filaments, as in species of *Audouinella*. The second is where there is differentiation between a principal filament and surrounding lateral filaments as in *Atractophora* or *Calosiphonia*. The aggregations of principal and lateral filaments are of regular shape and may undergo regular changes in shape as thalli age, as in *Gelidium*. It is convenient to refer to such an aggregation as an 'axis'. The third level of differentiation is that found in complex thalli where there is not only differentiation between filaments in each axis but also differentiation between axes. Differentiation between axes, as in *Bonnemaisonia* or *Plocamium*, produces regularly and elaborately branched thalli and contributes much to their aesthetic appeal. For both axes and their constituent filaments, this differentiation involves control of growth and it is necessary to have two systems of terminology referring to and discriminating between filaments and axes. In terms of behaviour, an axis is an integrated aggregation of filaments and represents more than their summation. It is convenient to refer to the principal axial filament or filaments as being of unlimited growth, while the surrounding lateral filaments are of limited growth. In discussions of differentiation between axes, these can be described as determinate or indeterminate depending upon whether their growth is restricted or not.

Florideophycean erect thalli of primary heterotrichous construction fall into two major categories, depending upon the number of filaments of unlimited growth in each axis (Fig. 3). As indicated previously, the situation in which a

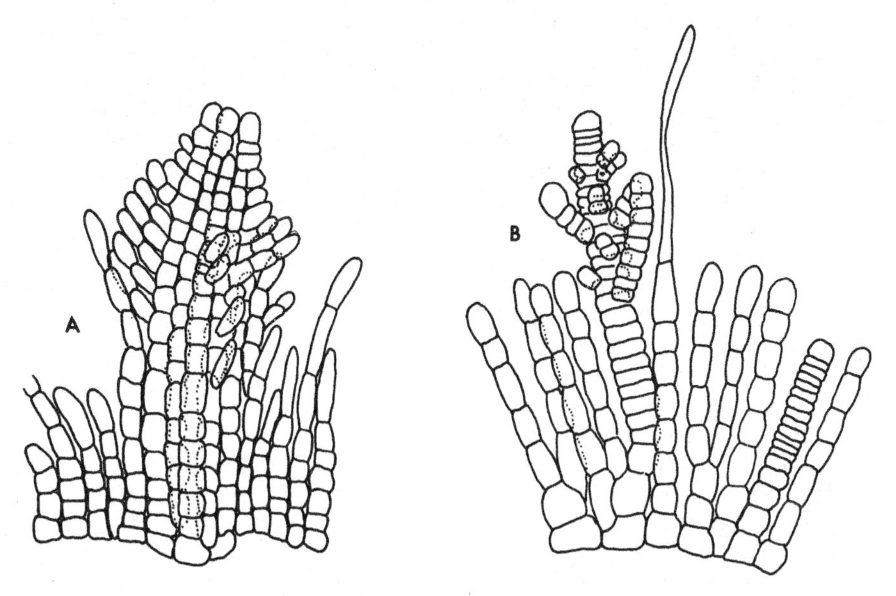

Fig. 3 Origin of the erect fronds from prostrate bases
A. *Platoma bairdii*, with multiaxial erect frond (after Kuckuck) ×535; B. *Gloiosiphonia capillaris*, with uniaxial erect frond (after Oltmanns) ×350.

single axial filament occurs is referred to as uniaxial construction, while that in which several occur is said to be multiaxial. The two categories provide a useful characterization, sometimes of systematic importance, although distinction is not as easy as some authors have indicated. The final form of the thallus is usually no indication of construction. One critical aspect related to the differences between multiaxial and uniaxial organization is frequently overlooked. The apical meristem of an axis of multiaxial construction contains a number of apical cells. This meristem can divide into two parts thereby undergoing truly dichotomous branching. The apical meristem of an axis of uniaxial construction contains a single apical cell, which *cannot* separate into two equal and equivalent parts. Thalli of uniaxial construction frequently have a dichotomous aspect, as in most species of *Ceramium,* the pattern of development involving the formation of a *lateral* axis and then the rotation of both lateral and parent axes to give the dichotomous appearance. This rotation is termed 'evection'. The mode of branching of a thallus of dichotomous aspect need not be, and often is not, true dichotomy.

Audouinella daviesii is a thallus of uniaxial construction, while the freshwater *Batrachospermum* is a more complex example of the same type, with the thallus differentiated into filaments of limited and unlimited growth. The apical cell of the filament of unlimited growth divides transversely, each segment giving rise to four primordia which function as the apical cells of filaments of limited growth. The filaments of limited growth form a loose aggregation, radiating from the axial filament of unlimited growth. In addition some grow downward in proximity to the axial filament, the degree of investment depending on the relative growth of the two types of filament. The down-growing filaments of limited growth will be referred to as rhizoids in the present treatment. The thallus of *Batrachospermum* is radial in symmetry, the filaments of limited growth being of equivalent development. In many genera there is a bilateral symmetry. Two primordia are cut off on the lateral flanks of the segment formed by the division of the apical cell and these then give rise to filaments of limited growth of some length. The second pair of primordia produce much shorter filaments which contribute to the depth of the thallus. Thus, the thallus is of greater breadth than depth. In compact thalli, the apical cells aggregate to form the superficial layer of cells, so that the layer of cells over the surface of the thallus is not an inert 'epidermis' but a region of apical cells, dividing actively or capable of resuming meristematic activity. This aspect of meristematic organization and the ease with which meristematic activity can be resumed accounts for the frequency with which proliferations can arise and the marked effects of animal grazing and fungal attack. The *Batrachospermum* type of thallus organization can give rise to tubular thalli, as in *Dumontia* or *Gloiosiphonia.* The innermost cells of the filaments of limited growth elongate in a radial direction, while the outer cells aggregate to form a continuous tube. The tube is traversed by the axial filament, with each cell of the latter attached to the tube by the elongated innermost cells.

A more elaborate pattern of segmentation occurs in thalli where the apical cell divides obliquely rather than transversely. The segments formed are wedge-shaped and each segment gives rise to a single filament of limited growth. The orientation of successive filaments of limited growth is determined by the obliquity of the divisions. In *Cystoclonium* this is alternate while in *Plocamium* the apical cells of a major axis may segment in three planes but only in two planes in smaller axes. The degree and orientation of the obliquity of the apical cell divisions may produce spiral arrangements of some complexity, as in *Naccaria* and *Bonnemaisonia*.

Distinguishing regions such as 'medulla' and 'cortex' may be helpful for purposes of identification but this is not a fundamental difference. Developmentally, the tissue which is in the inner part of the thallus represents the axial filament together with the products of division of the apical cells of the filaments of limited growth at an early stage of development. The tissue which is now immediately behind the apical cells of those filaments was produced by the division of the same apical cells but at a later time. Thus, much of the tissue which is now 'medulla' was once 'cortex'.

Turning next to a consideration of florideophycean thalli of multiaxial organization, the thallus of *Nemalion* represents the classic example of this type. The filaments are loosely aggregated, embedded in mucilage and examination is relatively easy. Each axis contains up to 250 filaments of unlimited growth which form a compact mass from which the filaments of limited growth radiate. The first-formed cells of the filaments of limited growth elongate considerably. The last formed, lying immediately behind the apical cells from which they have been formed, form a compact cell mass at the surface of the thallus. Adjacent cell masses pack together to form an almost complete surface layer. *Nemalion* shows radial symmetry, as do many other multiaxial Florideophyceae in which the structure is more compact and cartilaginous, such as *Ahnfeltia* or *Polyides*. Some multiaxial thalli are either compressed or markedly flattened, particularly in many genera referred to the Cryptonemiales, Gigartinales and Rhodymeniales (*Chondrus, Gigartina, Halarachnion, Kallymenia, Rhodymenia*). The external appearance of many foliose thalli is very similar. The internal structure, however, as seen in section or squash is usually characteristic. Features of importance include cell shape and size, cell wall thickness and the extent to which an intercellular mucilaginous matrix occurs; patterns of cell rows and special arrangements of cells developing as a result of the formation of secondary pit connections can also be diagnostic.

Some multiaxial thalli are calcified. In articulated coralline algae, the thalli are composed of calcified portions (the intergenicula) separated by flexible non-calcified joints (the genicula). In *Corallina*, the axes contain 100–250 axial filaments, composed of elongate cells, and the filaments of limited growth arise from these. At a geniculum, the thallus consists only of the filaments of unlimited growth, the cells of which may be particularly elongate. The deposition of calcium carbonate takes place in the cell wall of the filaments of limited growth so that the geniculum remains devoid of calcification.

Two types of tubular organization are found in Florideophyceae with multiaxial thalli. The first, which occurs in *Scinaia,* is analogous to that described previously for *Dumontia.* The mass of axial filaments in *Scinaia* forms a central core. This becomes separated from the compact outermost tissue formed from the terminal portions of the filaments of limited growth by the elongation of the lowermost cells of the latter. A second type of tubular thallus based upon a multiaxial organization occurs in the Rhodymeniales. In *Lomentaria* and *Chylocladia,* the filaments of unlimited growth do not form a core, as in *Nemalion,* but are themselves arranged as a hollow tube. The tube is unconstricted in certain species *(Lomentaria clavellosa)* whereas in others it is constricted at intervals, in some cases *(Chylocladia verticillata, Champia parvula)* being associated with plates of tissue formed at intervals across the cavity.

All the examples which have been considered exhibit some degree of heterotrichy. The remaining genera to be considered, referable to the Ceramiales, are characterized by the absence of heterotrichy. Species of *Callithamnion* are probably the most simple from the morphological point of view. Thalli are branched, the principal axes and lateral branches are uniseriate, the lateral branches arising alternately from every cell in either distichous or spiral arrangements of varying degrees of complexity. There is no morphological differentiation between filaments of limited and unlimited growth. *Antithamnion* is similar to *Callithamnion,* but the branching is opposite as opposed to alternate. *Ptilota* is an excellent example in which to consider the interactions between filaments and axes. Every cell of the principal axes gives rise to 4 primordia. Of these, each primordium of an opposite pair on alternate cells of a principal axis forms a lateral filament of unlimited growth; the remaining primordia give rise only to filaments of limited growth. The two lateral filaments of unlimited growth formed by the two opposite primordia of a pair develop unequally so that one is longer than the other when growth ceases. These long and short lateral axes are arranged alternately and each is of equal development for considerable distances, so that the outline of the frond is highly regular. A similar series of interactions to those observed in *Ptilota* can also be demonstrated in the thalli of various species of *Ceramium.* Each segment formed by the division of the apical cell of a principal axis in *Ceramium* gives rise to a number of primordia. Each primordium gives rise to filaments of limited growth, usually four in number, of which two grow upwards and two grow downwards. The upward- and downward-growing filaments of limited growth aggregate to form a compact ring of tissue. Adjacent rings of tissue may remain discrete, or they may meet or even fuse to form a complete investment to the filament. Although the thallus of *Ceramium* has the appearance of being dichotomously branched, the actual mode of branching in this or any other uniaxial thallus cannot be dichotomous (see p. 10).

The genera *Callithamnion, Antithamnion, Ptilota* and *Ceramium* are all representatives of the family Ceramiaceae. By comparison, in the Delesseriaceae, the structure resembles closely a thallus such as that of *Antithamnion* with the principal axis, lateral axes and filaments of limited growth adhering to form a flat sheet of tissue. *Hypoglossum* is one of the best examples of this type of organization. Each principal axis possesses a single apical cell, which divides transversely, each product giving rise to four primordia. The two first-formed primordia develop laterally and a second pair of primordia is produced on the upper and lower surfaces. The lateral primordia divide obliquely to produce laterals of first and second order. The apical cells of these laterals reach to the periphery of the thallus so that growth must be co-ordinated. The principal axial filament, together with the second pair of primordia and the products of division of the latter, aggregate to produce the midrib of the thallus. In *Delesseria* the general pattern of development is similar except that the segments formed by the division of the principal apical cell of an axis undergo intercalary cell division of a highly organized type. Cell divisions in thalli of *Cryptopleura* depart even further from the regular pattern exemplified by *Antithamnion.* Although obviously uniaxial in the sporeling stage, the filaments of limited and unlimited growth are not morphologically differentiated as in *Hypoglossum* but merely exhibit physiological differences as in *Callithamnion.* The result is a thallus which may be interpreted as a 'webbed' structure, where filaments of limited and unlimited growth are not discernible in the mature state.

The Rhodomelaceae is a large family, all genera of which have a basic construction similar to that of *Polysiphonia.* The dome-shaped principal apical cell of each axis divides transversely to form segments. Each of these divides again immediately to form a ring of primordia. In mature thalli of all genera of the Rhodomelaceae, each axial cell is surrounded by a ring of cells which are of the same length as itself, usually referred to as pericentral cells, although this term is applicable to any lateral primordia formed from an axial cell whether arranged in a ring, as in *Polysiphonia* or *Ceramium,* or formed singly, as in *Cystoclonium.* In most genera of the Rhodomelaceae, lateral branches of two kinds are formed. Some have pericentral cells like the parent axis, whereas others are uniseriate hair-like structures of limited growth. The latter are the trichoblasts. These are usually ephemeral and composed of colourless cells, although in *Brongniartella* they contain chloroplasts and are relatively long-lived. The trichoblasts bear the gametangia and in some species of *Polysiphonia* only gametangial-bearing trichoblasts are formed although this is not the case with all species. In some species of *Polysiphonia,* downgrowing filaments develop from the basal poles of the pericentral cells and invest the axis with a heavy filamentous covering. Despite the universal *Polysiphonia*-like construction of all thalli in members of the Rhodomelaceae, outward evidence of this may be lost at an early stage of development as in *Chondria* or *Laurencia.* A modification, found in many of the Rhodomelaceae, relates to the bilateral arrangement of axes. The tier-like arrangement of cells characteristic of *Polysiphonia* occurs in these genera but the axes are arranged in specialized bilateral patterns. In some, the basic structure is indistinguishable from that of *Polysiphonia* although in more advanced forms the thalli are of remarkable complexity. Special arrangements of branches produced on every cell, every second, third, or fourth cell produce thalli of remarkable beauty and symmetry. A further specialization involves a process akin to the 'webbing'. This can cause some of the pericentral cells to be absent on the adaxial sides of the lateral axes in the area of fusion, as in *Pterosiphonia complanata*.

Few genera are assigned to the Dasyaceae, the fourth family of the Ceramiales. For many years these were included in the Rhodomelaceae, but they differ in that their basic pattern of development is sympodial. The main axis of the thallus is made up of the basal portions of successive laterals. In *Dasya*, each lateral contributes only a single cell to the principal axis whereas in other genera, continuation of the principal axis is by a lateral formed on the second or third cell. The principal axis becomes polysiphonous. In *Heterosiphonia* the lowermost segments, usually the lowest three, also develop in this way, the apical portion of each lateral filament remaining uniseriate.

Regeneration

The capacity of all red algae to regenerate is virtually unlimited and the ecological and biological implications of this are considerable. Fragments which have overwintered, or survived some other adverse period buried under sand or debris, can produce new thalli when conditions improve. This is the most significant reason for the persistence of red algae.

In *Griffithsia corallinoides* cells may be separated one from another simply by putting a thallus into a bottle half-filled with seawater and shaking vigorously for a few minutes and it has been found that a complete thallus can regenerate from a single cell. Regeneration in thalli of more elaborate morphology is more complex. In material of a species with well-differentiated axes, the principles of hierarchical organization will operate and the later appropriate lateral axis will take over the functions of the axis whose apex has been damaged or destroyed. In older material there may not be a functional axis remaining, and new apices arise through the conversion of the apical cell or cells of one or more filaments of limited growth into filaments of unlimited growth. Filaments of limited growth will regenerate rapidly by re-induced divisions of adjacent apical cells. This can be seen clearly in flat membranaceous thalli of some Delesseriaceae, where perforations of the lamina will heal rapidly by the reactiviated divisions of surrounding cells.

Sporangia may regenerate once the original spores have been shed. This is a relatively simple process in Bangiophyceae: the mother cell, or an adjacent cell, expands to fill the space which is now available and forms a new sporangium. The process is a little more complex in Florideophyceae. When the initial spores have been shed, cytoplasm protrudes through the pit connection. The protrusion is cut off to form an apical cell which develops into a new sporangium. Regeneration may occur several times in a given sporangium.

Galls and tumour-like growths

Gall-like proliferations are more frequent on members of the Florideophyceae than Bangiophyceae. They develop as irregular masses of cells which may be visible to the naked eye. Relatively few galls occur in Rhodophyta compared with the dicotyledonous Phanerogams which are remarkable for the abundance and diversity of their galls.

Various causal agencies are known. In some cases a bacterium has been isolated which on re-innoculation causes the development of new galls. The organism responsible for 'crown-gall' disease in higher plants *(Agrobacterium tumefaciens)* will induce tumours in many marine algae. Fungi and endophytic algae have also been reported as causing distortion in many red algae. The production of galls by fungi is rare whereas endophytic algae frequently cause distortions of red algal tissues. A species of '*Chlorochytrium*' causes distortions of some species of *Porphyra* and the Delesseriaceae. Species of *Streblonema* cause distortions in many Delesseriaceae, *Gelidium, Pterocladia* and *Furcellaria.* Clusters of diatoms which develop internally can also give rise to tissue distortion. As with higher plants, galls may be produced by animal infestation. In general, two types of animal are involved – nematodes or harpacticoid copepods. Nematodes have been detected in the galls of *Furcellaria* and *Chondrus,* while harpacticoid copepods have been reported in the galls of *Palmaria, Rhodophyllis, Stenogramme, Polyneura, Cryptopleura* and *Nitophyllum.* Some galls are caused by chemical factors in the environment and it could well be that many reports of galls for which no cause can be identified are produced in this way. Viral infections have been suggested as the agent in such cases although there is no experimental evidence. Acarina and insects, which are responsible for most galls in Phanerogams, have never been shown to give rise to galls in Rhodophyta or other algae.

Gall-like proliferations have often been misunderstood. The alleged 'cystocarps' of some species of *Ceramium* are galls and not reproductive structures. Several supposed 'parasitic' red algae are nothing more than tumour-like proliferations of the host species, as with *Choreocolax cystoclonii* reported on *Cystoclonium.* Some reported instances of 'parasporangia' could well be based on disorganized cell growth such as occurs in galls.

Rhizoids, tendrils and attachment structures

Most thalli of the Florideophyceae and Bangiophyceae are attached to the substrate by rhizoidal structures. In some cases, rhizoids are lacking and the thallus is attached directly by some form of 'cementation' which can be compared with the cohesions between filaments found in all complex thalli. Attaching rhizoids are single-celled structures throughout the Bangiophyceae, although in *Porphyra* they may be produced in such quantity as to form a basal attachment of some size. Simple, attaching rhizoids are produced by most Florideophyceae. In many species of *Ceramium,* a single or several-celled clasping structure may form at the apex. Most rhizoids do not penetrate the substrate, although the attachment of *Polysiphonia lanosa*, an obligate epiphyte on *Ascophyllum* and other fucoids, consists of a swollen rhizoid which penetrates the tissue of the substrate species. Some Florideophyceae possess a different type of attachment structure which resembles a mass of rhizoids adhering laterally to form a multicellular structure. These may be formed laterally, as in *Gelidium* or some Delesseriaceae, or terminally by the conversion of the apex of an axis, as in *Plocamium*. Whether lateral or terminal, the attachment of prostrate axes at intervals produces an organization similar to stolons of higher plants.

Cell and tissue adhesions and fusions

Secondary pit connections are probably the best-known type of cell fusion in the Florideophyceae. They develop between cells not of kindred origin. Formation of a secondary pit connection may involve the production of a single-celled lateral with a normal pit connection. The single cell fuses completely with an adjacent cell so that the original pit connection is left between two cells which are not of common origin. Such pit connections have been reported only once in the Ceramiaceae; they are absent in simpler members of the Cryptonemiales and Gigartinales and throughout the Nemaliales with the exception of the Gelidiaceae. Secondary pit connections of a different type occur in the Corallinaceae. Direct cellular communication develops between two cells through the dissolution of the cell walls at the point of contact, with a plug-like pit connection laid down in the channel of communication. Cell fusions not associated with secondary pit connections also occur extensively in the Corallinaceae and have been reported elsewhere in Bangiophyceae and Florideophyceae. The filaments of the *Conchocelis* phases of *Bangia* and *Porphyra* form a mass with abundant cell fusions. In *Ahnfeltia plicata, Palmaria palmata, Pleonosporium borreri, Bornetia secundiflora, Corynospora pedicellata* and several crustose thalli, similar fusions occur to a varying degree.

The organization of many Florideophycean thalli depends upon the adhesion of the filaments of which they are composed. This adhesion can break down, particularly in cultures which are stale or ageing. The thallus is reduced to a mass of branched filaments reminiscent of species of *Audouinella*. In complex multicellular thalli, contact between axes may result in permanent adhesion. Attachments may arise when axes of two species come into contact and there is some evidence to indicate that secondary pit connections can develop between axes of different species.

Curved laterals which, after contact with the same or another alga, form one or more firm coils around it, have been reported in diverse Florideophyceae. Such laterals of *Bonnemaisonia hamifera* rarely curve more than a half turn while those of *Cystoclonium* or *Calliblepharis* may make four or five coils. The short reflexed spines of *Asparagopsis armata* do not encircle objects with which they come into contact although the tip may form an attachment structure like that in *Plocamium*.

REPRODUCTION IN FLORIDEOPHYCEAE

All reproductive bodies in the Rhodophyta are devoid of flagella. Release of zoospores from fungal infections was probably the cause of early reports of flagellated spores in red algae. The nuclear phenomena in many sporangia and the products of germination of various spores are wholly unknown. The role of a spore may be a matter for conjecture, supported by little real evidence. One particular point which should be made clear concerns the use of the term 'sexual'. In many treatments, the term is restricted to the process of syngamy. If one accepts the genetic definition of sex, meiosis is as much a part of sex as syngamy, so that sporangia in which meiosis occurs should not be described as 'asexual' structures.

All Florideophyceae are characterized by a highly specialized type of oogamous reproduction. The zygote is retained on the female plant, giving rise to a complex post-fertilization development known as the carposporophyte. Various sporangia occur in this class, their terminology derived either from the number of spores produced or from the name of the genus for which they are characteristic.

Gametangia and gamete formation

Sexuality in the Florideophyceae and the nature of the gamete-producing reproductive organs was established by Bornet and Thuret (1867, 1876–80). In a series of superb studies, these investigators first demonstrated the structure and disposition of male and female gametangia and the process of fertilization.

Carpogonia The carpogonium is the female organ. It consists (Fig. 4) of an inflated base which contains the reproductive nucleus and an elongate process, the trichogyne, which is the receptive area for male gametes. The carpogonium is formed by the transformation of the apical cell of a filament which may contain a specified number of cells, often three or four, and arises terminally or laterally as an adventitious and specialized structure. In such cases, the cells are often different in size and shape from vegetative cells and usually devoid of chloroplasts. This filament associated with the carpogonium is usually referred to as the carpogonial branch. In many members of the Nemaliales, the degree of specialization of the cells associated with the carpogonium is variable. Chloroplasts diminish in number although there are usually some present even in the carpogonium. The absence of a change in appearance between the vegetative cells and the cells associated with the carpogonium means that in these Nemaliales it is sometimes difficult to define the carpogonial branch. The carpogonium in certain genera is sessile or intercalary. Sessile carpogonia are formed from a lateral apical cell. The transformation may even occur before such a lateral apical cell has separated completely and the resulting carpogonium is then an intercalary structure.

Carpogonia are produced in growing areas since a carpogonium is formed from an active apical cell. In some cases there may be a resurgence of activity in apical cells in mature areas, forming a superficial tissue mass in an area where cell division has normally ceased. Observations on species of *Antithamnion* and *Pleonosporium* indicate that development of the carpogonium is completed in about 14 to 18 hours. Whether this time sequence applies in other genera is unknown.

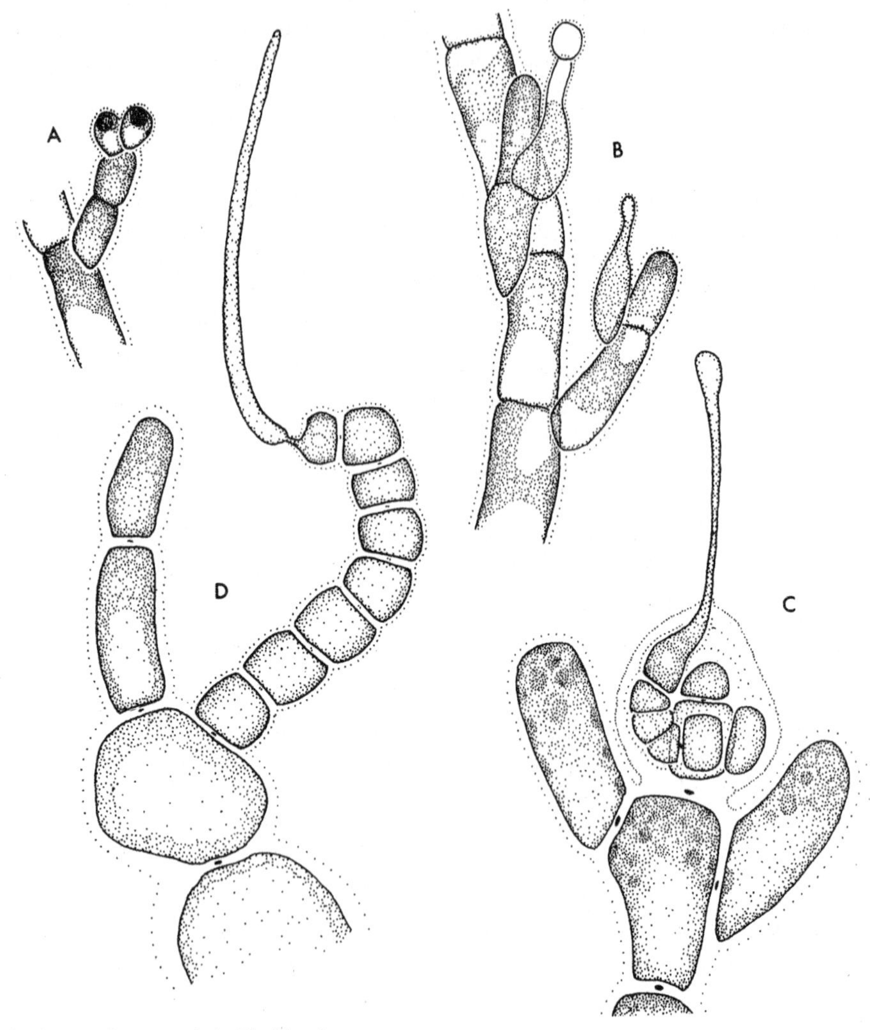

Fig. 4 Gametangia in Florideophyceae

A. *Audouinella thuretii*, spermatangia ×1250; B. *A. thuretii*, carpogonia, both fertilized (upper) and prefertilization (lower) ×1250; C. *Ptilothamnion pluma*, carpogonial branch ×1250; D. *Dudresnaya verticillata*, carpogonial branch ×1250.

The size, shape and structure of the trichogyne and its persistence vary enormously. In many Gigartinales the trichogyne is short, about 5 μm in length, whereas in many Ceramiales it is long, in excess of 100 μm. The trichogyne is cylindrical, although coiled or twisted in compact thalli. The trichogyne is an ephemeral structure, whose survival must be measured in hours. The wall is often thick and the frequency with which detritus adheres to it indicates that it is mucilaginous.

Spermatangia The male gametes, or spermatia, are spherical or ovoid, 2–5 μm in diameter, formed singly within the gametangia, or spermatangia. The spermatangia are produced from spermatangial mother cells which may or may not be distinguishable from vegetative cells (Fig. 4A). The spermatangial mother cells may have reduced chloroplasts or even be colourless, whilst they and the derivative spermatangia are usually uninucleate irrespective of the nuclear constitution of the normal vegetative cells. From two to five spermatangia are produced from the subterminal portions of the mother cell. The formation of spermatangia in rows has been reported in several genera. Young spermatangia are slightly elongate and measure 3–7 μm in diameter, with a relatively thick wall, often more than 1 μm thick. Spermatangia are usually colourless, although chloroplasts have been reported in some Nemaliales. Patches of spermatangia may be sufficiently large for the colour differences to be visible to the naked eye. When spermatangia are produced singly they may be impossible to detect without sectioning. The spermatangia are overlaid by wall material. With the progressive release of spermatia, the thickness of this overlying layer increases. The greater depth of wall may be sufficient to indicate the occurrence of spermatangia. In some cases, spermatangia may proliferate within the empty wall remaining after the spermatium has been shed.

Fertilization The trichogyne is mucilaginous and the spermatia make contact and adhere to it. An open connection is established between the cytoplasm of the trichogyne and the spermatium. A delimiting wall is present in the spermatium by this time despite its condition at release. Following fertilization, the trichogyne separates from the basal part of the carpogonium. It has been suggested that fertilization may not precede further development in certain Florideophyceae.

The carposporophyte and its development

The Florideophyceae are characterized by the retention of the zygote on the female plant where it gives rise to a post-fertilization development, the carposporophyte. Interpretation of this structure has changed during the past two centuries but there are still some uncertainties remaining. It was appreciated at a very early date by Stackhouse (1801) and Turner (1802) that in those algae now referred to the Florideophyceae the spores, or 'seeds' as they were then called, were produced in two very different ways. In one, the spores were formed in groups of four, whereas in the other the spores were produced in a mass which might or might not possess a surrounding envelope. Many different terms were applied to the variants of the latter and most of these are no longer in use, but the term 'cystocarp', first applied by Agardh in 1844 to spore masses surrounded by a definite envelope, is still widely used.

Following the discovery of the process of fertilization by Bornet and Thuret, it became clear that these 'cystocarpic' spore masses resulted from gametic fusion. Details of the development were first elucidated by Schmitz (1883) although his interpretations were not entirely correct. The most important discoveries made by Schmitz were:

1. that when an envelope surrounds the spore mass, the spore mass is formed from the zygote, whilst the envelope is produced by an out-growth of the female plant,
2. that the zygote may develop directly into the spore mass or indirectly through a complex process involving further cell fusions.

Schmitz assumed that the further cell fusions were equivalent to the preceding gametic fusion. The correct interpretation was provided somewhat later by Oltmanns (1898) who showed that these further fusions involved nuclear transfer but not nuclear fusion. Oltmanns regarded the post-fertilization growth as equivalent to a plant producing tetrasporangia or gametangia in that he called it the sporophyte. The word 'carposporophyte' was coined by Janet (1914) and its use together with the concept of it being a phase or generation equivalent to the gametangial and tetrasporangial plants have now been accepted almost universally. The carposporophyte is formed from one or more primordia, the gonimoblast. The carposporophyte terminates with the production of carposporangia, which may be formed only from the apical cells of the carposporophyte or, in addition, from intercalary cells. Each carposporangium usually liberates a single spore although in a few examples four spores are produced. Liberation takes place through the rupture of the sporangial wall, and regeneration has been reported in several cases. It would be inappropriate in the present work to discuss the many variations in carposporophyte development which occur throughout the Florideophyceae. For those interested in such details, the accounts by Fritsch (1945), Drew (1954), and Kylin (1956) should be consulted. However, in a floristic treatment such as this, reference will be made on many occasions to the mature product of carposporophyte development because it is often of considerable taxonomic value. The following brief survey is intended merely to explain those structures to which reference is made in the text.

The discovery that the zygote, or its derivative product, develops into the sporangial mass, and that the envelope is formed as an outgrowth of the female plant, has been confirmed by virtually all subsequent investigations. In most cases, the mature carposporophyte is represented by a mass of sporangia together with a variable number of sterile cells. There is variation in the number of sporangia, the size of the carposporophyte and of the sporangia, and whether only terminal cells convert to sporangia or intercalary cells are also involved. There is some variation in location, but this reflects the positions at which carpogonia and/or auxiliary cells are formed and the extent to which vegetative growth continues after gamete fusion and development of the carposporophyte. The nature of the envelope when present is of considerable systematic importance. Some sort of protection around the carposporophyte is present in almost all Florideophyceae. The simplest form, occurring in members of the Nemaliales and Ceramiaceae with obviously filamentous thalli, consists of a cluster of one or more involucral filaments or axes which arise during development of the carposporophyte and loosely encircle it. In certain Nemaliales, a cluster of such filaments is present even when the carposporophyte is embedded within the thallus. More elaborate protection occurs in the Dasyaceae, Delesseriaceae, Rhodomelaceae and Rhodymeniales where the protective filaments are laterally adherent to form a complete envelope. This is usually provided with an aperture through which carpospores are discharged. The most substantial coverage occurs in Gelidiaceae and genera of the Cryptonemiales and Gigartinales with cartilaginous thalli. The carposporophyte is embedded in the female gametangial plant and enveloped by a very substantial structure.

The term 'cystocarp' was first applied to the structure seen in the Rhodomelaceae where the carposporangial mass is surrounded by a flask-shaped envelope. Subsequently, the term was applied throughout the Florideophyceae irrespective of the presence of a protective envelope. In those cases where a protective envelope is absent the term cystocarp refers to the carposporophyte alone. When an envelope is present, it applies to a compound structure composed partly of the carposporophyte and partly of female gametangial tissue. Because of these two different applications, several investigators have objected to its use. From a pragmatic standpoint, there is still value in the term providing it is understood that it may be applied in ways which are not strictly comparable. For this reason, it should be appreciated that the term is used in the present treatment in a general sense. Where reference to the carposporophyte alone is intended, that term will be used while the term 'pericarp' applies only to the protective envelope.

Other sporangia

Tetrasporangia Tetrasporangia, each of which produces four spores, are the most widely occurring sporangia found in the Florideophyceae. Their size varies from about 10 μm in members of the Phyllophoraceae to 600 μm in some Corallinaceae. Occasionally, tetrasporangia may develop in an intercalary position *(Petrocelis)* although, in most cases, they are formed terminally or laterally on filaments of limited growth. Where the aggregation of filaments is minimal the tetrasporangia are free, whereas in more compact thalli the tetrasporangia are partially or completely embedded in the cortex. Tetrasporangia may be scattered irregularly over the surface of the thallus or restricted to particular areas, such as marginal proliferations *(Cryptopleura)* or specialized adventitious axes (*Plocamium, Dasya,* etc.). In certain genera of the Champiaceae the tetrasporangia are formed in sori which infold as sunken pits.

Tetrasporangia are of three types, cruciate (e.g. *Chondrus crispus*), zonate (e.g. *Furcellaria lumbricalis*), or tetrahedral (e.g. *Bonnemaisonia hamifera*), depending on the final arrangement of the four spores. Meiosis has been demonstrated in various tetrasporangia although apomeiosis might occur in one or two cases. In tetrahedral tetrasporangia, two nuclear divisions occur prior to the simultaneous development of invagination furrows. In most cruciate and some zonate tetrasporangia the first transverse cleavage occurs before the second nuclear division, whereas in other zonate tetrasporangia the quadrinucleate condition is established before the first cleavage. The time interval between initiation of the primordium and the release of spores varies enormously from species to species. For some, the process is completed in 12–14 days, with sporangia being produced for a considerable period, whereas in *Furcellaria,* the initials are formed in April with meiosis in early November and release of tetraspores taking place almost simultaneously throughout the plant in late December. Thus, tetrasporangia may be borne on a plant for a considerable period, but be functional only for a short time.

Dehiscence takes place by a regular splitting of the wall, or by a more irregular breakdown, or by the detachment of a lid from the apex. Following release of spores from a tetrasporangium, the mother cell may proliferate a new sporangium into the old empty wall, a process which may occur more than once.

Polysporangia and parasporangia Sporangia containing more than four spores occur in various Florideophyceae, and the terms 'polysporangium' and 'parasporangium' have been applied indiscriminately. Drew (1939) suggested that sporangia containing more than four spores are of two categories:

a. those which are homologous with tetrasporangia; meiosis occurs during spore formation and the structures occur in place of tetrasporangia,

b. those which are not homologous with tetrasporangia; meiosis does not occur during spore formation and the structures do not replace tetrasporangia.

She suggested that the term 'polysporangium' should be applied to the first type and 'parasporangium' to the second.

Sporangia of the first type occur in Britain in *Pleonosporium borreri* and certain Champiaceae. It is unfortunate that nothing is known of the cytology of these polysporangia. Several sporangia exemplify Drew's concept of the parasporangium. Their origin is the result of cellular proliferation and separation and gives rise to a cluster of 'spores' of various numbers and sizes, ranging from 6 or 7 in the smallest parasporangia of *Plumaria elegans* to several hundred in some examples in *Ceramium.* Cytological observations are available for *Plumaria,* where the divisions are mitotic.

Bisporangia Sporangia containing two spores have been reported in the Corallinaceae and Ceramiaceae. Bisporangia occur as the only form of reproduction or mixed with tetrasporangia. There are few reports of their occurence on carpogonial or spermatangial plants. The impression (Fritsch, 1945) that bisporangia are homologous with tetrasporangia may have been strengthened by the delay between the first and subsequent divisions in many cruciate and some zonate tetrasporangia. Two-celled structures appear on what are tetrasporangial thalli but it is not known whether spores are released from these incompletely-developed tetrasporangia.

Bisporangia are of two types, the mature spores being either uninucleate or binucleate. Suneson (1950) showed in *Dermatolithon litorale,* for which only plants bearing bisporangia are known, that two nuclei are formed by mitosis and these pass to the spores, each of which is uninucleate. Suneson also showed that three types of sporangia were produced in Swedish material of *D. corallinae:* quadrinucleate tetrasporangia, quadrinucleate bisporangia and binucleate bisporangia, with meiosis occurring in the first two types but not the last. The contradictory reports of bisporangia indicate that caution is needed in their interpretation. In *Crouania attenuata* it appears that bisporangia are formed during the winter with tetrasporangia produced on the same plants during the summer.

Monosporangia Various types of sporangium occur from which a single spore is released. Monosporangia occur in *Audouinella* and other Nemaliales, distinguished from spermatangia by their pigmented contents and larger size. Monospores of the Nemaliales are uninucleate and are liberated by rupture of the sporangial wall. Regeneration occurs by proliferation from the cell below and this can occur many times in one sporangium. The monosporangia of *Corynospora* differ in that they are multinucleate structures, formed by the transformation of the apical cell of a two-celled branchlet. In *Ahnfeltia,* the monosporangia are formed in tissue masses once considered as a 'parasitic' Florideophycean alga, *Sterrocolax decipiens.* Monosporangia in terminal rows occur in species of *Seirospora.* These structures are frequently referred to as 'seirosporangia', but are simply monosporangia.

Monosporangia have been detected in culture in species where they have not been reported in the field. These are little-differentiated and release consists of the loss from a cell of its contents.

REPRODUCTION IN BANGIOPHYCEAE

In many Bangiophyceae, a major problem arises in defining a spore, because of the lack of clear distinction between a vegetative cell and a spore or spore-producing cell. In addition, recognition of gametes is even more uncertain. As a result of these two features, knowledge of reproduction in the Bangiophyceae is much less certain than for the Florideophyceae.

Sporangia and spore formation

Three types of spores have been recognized in the Bangiophyceae, according to the numbers produced and the type of mother cell from which they are formed (Drew, 1956):

Type 1: monospores produced by differentiated sporangia;
Type 2: monospores produced from undifferentiated cells;
Type 3: spores produced by the successive divisions of a mother cell.

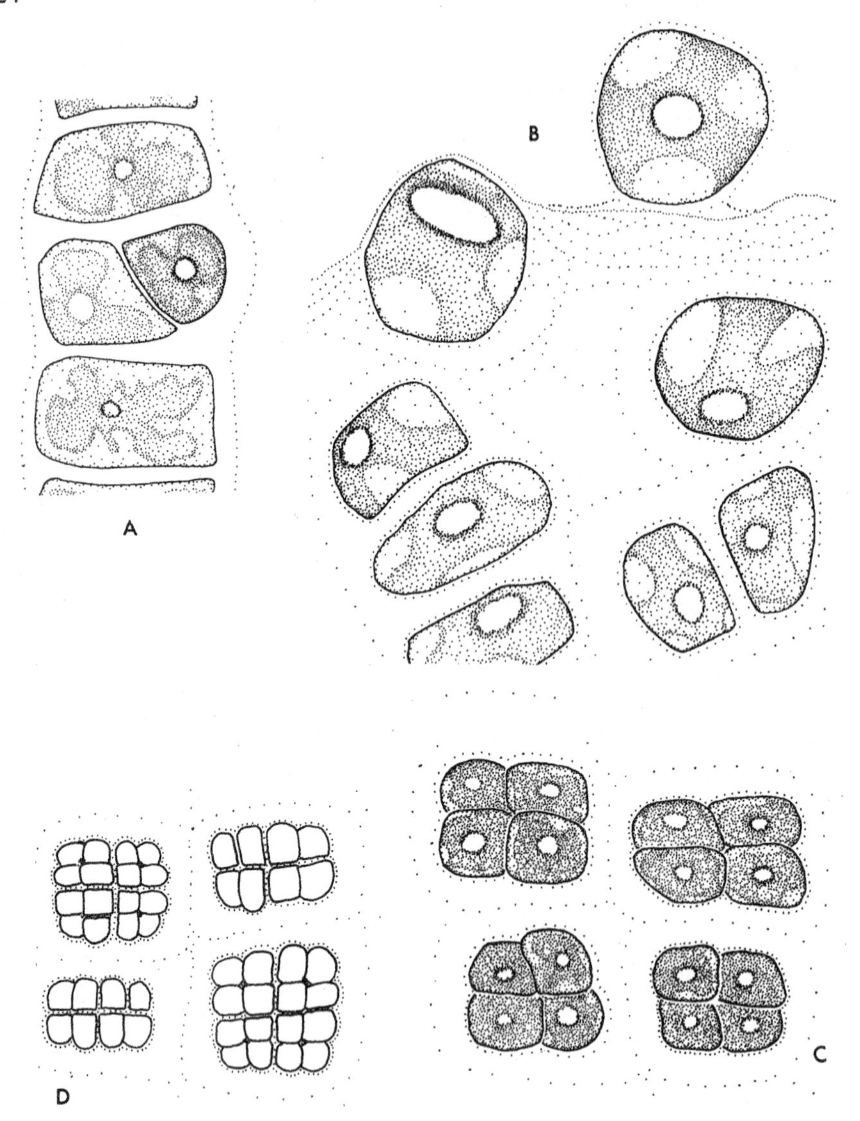

Fig. 5 Spore production in Bangiophyceae

A. *Erythrotrichia carnea*, formation of (Type 1) monosporangia ×1250; B. *Porphyra* sp., formation of (Type 2) monosporangia ×1250; C. *Porphyra* sp., formation of (Type 3) sporangia, or α-sporangia, often regarded as 'carposporangia' ×1250; D. *Porphyra* sp., formation of (Type 3) sporangia, or -sporangia, often regarded as 'spermatangia' ×1250.

The two types of monospore were distinguished by the degree of differentiation. If every cell division prior to the release of a spore was identical with those characteristic of vegetative growth, the spore was said to be of Type 2 (Fig. 5B). Such a spore represents the release from a multicellular body either of a whole cell or the contents of a cell. Type 1 spores were those where cell function changed at the formation of a sporangium, a specialized portion of a cell being cut off by a division different from those by which vegetative growth was achieved (Fig. 5A). A spore was then released from this specially cut off portion. It has been customary to regard monospores of either Types 1 or 2 as not being of sexual origin or related to a sexual process. Because of the absence of differentiation prior to the formation of Type 2 monospores, they have been frequently overlooked because they can be detected only by observation of living material. Type 1 monospores are formed in differentiated sporangia. Following nuclear division, an oblique wall is laid down in the apical pole of a cell to divide it into two unequal portions. Each portion contains a chloroplast and pyrenoid. The smaller portion is the sporangium which releases its contained spore; the larger portion may then expand and divide again to form a second sporangium. Unlike the Florideophyceae, where the remnants of old sporangial walls persist, no traces remain in the regeneration of a Type 1 sporangium in the Bangiophyceae. Spores formed in differentiated sporangia are liberated by rupture of the wall and it has been suggested that expulsion is violent. Although it has been inferred that Type 1 spores are not of sexual origin and do not participate in a sexual process, there are several examples contrary to this hypothesis. There is evidence of a Type 1 spore functioning as a male gamete, as a post-fertilization spore, or as an 'asexual' spore (Dixon, 1973).

The third type of spore formation (Type 3 of Drew) occurs through the successive divisions of a mother cell. Such divisions, by walls laid down in two or three planes at right angles, have been described many times in *Porphyra* and *Bangia.* The volume of the mother cell increases slightly during the process of division while the original cell wall remains distinct so that the mature spores are arranged in 'packets'. Two different forms of Type 3 spores may be produced. In one case, each mother cell divides two or three times to give four or eight products, the products remaining pigmented and being released in this condition (Fig. 5C). The function of the mother cells and the products of division has been the subject of much controversy. The mother cell has been considered as a gamete which is fertilized although the spores have also been so regarded. Most authors have considered the spores to represent the product of some form of sexual reproduction and referred to them as 'carpospores'. In the second case, the mother cell divides to produce a much larger number of products (Fig. 5D), usually 64 in most species of *Bangia* and *Porphyra,* although in the latter there may be as many as 128 or as few as sixteen or even eight in some of the distromatic species. However many products are formed, pigment is lost by the chloroplast which also decreases in size although it may be recognizable in the liberated product. As the products are formed from large numbers of mother cells, areas of the thallus become pale yellow or greenish. Because of the greater number of divisions and the minimal increase in size of the mother cell during division the resulting colourless bodies are much smaller than the pigmented type discussed previously. The products are liberated in vast numbers by the dissolution of the cell walls. Virtually all authors have considered these smaller, non-pigmented bodies to be male gametes and they have been referred to as 'spermatia' in consequence. Although these supposed 'carpospores' and 'spermatia' of *Porphyra* and *Bangia* can be obtained in vast quantities with ease, the evidence on which such interpretations are based is virtually non-existent. About all that can be stated with certainty is that *Porphyra* and *Bangia* produce spores of two different sizes by the repeated division of mother cells. Obviously one should not continue to refer to these as 'carpospores' or 'spermatia'. Conway (1964) has suggested that it would be more appropriate to refer to these simply as large spores, or α-spores, and small spores, β-spores. Despite the variation in number of divisions in both cases, and the resulting variation in size, the two classes are quite distinct and can usually be recognized on the basis of size and colour.

In summary, it can be stated that unicellular bodies of different shapes and sizes are produced in the Bangiophyceae in one or more of three ways. The three methods of formation outlined by Drew provide a useful categorization but there is little real understanding of the origin, behaviour and function of the spores.

Evidence for sexual reproduction in the Bangiophyceae

Evidence for sexual reproduction consists in the recognition of gametes and observation of their fusion. Subsidiary evidence may be obtained from cytological studies indicating counts of the number of nuclei or chromosomes. Sexual reproduction has been reported for some Bangiophyceae found in Britain: *Bangia, Erythrocladia, Erythropeltis, Erythrotrichia, Porphyra,* although the evidence on which these reports are based is incomplete and often conflicting.

Considering the gametes, the descriptions may be grouped into four types:

1. The fusion of two released cells has been described (Fig. 6A). Of the British marine Bangiophyceae, such gamete fusion has been reported in *Porphyra* and *Bangia*. However, the occurrence of spore-like structures with protuberances has been shown to be indicative of division rather than fusion so that these reports must be regarded with suspicion.

2. The fusion of adjacent cells in a thallus has been reported in *Erythrotrichia* (Fig. 6B).

3. The fusion of a released cell with an undifferentiated cell of the thallus has been reported in *Bangia* and *Porphyra* (Fig. 6C). Contact is said to be by means of a tube which develops and penetrates the undifferentiated cell. This development is very like fungal infection.

4. The fusion of a released cell with a cell of the thallus which has differentiated to produce a protuberance (Fig. 6D) has been reported in *Porphyra,* occasionally in *Bangia,* and also in *Erythrotrichia, Erythrocladia* and *Erythropeltis.*

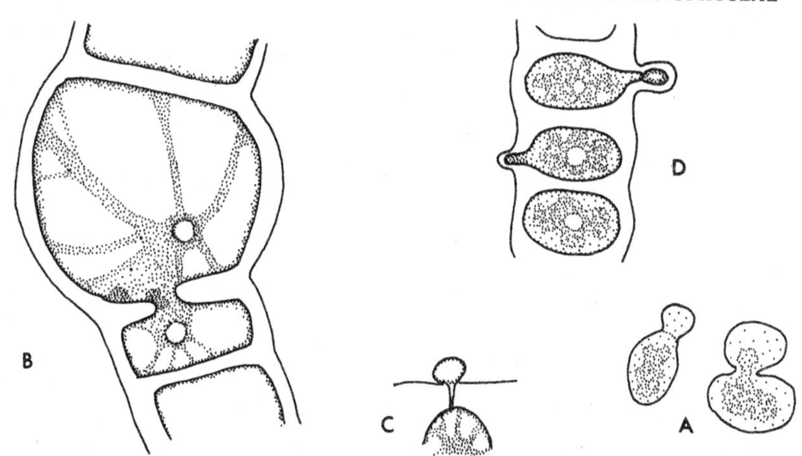

Fig. 6 Alleged modes of gamete fusion in Bangiophyceae
A. *Porphyra* sp., fusion of two released cells (after Goebel) ×700; B, *Erythrotrichia carnea,* fusion of two adjacent cells (after Heerebout) ×1325; C. *Porphyra* sp., fusion of a released cell with an undifferentiated cell (after Berthold) ×200; D. *Erythrotrichia* sp., fusion of a released cell with a differentiated cell (after Berthold) ×200.

Evidence of a relationship between the released cells reported as participating in fusion and the alleged gametes described on purely morphological grounds is minimal. Cytological evidence of fertilization is scanty and contradictory even in *Bangia* and *Porphyra* which have been well studied by comparison with other Bangiophyceae. At the nuclear level, observations of cells containing two nuclei are rare. Such binucleate cells have been reported for *Bangia* and *Porphyra* but it should be emphasised that bi- or tri-nucleate cells have been produced in the latter genus under the influence of discharges of industrial origin. It has been argued that meiosis occurs in *Porphyra* prior to the formation of 'carpospores', whereas others have claimed that such divisions are mitotic. A similar claim has been made for division of the mother cells in *Bangia atropurpurea.* In addition, some workers have indicated that the same chromosome numbers occur at all stages in both *Porphyra* and *Bangia.* The only other position in the life history of *Porphyra* where meiosis has been reported is in the fertile cell rows of the *Conchocelis* phase. Thus, in *Porphyra* and *Bangia* there is little agreement as to cytological events from which the occurrence or absence of fertilization might be deduced, and all that can be said is that doubling of the chromosome number may occur in certain circumstances although not indisputably under all conditions.

To summarize, there are claims for the formation and fusion of gametes and there are indications of the doubling of nuclei and subsequently of chromosomes, but for each piece of positive evidence there is an equally convincing piece of negative evidence.

LIFE HISTORIES

Life histories in the algae

A feature of plants is the development from a reproductive body of a morphological phase different from the phase on which it was formed. Nowhere is this displayed with greater diversity than in the algae. Despite innumerable treatments of algal life histories in recent years (Feldmann, 1952; Drew, 1955; Chapman & Chapman, 1961), there is still much confusion. First there is no agreement as to the terminology to be applied to the phenomenon as a whole or to the constituent parts, or to the different life histories. Secondly, the evidence accepted as 'proof' of a life history is often dubious. Thirdly, many of the anomalies may be explained by the fact that behaviour, even in the same species, can be very flexible. Under different circumstances or in different geographical localities the sequence of phases can differ considerably.

Considering terminology, the phenomenon under discussion will be termed 'life history' in the present treatment. There are two other terms – 'alternation of generations' and 'life cycle' – both used widely. The term 'alternation of generations' represents a mis-translation of "*Generationswechsel*", which does not mean 'alternation of generations', but rather '*change* of generations'. The latter would be much more acceptable for the algae. The term 'alternation of generations' is more representative of the higher cryptogams, where there is an almost obligatory alternation of spore-producing and gamete-producing phases. In the algae there are several life histories similar to this strict 'Hofmeisterian cycle', but there are many more which are not. It makes little sense to consider algal life histories from the standpoint of the land flora, because only one of many types of life history found in the algae has survived terrestrial colonization. A further objection is that many Florideophyceae have life histories with more than two phases and one cannot have an alternation of three objects. The term 'life cycle' implies that the sequence of phases consists of a single, obligate cycle. In some algae the life history is not obviously cyclic, while in many it would be better described as polycyclic. For these reasons, 'life cycle' is hardly suitable for general use, even if applicable to the higher cryptogams. The term 'life history' is best, in that it is clear, precise and devoid of unnecessary or unwanted connotations.

Another problem concerns the names for the constituent parts. Here again, difficulties arise from attempts to apply to the algae a terminology developed for the higher cryptogams, where there is almost complete coincidence between the somatic phase and the nuclear state. The gamete-producing phase is haploid, the spore-producing phase is diploid, and the terms 'gametophyte' and 'sporophyte' have been applied to these. In the algae, the coincidence between somatic phase and nuclear state breaks down extensively. In *Fucus*, gametes are produced on a diploid plant and there have been disputes as to whether this should be considered as a gametophyte or a sporophyte. In most Florideophyceae, a diploid carpospore-forming phase is retained on the haploid female gamete-forming phase, and there is a second distinct diploid spore-forming phase producing tetraspores.

Further, there are those algae where the life history exhibits several similar morphological phases of different ploidy levels, each of which may produce reproductive structures of two different types. It seems best to avoid terms such as 'gametophyte' and 'sporophyte' as they are not applicable in the sense in which they were originally intended. The simplest procedure, which will be adopted in the present treatment, is to name each morphological phase in terms of the reproductive structure produced, so that one may speak of 'gametangial phase' or 'tetrasporangial phase'.

A third problem concerns the discrimination between different algal life histories. As the known diversity of algal life histories has increased, so has this problem. Most schemes have used a terminology which attempted to be descriptive. The objections to the most commonly used systems using such terms as haplobiontic/diplobiontic, haplontic/diplontic/diplohaplontic or references to both morphological and cytological phases have been dealt with elsewhere (Dixon, 1973). It was suggested that a 'type' system was simpler, more flexible and capable of expansion to accommodate new data.

The life history of an alga may be defined as the recurring sequence of morphological and cytological phases. A morphological phase may be defined as a state of an organism recognizable by a constant characteristic morphological appearance irrespective of chromosome number. A morphological phase may be considered as beginning with the single cell from which it arose and ending with a single cell, the reproductive body which it produces. In Florideophyceae, virtually all authors have considered the carposporophyte as equivalent to the structures which produce gametangia or tetrasporangia. The carposporophyte begins with a single-cell (the zygote) and terminates with the production of one or more carpospores and fulfills the requisites for a morphological phase; it is best that it be so interpreted. A cytological phase may be defined as that state of an organism characterized by mitotic divisions all showing the same chromosome number.

In order to establish a life history certain specific information must be available. The first set of data concerns a knowledge of the kinds of morphological entity and reproductive body representing the species under consideration. This may or may not be easy to obtain. There are some taxa for which the morphological units are similar, although in others they are dissimilar. The second set of data concerns the germination of each reproductive body and its growth to maturity, which helps to associate morphological units of unlike appearance. The third set of data relates to the cytological phases. It demands a knowledge of the chromosome number of each morphological entity as well as determination of the positions of syngamy and meiosis. These basic requirements are sufficient to establish the life history; the difficulty is that for most red algae one or more is lacking and the number for which all data are available is extremely small. Lack of data has resulted in assumptions being made and constant repetition of these obscures the fact that one is dealing only with an assumption. It is essential that conclusions in the laboratory are confirmed in the field, since differences have been demonstrated.

Life histories in the Florideophyceae

It was appreciated by the early phycologists that different types of reproductive structures occur on different plants in what are now considered members of the Florideophyceae. During the nineteenth century, Schmitz elucidated the nature of the carposporophyte and its relationship to the cystocarp. Strasburger published his generalization on the periodic reduction of chromosome numbers and this changed the outlook on life histories, because the significance of syngamy and meiosis was now apparent.

For the red algae, Yamanouchi (1906) showed that meiosis occurred in the tetrasporangia of *Polysiphonia flexicaulis* (as *P. violacea).* The life history established for *Polysiphonia* on grounds of cytology has been substantiated by culture studies. There is now a large number of cases where it is known that the life history consists of a sequence of gametangial, carposporangial and tetrasporangial phases, the first being haploid and the last two diploid. In *Pleonosporium,* tetrasporangia are replaced by polysporangia and the same morphological sequence as in *Polysiphonia* has been proved by cultural studies although cytological evidence is lacking. Despite the abundant data confirming the occurrence of the *Polysiphonia* type of life history in many Florideophyceae, it is not universal. Evidence indicative of deviation includes the occurrence of gametangia and tetrasporangia on the same thallus, the germination of tetraspores to give tetrasporangial plants and the cytological demonstration for non-meiotic division in tetrasporangia.

Following the work on *Polysiphonia,* obvious targets for investigation were those genera, particularly of the Nemaliales, for which tetrasporangial plants had not been reported, or only doubtfully on rare occasions. Studies of *Scinaia* and *Nemalion,* reinforced by later investigations of *Asparagopsis* and *Bonnemaisonia,* led to the concept that meiosis followed immediately after syngamy and that gametangial and carposporangial phases were haploid. The general idea developed that the Nemaliales was set aside from other Florideophyceae through this absence of tetrasporangial phases. More exact cytological investigations and determination of the sequence of morphological phases through culture studies have shown that this view is no longer tenable. The new data indicate that in the Nemaliales the life history consists of the same sequence of gametangial, carposporangial and tetrasporangial phases as in *Polysiphonia* but with the difference that the gametangial and tetrasporangial phases are morphologically dissimilar. In addition, there is evidence indicating that the life histories of genera in Gigartinales and Cryptonemiales for which tetrasporangial phases had previously not been detected are also of this pattern. There are still many outstanding questions regarding the identity of the tetrasporangial phases, determination of the products of spore germination and, particularly, cytological details.

For several genera and species, the product of germination of a carpospore resembles a species of *Audouinella*. Other growths are more organized, forming prostrate systems of varying degrees of aggregation. The more loosely associated germlings have been described as the '*Naccaria* type' while the more compact have been termed 'discoid'. This categorization is not fundamental because there can be variation in the compactness of germlings even in the same species.

Despite the evidence for a range of morphological phases in life histories, there is still uncertainty as to the nature of these phases in some cases. In several instances it has been shown that the product of germination of a carpospore gives rise to tetrasporangia, and the release from these of tetraspores has been demonstrated, the tetraspores germinated and their development followed to such an extent that the product may be identified and, in a few cases, followed to the point where it is reproductively mature. On the other hand, there are several instances where the plant formed by the germination of carpospores gives rise directly to what is identifiable on morphological, if not reproductive, grounds as the gametangial plant. Thus, there are two possible pathways of development and both have been demonstrated for certain species. Interpretation would be facilitated if detailed cytological data were available for examples of each method of development, particularly for species where both have been reported. A crucial question concerns the occurrence and position of meiosis. The studies of von Stosch (1965) and Ramus (1969a) on *Liagora farinosa* and *Pseudogloiophloea confusa* indicate that meiosis occurs in the tetrasporangia. For organisms where the plant identified as the gametangial phase arises directly as a bud on the product of germination of a carpospore, the only cytological data available are of *Lemanea mamillosa.* Magne (1967a) has claimed that meiosis occurs in the apical cell of the upright filament which produces the gametangial phase. Carpotetrasporangia were reported on the carposporophyte of *Liagora tetrasporifera* at the time when it was thought that life histories in the Florideophyceae could be considered on the basis of a *Polysiphonia* type and a further type, exemplified by *Nemalion,* where meiosis was thought to follow syngamy in the carpogonium. Such carpotetrasporangia are now known to occur in several Florideophyceae which occur in the British Isles.

Some species of the Phyllophoraceae possess the reproductive structures characteristic of *Polysiphonia* while others possess carpotetrasporangia reminiscent of *Liagora tetrasporifera.* For certain species such as *Phyllophora crispa, P. pseudoceranoides* and *Stenogramme interrupta,* normal tetrasporangia occur on one plant while gametangia, which ultimately give rise to carposporophytes with carposporangia, occur on another plant of similar appearance. The genus *Gymnogongrus* poses many problems. The thirty or so supposed species have been divided into two groups, those in which the gametangia produce an internal carposporophyte with carposporangia and those in which they give rise to external proliferations containing carpotetrasporangia. A species placed in one group may have its counterpart in the other, so similar in appearance as to be indistinguishable on the basis of vegetative morphology. The life histories of these supposed species are still incompletely known and, until this information is available, their taxonomic status must remain in doubt. New cultural and cytological evidence suggests that the generally accepted concept of the life history of *Phyllophora truncata,* as the sequence of a haploid gametangial phase

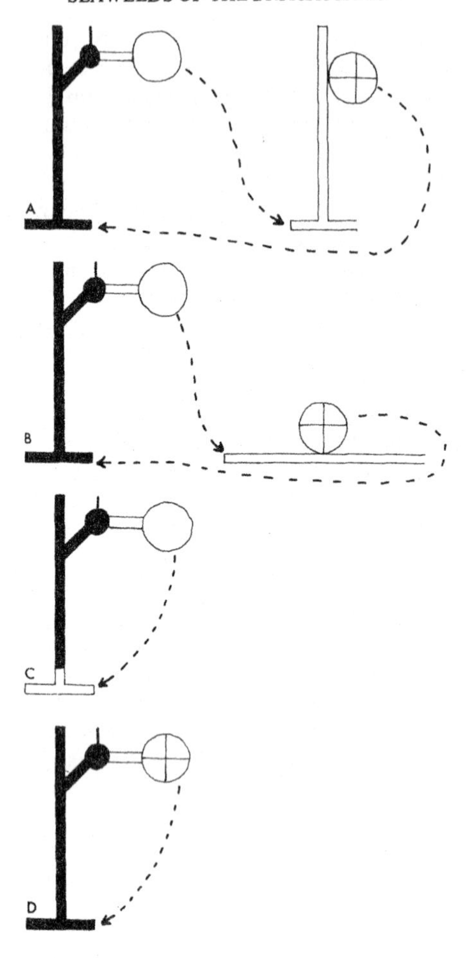

Fig. 7 Life histories in Florideophyceae
A. *Polysiphonia* type; B. *Bonnemaisonia* type; C. *Lemanea* type; D. *Liagora tetrasporifera* type.

and a diploid carpotetrasporangial phase may well be correct. The most peculiar of all Phyllophoraceae is *Ahnfeltia plicata,* in which the erect frond produces monosporangia with zonate tetrasporangia on the discoid base.

To summarize for the Florideophyceae, there appear to be four situations (Fig. 7) which can serve as a basis for comparison:

1. *Polysiphonia:* the life history consists of a sequence of gametangial, carposporangial and tetrasporangial phases, the first and last being morphologically

similar, the carposporangial phase developing on the female gametangial phase. Proof of such a life history has been obtained in numerous British species of the Florideophyceae and the indications are that it occurs in many others.

2. *Bonnemaisonia:* the life history consists of a sequence of gametangial, carposporangial and tetrasporangial phases, all three morphologically dissimilar, the carposporangial phase developing on the female gametangial phase. Proof of such a life history has been obtained in some British species and it probably occurs in others.

3. *Lemanea:* the life history consists of a sequence of gametangial and carposporangial phases which are morphologically dissimilar, with the carposporangial phase developing on the gametangial phase and with meiosis occurring somatically during the development of the latter. To date, this life history has been demonstrated only in the freshwater genus *Lemanea* but it may occur in some of the British marine representatives.

4. *Liagora tetrasporifera:* the life history consists of a sequence of gametangial and carpotetrasporangial phases which are morphologically dissimilar, with the carpotetrasporangial phase developing on the gametangial phase and meiosis occurring in the carpotetrasporangia. Of the British Florideophyceae, this life history has been demonstrated only in some Phyllophoraceae.

Life histories in the Bangiophyceae

If one accepts the criteria for the establishment of an algal life history listed above, there is not one member of the Bangiophyceae for which the complete life history is known.

The data are most complete for *Porphyra* and *Bangia*, for which there are three different accounts of the sequence of nuclear phases. It is thought by some that meiosis occurs immediately prior to the formation of the so-called 'carpospores' and that this is preceded by syngamy. It has also been claimed that syngamy occurs but that the subsequent divisions are mitotic. The first group of workers consider that the 'carpospores' are haploid whereas the second group regards them as diploid. There is the third group of workers who have been unable to obtain indications of syngamy or meiosis, with the same chromosome count occurring in all phases of the life history.

The occurrence of a *Conchocelis*-phase in the life histories of *Porphyra* and *Bangia* is now accepted and the idea that this represents a form of degeneration induced by culture has been completely abandoned. The formation and fate of reproductive bodies are governed by photoperiod in many species of *Bangia* and *Porphyra*. The so-called 'carpospores' appear to be formed only under 'long-day' conditions, with more than 12 hours light per 24 hours. Once formed, their manner of germination is controlled by the photoregime to which they are subjected. Under 'long-day' conditions the 'carpospores' germinate in a unipolar manner to give *Conchocelis* whereas under 'short-day' conditions the germination is bipolar and the germling develops into the 'leafy' phase. *Conchocelis* may form unicellular reproductive bodies which give rise to further *Conchocelis* phase.

Elongate cells which become divided are also produced on *Conchocelis*. These are the so-called 'fertile cell-rows' and the spores to which they give rise germinate to form the alternate phase in the life history. The formation of spores by *Conchocelis* is under strict photoperiodic control both in *Porphyra* and *Bangia*. Finally, there is also evidence for the occurrence of a phenomenon analogous to the 'direct development' described for members of the Florideophyceae.

In all other members of the Bangiophyceae, it is doubtful if even the sequence of morphological phases can be stated with any certainty. For species of *Erythrotrichia*, despite descriptions of syngamy similar to one of the methods described for *Porphyra*, there is no recent evidence for this. Similar comments must also be made about the very different account of gamete fusion which involved the fusion of adjacent cells. The only evidence is that monospores, on germination, reproduce the parental phase. For various species of *Erythrocladia*, little more is known of the life history than for *Erythrotrichia*. Syngamy has been described, while the product of spore germination is indistinguishable from the parent.

From this discussion, it appears that there is not one member of the Bangiophyceae for which the complete life history is known. The two critical aspects in need of resolution are the nature of the bodies loosely described as 'spores' or 'gametes' and the existence of a sexual process.

Theoretical and biological life histories

Knowledge of algal life histories has been derived almost entirely from laboratory studies. For macroscopic algae, the full biological significance of the life history cannot necessarily be extrapolated and some consideration of behaviour in the field is imperative. One consequence of laboratory orientation has been that attention has concentrated on reproduction by spores and gametes/zygotes. Such phenomena as vegetative propagation and perennation have been almost ignored.

Many species have perennial thalli. In some cases, macroscopic plants may be collected at any season of the year. Observation of marked plants of *Gelidium* and *Pterocladia* shows that some persist for five or even seven years. In other instances only a microscopic portion persists from one year to the next. Although an individual cell or axis may survive for only a short time, long term survival on a clonal basis can occur. In terms of occurrence, statements of 'absence' must be regarded critically, as reports of 'presence' tend to be based on visible macroscopic plants, the small fragments usually being overlooked, for obvious reasons. Detection of fragments beneath a covering of debris is impossible until the new frond develops at the beginning of the next growing season. The fragments which persist are of many different forms. The basal disc commonly perennates, while prostrate axes with their attachment at intervals provide for considerable perennation. For a plant to persist in this way it does not need to be particularly tough or cartilaginous. The occurrence of dense populations of *Porphyra* on the same sand-covered rock from year to year when other, adjacent, rocks are devoid of the alga is due to the persistence of fragments beneath the sand.

Considering next vegetative propagation, the facility with which fragments are detached and the ease with which reattachment can occur vary enormously. Reattachment is not always necessary and considerable growth can occur in loose-lying populations which develop in conditions where there is sufficient illumination but protection from disturbance by wave action. In bays, this may be at depths of 10 to 20 m. Such populations may be disturbed by the gales of autumn, but residual populations can persist in protected places throughout the year. The problem with vegetative propagation is not to show that it can occur, but rather to demonstrate its actual occurrence in the field.

Perennation or vegetative propagation can modify the occurrence of phases of a life history in the field, particularly if one phase is selectively influenced. There are many reports of species said to possess a *Polysiphonia*-type of life history, but where one phase is of greater frequency than the other or has a wider geographical distribution. In Europe, plants bearing tetrasporangia are frequently reported from further north than plants with gametangia, while at the northern limits of distribution most species are often sterile. Such observations raise questions as to the life history in these areas.

Investigation of a life history demands a study in the field and the laboratory of reproduction and its control, the production, viability and longevity of spores, perennation and vegetative propagation, using both cytological and cultural techniques. At the present time, the necessary data are almost non-existent or unreliable. It is usually assumed that the sequence of events demonstrated in a laboratory investigation represents the total expression of a life history, but this is merely the 'theoretical' life history. There may well be several expressions of life history possible for a species and to assume the existence of a single model is never justified. One can only reiterate the crucial need always to integrate the results of laboratory experimentation with field data.

ECONOMIC UTILIZATION

The earliest record of seaweed utilization is in a Chinese herbal of the third century B.C. while the use of brown algae as fertilizer was established in Europe by the twelfth century. There was extensive burning of kelp, first for the production of alkali in the eighteenth and subsequently for the manufacture of iodine in the nineteenth century. The present century marks the establishment of the real industrial use of algae. In the red algae, the major uses are as human food or as a raw material from which carbohydrates are extracted. The latter is by far the more important economically. There are a few reports of the use of marine red algae as animal fodder, while deposits of coralline algae are used for soil dressing. The following account will consider aspects of economic utilization only as they apply to the British Isles. For further details, see Chapman (1970), Levring *et al.* (1969) and Newton (1951).

Red algae as human food

In western Europe and on the Atlantic coast of North America, the two most important red algae used as food are *Palmaria palmata* and various species of *Porphyra*. The former is either eaten raw, fresh or dried, or cooked like spinach. There is still some collecting for personal use, but no commercial trade exists at the present time. Species of *Porphyra* have been eaten extensively in Ireland and Wales. There is still a considerable trade in South Wales where consumption is of the order of 200 tonnes (wet weight) per annum. The raw material is almost all imported from Cornwall, the Solway Firth, North Wales and Dunbar, with small quantities imported from Ireland when supplies from more accessible collecting grounds fail. The *Porphyra* is washed, boiled, minced and prepared for eating by warming in fat, sometimes being made into cakes with oatmeal. In Ireland, *Porphyra* is collected for personal consumption and for export. Preparation involves either frying with fat or converting to a pinkish jelly by heating the fronds in a saucepan with a little water and beating with a fork. Other species of red algae had some minimal use. *Laurencia pinnatifida* was probably the most widely used, as a condiment, 'pepper dulse'. It is said that *Dilsea carnosa* was used as a food species but the claim most likely represents an error in identification.

In addition to those which are or have been eaten whole, gelatinous extracts of other species have been used for many centuries. The use of *Chondrus crispus* and *Gigartina stellata* as the source of an edible gelatinous product is of considerable antiquity in Ireland and Scotland and the practice was introduced into North America early in the nineteenth century. The collected weed is washed, bleached and dried. It is added to milk and the carrageenan extracted by gentle heating. The blancmange-like product is insipid to most palates unless sugar, lemon rind or other flavouring is added.

Commercial extracts from red algae

Most of the procedures for the industrial extraction of agar, carrageenan, etc. were developed from the personal extraction processes. The characteristic polysaccharides of the red algae are mucilages which contain varying proportions of D- and L-galactose, 3,6-anhydro-D- and L-galactose, monomethylgalactoses and ester sulphates. Numerous polymers based on these units have now been characterized from various Rhodophyta although the commercial products fall into two classes, agar-polysaccharides and carrageenan-polysaccharides.

Agar Agar is the best known of algal products. The name is of Malaysian origin and was applied originally to the extract of *Eucheuma* although the product extracted from that genus is not the same as the substance to which the name is now applied. Agar is a phycocolloid of red algal origin, insoluble in cold but readily soluble in hot water with a 1·5 per cent solution being clear and forming a solid, elastic gel at 32–39°C, not dissolving again below 85°C. The properties of agar are often said to be a consequence of its being a mixture of two principal components, agarose and agaropectin. Agarose was considered to be neutral whereas agaropectin was charged. Recent results have shown that agar is a complex mixture of polysaccharides all having the same backbone structure but substituted to a variable degree with charged groups.

Agar is obtained from species of *Gelidium* and *Pterocladia* and other algae, often referred to as agarophytes. Preparation involves extraction of cleaned and bleached weed by boiling for several hours. The agar is decolourized, and purified by freezing.

The most important feature of agar is the high strength of the gel. Agar is used where material with minimal strength requires stiffening. The greatest use is in food preparation and the pharmaceutical trade. Agar is used in the canning of fish and meat, in the manufacture of processed cheese, mayonnaise, puddings, creams and jellies. Pharmaceutically, agar is used directly as a laxative and as an inert carrier for drug products where slow release is required, as a stabilizer for emulsions, and as a constituent of cosmetic preparations, ointments and lotions. The use of agar in bacteriology and mycology as a stiffening agent for growth media is still considerable.

Commercial production of agar was a monopoly of the Japanese before 1939 despite the development of agar industries in California and elsewhere. Demand in areas deprived of Japanese agar during the Second World War led to the development of industries in many countries, some of which have continued and prospered while others have since declined or disappeared. An attempt was made at this time in Ireland to develop an agar industry based on *Pterocladia capillacea* and various species of *Gelidium*. The initial standing crop was high so that the industry prospered in its first year, but then failed completely, due to inadequate crop replacement.

Carrageenan Carrageenan is the extract of 'carrageen', or 'Irish Moss', the trade names for a mixture of *Chondrus crispus* and various species of *Gigartina*, particularly *G. stellata*. Carrageenan is a mixture of two major components. These are the branched chain polymer of the kappa fraction and the linear chain of the lambda fraction. Kappa-carrageenan has a molecule with a backbone similar to that of agarose, although the units of which it is composed are different isomers. Commercial extraction is similar to that for agar although, unlike agar, carrageenan cannot be purified by freezing.

Carrageenans are used extensively for many of the same purposes as agar. Because of the lower gel strength, carrageenans are used less for stiffening purposes than agar although they are preferred for stabilization of emulsions in connection with paints, cosmetics and pharmaceutical preparations. For the stiffening of milk and dairy products, such as ice cream, carrageenans have supplanted agar completely in recent years and it is in this area that demand is greatest. One particular use is in 'instant' puddings, sauces and creams which do not require refrigeration.

Ireland is a major producer of carrageenan, although the greatest production takes place in Canada and the United States.

The use of coralline algae for soil dressing

Several species of crustose coralline algae occur in an unattached 'loose-lying' state, producing extensive beds in the shallow sublittoral, at depths ranging from 2 to 25 m below extreme low water. Plants are cast up from shallow water so that the local 'sand' is composed of crushed thalli of coralline algae. Living thalli are rose-pink to violet in colour but become bleached on being cast up. Calcareous material on the beaches may be recognized from a distance by the dazzling white appearance. Such beaches are called 'coral beaches' in western Ireland and southwest England and their frequency is indicated by the number of bays named 'Whitesand' or 'Whitestrand'. The coralline sand is used for liming acidic soils in Ireland. The rate at which material is cast up is adequate to replenish the quantity removed, probably because the use of 'coral sand' appears to have diminished during the present century due to the abandonment of the poorer land and the reduced costs of other sources of lime.

ECOLOGY AND DISTRIBUTION

The present treatment will provide a general survey of those factors which influence the occurrence and behaviour of marine Rhodophyta in the British Isles, together with some comments on geographical distribution, and brief notes on other aspects of ecological significance.

Ecological factors

Rhodophyta accepted as 'marine' occur in the sublittoral, the littoral and even into the maritime. For a discussion of the ecological terminology adopted in the present treatment, see p. 53.

Marine Rhodophyta of the sublittoral are less tolerant of extreme conditions than those of the littoral and within the littoral the tolerance limits are greater at the upper than the lower levels. There are problems involved in isolating the effects of any one ecological factor, in the field and in the laboratory. It is possible to give some indication of the ways in which species are influenced by particular environmental conditions although individual effects are rare. Factors to be considered fall into two categories, those which are physico-chemical in nature and those which are biological. Before considering specific parameters of the environment, a general account of climatic conditions in the British Isles is appropriate.

The climate of the British Isles Lewis (1964) has given a general summary of climatic conditions with respect to coastal ecology in the British Isles and a compilation of data from this source is shown in Table 1. The climate around the British Isles does not show any striking variation. In general, the north and west coasts have lowest summer temperatures, least sun, most cloud, and have rain most often. For more detailed accounts of the climate of the British Isles and of variations in sea and air temperatures, see Bilham (1938), Lumb (1961), and Meteorological Office (1953).

Temperature In considering the effects of temperature fluctuations on marine organisms it is necessary to consider both water temperatures and air temperatures for organisms which occur in the littoral.

The following summary of sea temperatures around the British Isles is based on Lewis (1964). Sea water temperatures fall progressively from south to north in the British Isles with widely-spaced isotherms and seasonal variations of the order of 7 to 12°C. Complications are produced by the almost land-locked Irish Sea, the southern North Sea and the English Channel, the nearness of south-east England to Europe, with its more continental climate, and circulation into the North Sea both from the English Channel and around the north of Scotland. The greatest seasonal range and the lowest winter mean sea temperatures occur off south-east England and in the eastern Irish Sea. Mean summer temperatures on the west coasts, except in the Irish Sea, show a change with latitude so that there is a difference of 2 to 4°C between south-west England and north-west Scotland. Temperatures in the North Sea may be influenced by a pocket of cold water but there is still a south to north gradient. In winter, the seawater isotherms around the British Isles run in a more north–south direction, so that mean temperatures off the west coasts (once again except in the Irish Sea) are several degrees higher than at the same latitude in the North Sea, while the mean seawater temperatures off north-west Scotland may be several degrees higher than in the southern North Sea although little different from those of the south-west coasts. In addition, there are local variations due to differences in coastal topography. Temperatures in shallow water may deviate by several degrees from those of the open sea whereas if the coast falls sharply into deep water the differences are minimal. Littoral tide pools exhibit marked temperature fluctuations in summer although in winter the deviation from open-sea temperatures is by no means as great.

The picture for the British Isles given by official weather reports indicates that air temperature isotherms tend to run east–west in summer and north–south in winter. The greatest seasonal range of mean temperature (12°C), with the highest summer mean temperatures and lowest winter mean temperatures, occur in southeast England. Minimal seasonal variation (7°C) in mean temperature occurs on the west and southwest coasts. Air temperatures are subject to greater and more rapid fluctuation than sea water temperatures. Loss of energy by back radiation on clear winter nights often results in air temperatures below 0°C although the southern and southwestern shores are relatively frost-free. Extremes of temperature have more severe environmental effects than are indicated by mean temperatures, particularly if the extremes last for several hours or even days. With respect to air temperatures, Lewis (1964) has stressed that little is known of air temperatures in proximity to rocky shores. Solar radiation raises the temperatures of rock so that the quantity of radiation has effects not indicated by official weather reports. Mean temperatures give the general pattern on a large-scale basis but are not the best parameter by which to relate the distribution of marine red algae to temperature and may even be misleading.

Comments about the effects of temperature in the field are often based on visual

TABLE 1

Climatic data for selected coastal stations of the British Isles
(Based on Lewis, 1964)

	Temperatures in °C								Sunshine		Fog	Rainfall	
	Monthly means Winter (Jan or Feb)		Summer (Jul or Aug)		Means of		Extremes recorded		Yearly mean per day (h)	Yearly % of possible sunshine	Days per year	Total amount (inches)	Days per year
	min.	max.	min.	max.	lowest annual	highest annual	min.	max.					
East Coast													
Deerness	2·2	6·1	10·0	14·4	–4·4	20·0	–13·3	24·4	3·1	25	12	35·5	215
Aberdeen	1·7	6·1	10·6	16·7	–9·4	24·4	–15·6	30·0	3·6	30	16	29·5	214
Tynemouth	2·8	6·7	12·2	18·3	–6·7	25·6	–14·4	31·1	3·9	32	42	24·5	179
Spurn Head	2·8	6·1	12·2	19·4	–4·4	26·7	–10·6	30·6	4·2	34	26	19·7	168
Gt Yarmouth	2·2	6·6	13·3	20·0	–6·1	25·6	–12·2	31·7	4·5	36	21	24·5	183
West Coast													
Stornoway	2·8	7·2	10·6	16·1	–6·1	22·2	–11·7	25·6	3·3	27	3	49·9	263
Rothesay	2·2	6·7	10·6	17·8	–6·1	26·1	–11·7	29·4	3·5	28	–	49·0	228
Malin Head	3·9	7·2	10·6	16·1	–2·8	23·3	–6·7	28·9	3·6	30	6	32·0	236
Valencia	4·4	8·9	12·8	17·8	–3·3	24·4	–6·7	27·2	3·7	31	3	55·6	252
Roche's Point	5·0	8·9	12·8	18·3	–2·8	22·8	–6·7	28·9	3·9	32	16	41·9	212
Irish Sea													
Douglas	2·8	7·2	11·7	17·2	–5·0	23·9	–11·7	27·2	4·3	35	17	41·2	204
Holyhead	4·4	7·8	12·8	17·2	–2·8	24·4	–8·3	30·0	4·3	35	12	34·9	201
St Ann's Head	4·4	7·8	12·8	17·2	–2·8	23·3	–7·2	27·8	4·3	35	29	35·2	201
South Coast													
Dungeness	2·8	7·2	13·3	20·0	–7·2	21·7	–12·8	28·3	4·9	40	44	24·4	169
Portland	4·4	7·8	14·4	18·3	–2·8	v23·9	–6·1	28·3	4·8	39	13	25·6	163
Plymouth	3·9	8·3	12·8	19·4	–3·9	26·1	–8·3	30·6	4·7	38	13	36·2	190
Scilly	6·1	9·4	13·3	18·3	–0·6	21·7	–3·9	27·8	4·7	38	27	31·9	207

observation. It is difficult to separate the effects of high temperature from those of related phenomena. Increase in temperature is a consequence of increased solar radiation but extremes of radiation may have destructive effects not related to temperature. The consequences of desiccation can be equally harmful. It is difficult to distinguish between the effect of temperature on the organism and secondary effects caused by reduction of oxygen or changes in the bacterial flora. Because intertidal species are both submersed and emersed one has to consider temperatures of air and water. Because these two media may be at different temperatures, the effects of shock on transfer from one to the other must be considered. The effects of extreme low temperatures are often more easily detectable than those of high temperatures and are most obvious when a hard frost occurs at a time when organisms of the lower littoral are emersed. Cold damage usually changes the colour of red algae to a bright yellow-orange.

Experimental investigation of the effects of temperature on marine red algae (see Gessner, 1970) necessitates a consideration of the same problems. There have been numerous investigations although most are comparative and the influences of other factors are not always excluded. Since the death point is dependent on temperature and length of exposure, a duration of 12 hours for experiments is practical and ecologically relevant. In general, temperature ranges from between –2° and +3°C to between 27° and 30°C characterize the tolerance of most sublittoral species although there are some deviations. Sublittoral species from the English Channel such as *Sphondylothamnion multifidum*, *Rhodophyllis divaricata*, and *Cryptopleura ramosa* were partially or completely killed at 5°C although capable of surviving at 27°C, while other species such as *Callophyllis laciniata*, *Calliblepharis ciliata*, *Plocamium cartilagineum*, *Polyneura hilliae* and *Heterosiphonia plumosa* died within 12 hours in sea water at 27°C, although surviving at a temperature of –2°C. Littoral algae have a greater tolerance to high temperatures than do sublittoral entities. *Bangia atropurpurea* survives when dried out daily on rocks which reach a temperature of 40°C, although when immersed in seawater it is not capable of tolerating temperatures over 35°C. This difference could be a reflection of deoxygenation or of bacterial activity. In order to study shock effects, several algae were grown at 14°C for a period of three months and then subjected for three hours to 32°C before being returned to the previous conditions. Photosynthesis and respiration were monitored continuously. The littoral species suffered a slight reduction in photosynthesis but the sublittoral species showed considerable depressions.

With respect to low temperatures, many algae are capable of surviving in the high Arctic completely embedded in ice. Littoral algae are much less sensitive to cold than sublittoral species. Swedish red algae are of two groups: those species such as *Delesseria sanguinea*, *Phycodrys rubens* and *Laurencia pinnatifida* which succumbed at temperatures of –3° to –5°C, and those which are more cold resistant, with death temperatures of –10° to –20°C, such as *Chondrus crispus*, *Nemalion helminthoides* and *Bangia atropurpurea*. Of thirty sublittoral species in the English Channel, some survived at –2°C, providing that ice formation did not take place, while most died at 1° to 3°C. Attempts to adapt algae to cold by maintaining them at −1° to +1°C for several months were not successful although in *Delesseria sanguinea* and a few other species a drop in the lethal temperature of the order of 1 to 2°C was detected.

Information obtained from such experiments is difficult to interpret and relate to conditions in the field. The data are sufficient to indicate that the tolerance of red algae varies considerably, and that littoral species have greater tolerance than those of the sublittoral.

Light Red algae are essentially photosynthetic organisms even though heterotrophic metabolism has been suggested in certain cases. The occurrence of complex mixtures of pigments in red algae has led to the idea that these organisms are distributed so that their absorption spectrum complements the spectral distribution of the light at a particular position. High intensities cause destruction of pigments or of the alga itself. Development and induction of reproductive structures (see p. 56) are not only dependent on the quality and quantity of light but also on the relative duration of light and dark periods.

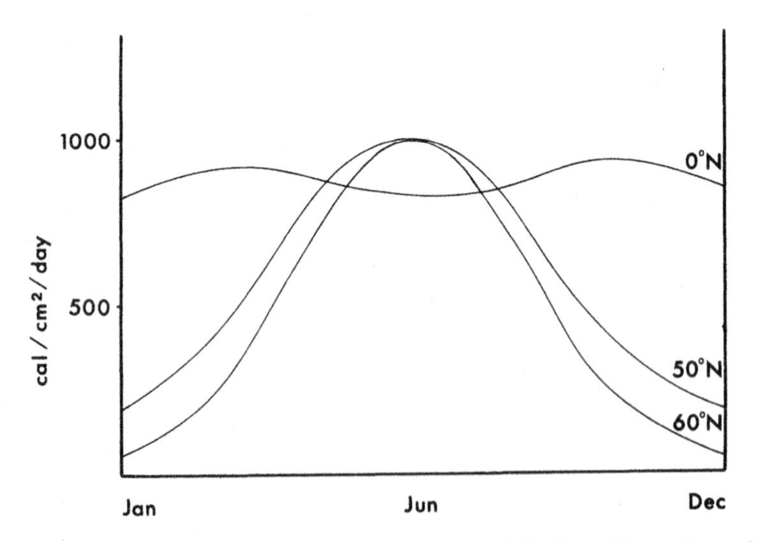

Fig. 8 Daily totals of solar radiation at 0°N, 50°N, and 60°N, at different times of the year, based upon a solar constant value of 1·94 g cal/cm²/year.

The British Isles lie between 50° and 60°N, with the Channel Islands and part of Shetland lying just outside these limits. Seasonal variations in light intensity and duration are shown in Fig. 8 and Table 2. The maximum daily totals of solar radiation in June are identical at 50° and 60°N although the minima in December are different, the daily total at 50°N being three times that at 60°N. The minima are about 20 per cent (50°N) or less than 10 per cent (60°N) of the summer maxima. Summer radiation maxima are higher than at the equator, due to longer daylength in high latitudes. The amount of effective radiation is reduced by the extent to which cloud cover occurs (see Table 1).

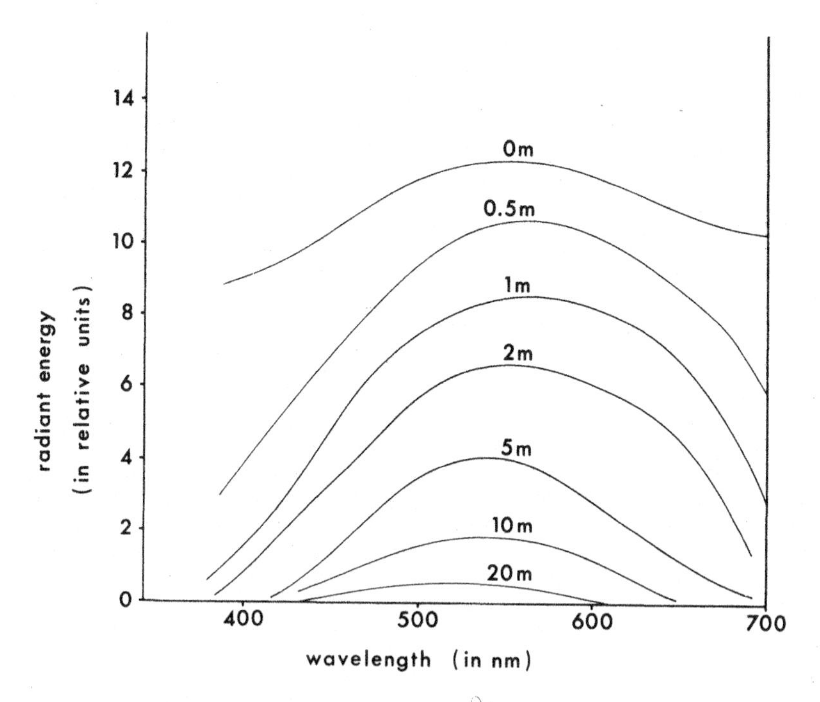

Fig. 9 Spectral distribution of radiant energy (in relative units), at noon, at various depths, measured on the west coast of Sweden (after Levring).

The conditions discussed obtain at the water surface, that is they apply to the illumination received by emergent algae. When light penetrates water the irradiance is reduced by absorption and scattering by the water itself, substances in solution and suspended silt particles. Benthic algae occur in shallow water, usually near to a coast, so that there is always some suspended material present compared with oceanic conditions and there is usually no light below 40 m. The spectral distribution of radiant energy at various depths under conditions usual on north temperate coasts is shown in Fig. 9. Red light is totally extinct at 10 m. Where the concentration of silt is high, the shorter wavelengths are absorbed to a greater extent, blue light becoming extinct at a depth of 2 m (Fig. 10) with virtually no light below 10 m.

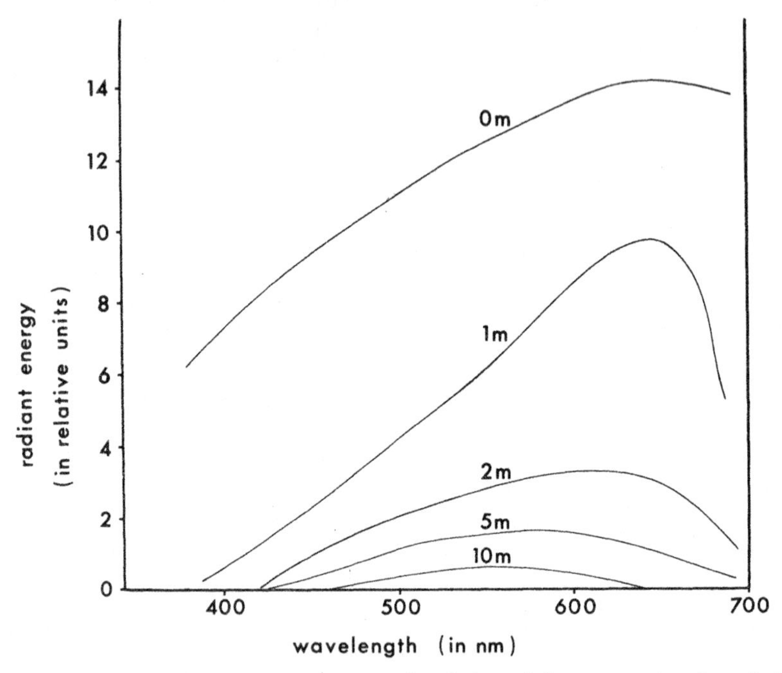

Fig. 10 Spectral distibution of radiant energy (in relative units), at noon, at various depths in water containing high levels of silt, measured on the west coast of Sweden (after Levring).

TABLE 2

Daylength variation in the British Isles

50°N	Summer maximum	16 h 20 min
	Winter minimum	8 h 4 min
60°N	Summer maximum	18 h 49 min
	Winter minimum	5 h 54 min

Based on official times for sunrise and sunset

For an alga, there are three critical levels of light intensity. The lowest is the intensity below which the alga ceases to grow and begins to degenerate; this represents the so-called compensation point. The intermediate critical level is the saturation point, the intensity above which no further increase in photosynthesis can be produced by increasing light intensity. The uppermost level is the intensity at which damage or death ensues. In some, the saturation point is identical with the level at which damage ensues, while in other cases it is not. None is a fixed point, in that adaptation can occur. The critical levels should be regarded as broad regions on the intensity scale.

The most recent discussion of the effects of light in marine ecology has been given by Hellebust (1970). It has been shown that littoral algae from the English Channel, such as *Plumaria elegans* and *Griffithsia flosculosa,* suffered no damage when exposed to direct sunlight for two hours, while sublittoral algae, such as *Polyneura hilliae, Antithamnion cruciatum,* and *Corynospora pedicellata* were killed. If intensities are high but sublethal, colour changes may be induced through long exposure, as in littoral algae by late summer. Red algae undergo changes in colour from the destruction of phycoerythrin. Chloroplast shape also changes with high light intensities. Cell death due to intense light may or may not be indicated by bleaching of the chloroplasts. Chloroplasts are still pigmented in *Griffithsia flosculosa* when vital staining shows the cells to be dead, while in *Ceramium ciliatum* the chloroplasts are bleached even though the cells are alive. Little is known of light requirements for growth in marine red algae. In most cases, the compensation point is about 50 lux whereas saturation values show considerable diversity, ranging from *Brongniartella byssoides* (220 lux) to *Plumaria elegans* (22 000 lux). Many filamentous Ceramiaceae have saturation values of the order of 500 lux beyond which damage ensues rapidly. Light requirements for growth and development may be modified by temperature and they may differ according to the age of the material.

Photosynthesis in algae is restricted to organisms which contain chlorophyll *a*. Pigments other than chlorophyll sensitize photosynthesis in brown, red and blue-green algae. Red algae have the highest photosynthetic activity in the middle of the visible spectrum. The theory of chromatic adaptation postulated that green, brown and red algae are distributed at different levels as a consequence of adaptation to the light spectra which they receive. An opposing view held that the zonation was due to the quantity of light; the demonstration that both wavelength and intensity of light are important reconciled these two views. It is difficult to apply the theory of chromatic adaptation to marine zonation. One cannot generalize about *all* red algae occurring in the sublittoral and *all* green algae occurring in the high littoral because there are too many obvious exceptions. Conditions in the sublittoral are those to which red algae are suited. Red light has been reduced by absorption and blue light diminished by dispersion; the light which remains is distributed in the middle of the spectrum, the region absorbed by phycoerythrin.

Several investigators have questioned whether benthic algae might be dependent on processes other than photosynthesis for organic materials. Such suggestions have been made for arctic seas and the vicinity of sewage outfalls. Sublittoral vegetation reaches its maximum development at a lower depth in the arctic, where there is no sun at all for several months, than in the boreal. Even though the arctic sublittoral at 100 m is lightless, numbers of algae occur including several red algae. The critical investigation to establish that heterotrophic nutrition is occurring where light intensity and temperature are low is still needed. Some algae occur in quantity in the vicinity of sewage outfalls. This may represent ability to survive in adversity, although the preponderance of coralline algae has suggested that they are capable of heterotrophic existence. Certain requirements must be met before uptake of organic materials can be accepted as a source of nutrient supply (North *et al.*, 1972). The alga must accumulate organic materials from very dilute solutions. The alga must incorporate the absorbed material into normal metabolic pathways, and it must be significant relative to material produced from CO_2 through photosynthesis. Exposure of red algae to solutions of glycine shows that accumulation and assimilation occur in certain cases (*Corallina, Gelidium,* etc.) although many algae (*Callophyllis, Cryptopleura*) have little or no uptake ability. The species with uptake and assimilation abilities are those which occur in waters receiving domestic sewage while those with little or no uptake ability are conspicuously absent.

Salinity Normal salinity of seawater around the British Isles is 34 parts per thousand. Changes occur in estuaries and the littoral at low tide due to dilution by rain or concentration by evaporation. Such fluctuations are significant in zonation and more local aspects of distribution. Species of the littoral have greater tolerance to increased or reduced salinities than those of the sublittoral. Most sublittoral red algae cannot withstand salinities below 15 parts per thousand and few littoral species can withstand salinities below 5 parts per thousand. Variations in salinity in estuaries are complex, each estuary showing different changes with season, wind and tidal conditions. There is rapid diminution in the number of red algae upstream from the mouths of estuaries and this has been regarded as a consequence of diminished salinity. In addition, suitable substrate is usually minimal and the level of suspended solids is high with deposition of mud and high turbidity. Almost all estuaries in the British Isles are associated with human habitation and discharges of domestic and industrial waste are considerable. In general, salinity is not a factor influencing large-scale distribution of marine Rhodophyta, and no attempt will be made to discuss this subject in any further detail here. A summary has been given by Gessner & Schramm (1971).

Substrate Marine Rhodophyta occur on various substrates, organic and inorganic. The present discussion will be restricted to inorganic substrates; organic relationships are treated later (p. 48).

An inorganic substrate serves as an attachment surface from which no materials are obtained directly. The nature of this surface in terms of hardness, texture, topography, etc. appears to be very critical. Most species possess some degree of preference regarding substrate. The most sterile shores are those where large shingle or small cobbles occur in the littoral. This sterility applies to all organisms, not merely to red algae and is due to the abrasion by the substrate in motion and the speed of drainage during emergence. Few marine Rhodophyta occur on sand or mud in the littoral because such substrates are unstable. The association of red algae with sand-covered rocks reflects the capacity to survive burial by sand, development having commenced on underlying rock. Considerable populations can develop on more unstable substrates in the sublittoral providing that these are below the effect of wave action.

Despite the obvious importance of substrate, experimental work evaluating this is almost non-existent. One study (Linskens, 1966) showed that spores of some Rhodophyta develop on smooth surfaces while others prefer a rough surface.

Substrate can vary over short distances and has effects on local aspects of distribution. Large-scale distribution may be affected if unfavourable geological conditions prevail for some distance. In England, Lewis (1964) has indicated various organisms absent from shores to the south of Flamborough Head and east of the Isle of Wight. For red algae, the area of minimal representation lies between Flamborough Head and Herne Bay, where the substrata are post-Cretaceous overlain by boulder clay. Turbidity is high while the substrate is extremely soft. Sublittoral populations develop on moraines of harder material. Similar unsuitable substrates exist in the Irish Sea between Colwyn Bay and Dumfries and between Wicklow and Newry.

Biological interactions

Interactions of all kinds occur between spores, cells and tissues of red algae as well as between red algae and other organisms. The amount of information available is slight and even non-existent in some cases.

Interactions between spores and tissues A simple type of interaction occurs between spores. If spores are distributed sparsely, development may be slow or cease at an early stage. Development is more rapid when spores are clustered, and the basal discs formed from several spores may coalesce. The coalesced discs function as a single organism and there is an acceleration of development. If spore aggregation is necessary for the formation of erect fronds, low spore settlement will result only in the production of basal discs and the occurrence of the species is not likely to be reported. In a species where gametangial plants are dioecious, such as *Gracilaria verrucosa,* two spores from each tetrasporangium produce spermatangial plants and two carpogonial so that coalescence may involve spores of different genetical make-up.

Coalescence between tissues of the same or different species can occur where incurved laterals are formed (see p. 16). The degree of incurvature varies considerably; those of *Bonnemaisonia hamifera* rarely curve more than half a turn, whereas in *Cystoclonium purpureum* and *Calliblepharis jubata* the tendrils are long and may make 4 to 5 coils around an object. In *Cystoclonium,* the tendrils attach strongly when an axis of the same species is enrolled whereas in other species the degree of enrollment is not stimulated by an axis of the same species but strongly by other materials. The reflexed laterals of *Asparagopsis armata* do not encircle objects with which they come in contact although the tip of the spine forms a terminal attachment structure. Many species lacking tendril-like laterals undergo fusions with axes of the same or another species. Sometimes rhizoids develop but reciprocal development from both participants occurs only when axes of the same species are involved. In many cases, contact may involve secondary pit connection formation and this can occur between axes of different species.

One type of tissue fusion occurs when tetraspores germinate *in situ* and the product fuses with the parent plant, making it difficult to distinguish between the two. The germling may form gametangia and some records of different reproductive structures on the same plant may result from such *in situ* development. When tetraspores germinate *in situ,* one to four spores develop and each usually gives a distinct product. In *Champia, Cystoclonium, Gracilaria* and *Lomentaria,* however, the four spores in a tetrasporangium may form a single compound individual.

Epiphytism and parasitism A significant form of interaction involves a relationship with a living substrate. Such algae have been termed 'epiphytic', 'epizooic', or 'parasitic'. Unfortunately, experimental investigation has been undertaken only rarely. It has been customary to regard cases where the relationship is superficial as epiphytic or epizooic. Even with such a relationship, substances may move from the substrate species to the attached species or in the opposite direction (Linskens, 1963; Harlin, 1973). The balance between movement in the two directions varies from example to example. Some species characteristically bear abundant epiphytes while others are always devoid of them. Differences in epiphytic development may reflect the surface tension and/or chemical composition of the host.

Some species occur as epiphytes or endophytes on a single substrate species, but such reports are often based on limited collections. Some endophytes will grow *in vitro* away from the substrate species. *Audouinella asparagopsis* and *A. infestans,* reported from red algae other than *Heterosiphonia plumosa,* germinated on that species but failed to penetrate, while spores of *Audouinella endophytica* obtained from *Heterosiphonia* produced endozoic filamentous growths within a hydroid organism (Boney, 1972). Such studies indicate the need for investigation of host specificity and until more information is available its use as a taxonomic criterion is open to question. There appears to be a complete gradient of relationships from facultative epiphytism through obligate epiphytism to interactions of a more specialized nature. One species of an intermediate category is *Polysiphonia lanosa,* which occurs on *Ascophyllum* and other Fucàceae. The initial primary rhizoid is bulbous, entering the substrate species through the cuticle, sites of damage or the apices. On the basis of histochemical staining, the bulbous rhizoid withdraws material from the substrate species, although there is no evidence of haustorial contact. Reports of *Polysiphonia lanosa* on rock are partly based on specimens growing on the basal crust of *Ascophyllum* which resembles the underlying rock. Plants of *Polysiphonia lanosa* also form normal rhizoids at a distance from the primary bulbous rhizoid and attach directly to a rock substrate.

Florideophyceae which penetrate deeply into the substrate species and which exhibit reductions of morphology or pigmentation provide better indications of 'parasitism'. *Ceratocolax hartzii* occurs on *Phyllophora truncata* as small pinkish tufts. The *Ceratocolax* spreads below the cuticle or extensively through the medulla of *P. truncata* causing tissue breakdown. *Gonimophyllum buffhamii* occurs on *Cryptopleura ramosa* and possesses pale pink fronds (up to 10 mm long) which arise in tufts from the substrate species with some hypertrophy. Some photosynthetic pigmentation is produced both in *Ceratocolax* and *Gonimophyllum* although the quantity is less than normal for a red alga. Photosynthetic pigments are lacking in *Harveyella mirabilis* and *Holmsella pachyderma* which occur on *Rhodomela confervoides* and *Gracilaria verrucosa*, respectively. In both, filaments penetrate the substrate species and secondary pit connection formation occurs between the substrate species and the 'parasite'. The greatest morphological reduction occurs in *Choreonema*. This is colourless and largely endophytic, only the reproductive stages occurring above the surface of the substrate species.

The parasitic habit in these Florideophyceae is generally assumed although the evidence is rather weak. Many of the 'parasites' are associated with species of close affinity, usually of the same family. Such affinity suggests that examples might provide material for studies of DNA and enzyme relations of substrate and associated species, whether or not 'parasite' is the correct term to be applied. The close affinity may mean that the organism described initially as a 'parasite' is only a gall, e.g. *Choreocolax cystoclonii* on *Cystoclonium*.

Relations between marine Rhodophyta and fungi By comparison with terrestrial fungi which have been studied since Theophrastus and Aristotle because of economic and culinary properties, saltwater fungi are virtually unknown. Most associations between marine Rhodophyta and fungi in the British Isles involve Phycomycetes or Ascomycetes.

Marine Phycomycetes are single-celled or non-cellular organisms. A taxonomic discussion has been given by Sparrow (1960), while ecological aspects have been discussed by Johnson and Sparrow (1961). The single-celled fungi can be detected when mature by the presence of sporangia. Earlier, the fungus can be detected by the difference in refractive index between normal and infected cells. Sites of infection are often apical cells. When the principal apical cell is infected, this can cause regeneration of new axes. *Olpidiopsis feldmannii* infections of the *Falkenbergia* phase of *Asparagopsis armata* (Fig. 11) can change the morphology of the host species. Sporangia are also frequently infected. *Petersenia lobata* is one species reported many times in sporangia of the Ceramiaceae. Non-septate filamentous Phycomycetes are less common than the single-celled types. One species of *Pythium* (Fig. 11) is pathogenic to *Porphyra* spp. in Britain although not as frequently as the species responsible for destruction of cultivated pop-

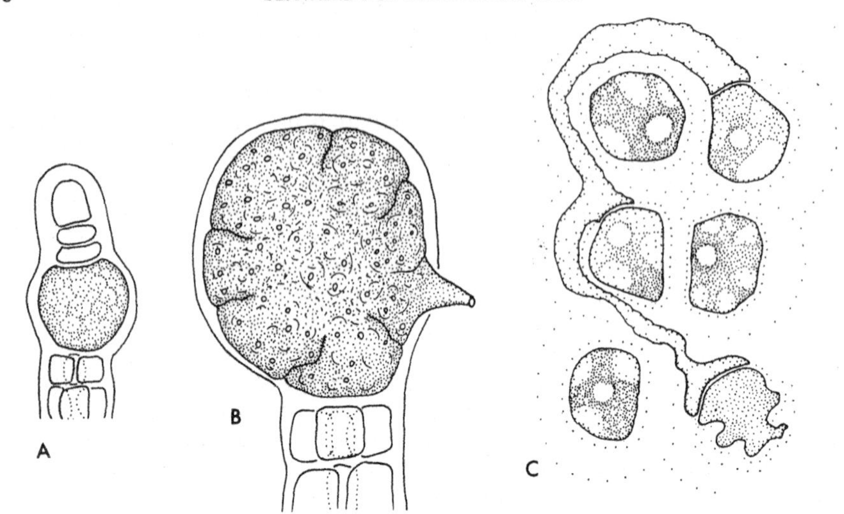

Fig. 11 Fungi associated with red algae
A. *Olpidiopsis feldmannii* in the *Falkenbergia* phase of *Asparagopsis armata,* at an early stage of infection ×1000; B. Later stage, with sporangial formation ×1000; C. *Porphyra* sp., infected with *Pythium* ×1000.

ulations in Japan. Mucorales occur in Britain on cast-up, decaying algae but such infections represent only terrestrial species with high salt tolerance rather than truly marine representatives.

It is difficult to determine the relationship between a red alga and a fungus. The occurrence of a fungus in a dead or dying specimen does not necessarily indicate the cause of death. The fungus may be truly pathogenic as in the case of the *Pythium* infection of *Porphyra,* or it may behave as a parasite or saprophyte at different stages of development. *Petersenia lobata* is an active parasite during initial infection and early development, although destruction of the host is caused by secondary bacterial infection. Other Phycomycetes appear to have little or no effect on red algal host species. *Thraustochytrium* can be obtained from any red alga or from any substrate, living or non-living, organic or inorganic, removed from the sea.

Marine Ascomycetes include the yeasts, which occur as single cells or as chains of cells, and those fungi with multicellular, septate mycelia which form pseudoparenchymatous masses. Although virtually impossible to detect by direct observation, yeasts can be obtained by culturing the surface washings of red algal thalli. Multicellular fungi with septate mycelia which occur on marine red algae are all Pyrenomycetes. Identification requires the presence of fruiting structures called ascocarps, which are superficial dark-brown or black structures about 1 mm in diameter. Pyrenomycetes occur on various marine Rhodophyta in the British Isles although they are not common. Their occurrence has often been overlooked due to the similarity between the fungal ascocarp and the cystocarp of the alga. The status of Pyrenomycete fungi is difficult to ascertain. Many have been described as parasitic but the fungi are very weak pathogens or facultative parasites capable of living both on live and dead algal thalli. In Phaeophyta, Pyrenomycete fungi may produce galls although there is no evidence for this in Rhodophyta.

It has been considered (Johnson & Sparrow, 1961) that fungi may kill some marine algae although most destroy only weak or moribund specimens.

Relations between marine Rhodophyta and bacteria Bacteria occur in marine Rhodophyta, both superficially and internally. The present discussion will be restricted to aspects relevant to a Flora.

Bacteria most obvious during examination of Rhodophyta are whisker-like filaments, 2–5 μm in diameter, referred to *Leucothrix*. The filaments are about 100 μm in length and cross-walls are just visible in living material. The filaments often wave from side to side although they never appear to glide like blue-green algae. The filamentous bacterium found in rotting masses of seaweed is not *Leucothrix* but *Beggiatoa*, which is more like a colourless *Oscillatoria* and has filaments that glide. *Beggiatoa* is present only when decomposition has led to H_2S production.

Galls of bacterial origin occur in some Rhodophyta. In *Chondrus crispus*, for example, a bacterium has been isolated and grown in culture, and has produced galls on inoculation. The bacterium involved is similar to, if not identical with, *Agrobacterium tumefaciens*, the cause of 'crown-gall' disease in higher plants.

Marine Rhodophyta derive most, if not all, of their B-vitamin supplies from associated bacteria. B-vitamins (particularly B_{12}) must be added to axenic cultures if normal growth rates and morphologies are to be maintained (Provasoli, 1964).

Relations between Rhodophyta and viruses The suggestion that viruses might be the cause of otherwise inexplicable disappearances of algae, particularly in freshwater, was made by Krauss (1961). There are numerous examples of virus pathogenicity in blue-green algae (Fogg *et al.*, 1973). In Rhodophyta, viruses have been identified by means of electron microscopy (Lee, 1971) and it is obvious that they are widespread. However, nothing is known of their effects on growth or reproduction in red algae and experimental transfer from an infected to an uninfected specimen has not been achieved.

Zonation

By far the most significant forms of interaction in the marine environment are those which give rise to the phenomena of zonation. The present discussion is intended only as an explanation for the terminology used in our statements of position on the shore where species occur. It is unfortunate that none of the schemes of terminology proposed in recent years (Stephenson & Stephenson, 1949, 1972; Lewis, 1964) is directly applicable here. As will be seen, however, the scheme used represents little more than a simplification of terminology from that proposed by Lewis (1964).

The stratification of organisms at the edge of the sea has been known for many years. The earliest workers, at the beginning of the nineteenth century, developed schemes based on predominant organisms. Once accurate tidal data became available the importance of tidal oscillation began to be appreciated. Zonation is influenced by many factors in addition to tidal movement, however, and the location of the zones with respect to tidal position depends upon the degree of exposure to wave action. Three major biological zones occur on rocky shores around the world and, in the British Isles, these are characterized by:

a. Littorinids and blackening organisms such as Cyanophyta and species of the lichen genus *Verrucaria.*

b. Barnacles, limpets and fucoid algae.

c. Laminarians.

The simplest procedure for defining position on the shore would be to cite the biological zone in which a species occurs. Unfortunately, this becomes too complicated because some of the accepted 'marker species' for zones have peculiarities of occurrence and distribution. In Britain, tidal data are well-founded and widely available; one could cite position on the shore in relation to tide height as has been done in other places (Smith, 1944). On a small-scale basis in areas where tidal amplitude is minimal this procedure is satisfactory. But because of the modifying effects of wave action on the relationships between zones and tidal position (Fig. 12), such statements of position in relation to tide height are of little value for the British Isles, where tidal amplitudes are as great as anywhere in the world and where variation in exposure to wave action is considerable. Even general statements based on tidal data, involving such terms as 'midtidal' or 'subtidal' are meaningless for large scale comparison and should be avoided. We have therefore divided the marine algal populations into 'littoral' and 'sublittoral' components. The former comprises those algae which are adapted to or require alternating exposure to air and wetting by immersion or by splash and spray. The sublittoral components are those which are totally or almost totally immersed, with only the uppermost part ever uncovered for short periods.

The boundary between the littoral and the sublittoral has been placed at the upper limit of Laminarians. The depth to which algae can penetrate is determined by the extent of light penetration. The absolute limit, anywhere in the world, appears to be of the order of 100 m although in most places it is less than this. In coastal waters around the British Isles, light penetration is reduced by high turbidity and one rarely finds marine Rhodophyta at a depth greater than 30 m.

The littoral zone (as used here) extends from the upper limit of Laminarians to the highest point of occurrence of *Littorina* and *Verrucaria.* For more precise definition, we are accepting three subdivisions of the littoral zone. The first comprises such organisms as littorinids and *Verrucaria* which are the uppermost marine organisms; this subdivision we refer to as the 'upper littoral'. The term 'supralittoral' of Stephenson & Stephenson (1949) corresponds approximately to

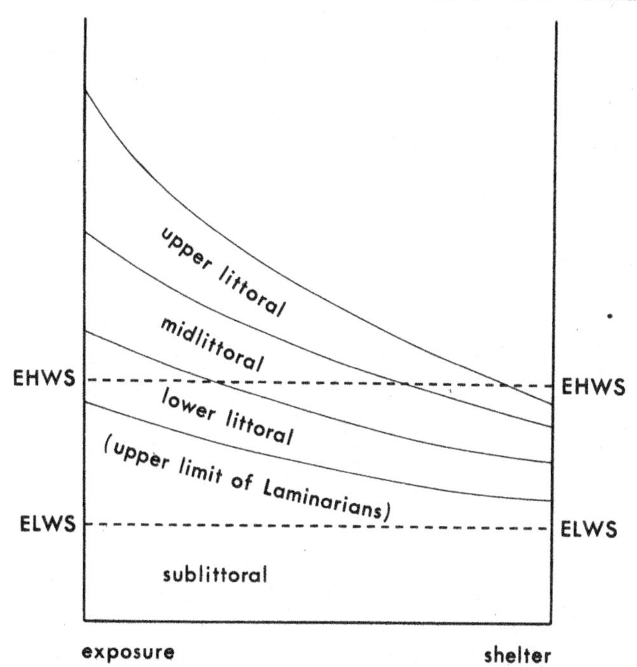

Fig. 12 The terminology for zonation used in the text. The position of the highest spring tide is indicated by EHWS; the position of the lowest spring tide is indicated by ELWS. Based on Lewis (1964).

our 'upper littoral' although the former term should be avoided because it is based on a misconception: organisms of the upper littoral lie within tidal limits on sheltered shores and are above tidal limits only on shores exposed to wave action (Fig. 12). The midlittoral and lower littoral contain those organisms such as barnacles, limpets and fucoid algae which Lewis (1964) terms the eulittoral. Precise definition of the boundary between the mid and lower littoral is not possible as its position varies from shore to shore and our distinction is largely one of convenience.

Environmental control of growth and reproduction

The variation in thallus form and reproductive behaviour under different environmental conditions creates a major source of difficulty in algal taxonomy. Much early taxonomic work was undertaken by botanists who regarded any morphological variant as a distinct taxon. The causes of morphological variation are now regarded more critically although even today only the mature form is considered in many cases. Comparisons must be based not only on mature form but on the pattern of development by which this comes into being. Thus, a critical appreciation of morphogenesis, morphogenetic processes and their environmental control is indispensable.

In the Bangiophyceae, it has been shown that certain effects result from particular conditions but experimental investigation is difficult and the results have little relevance to taxonomic problems. In the Florideophyceae, many encrusting thalli grow slowly and it is difficult to monitor growth in living material. The erect fronds have proved more amenable to field investigation and experimental study.

It has been shown (Dixon, 1966) that external form can be assessed in terms of:

a. disposition of axes,
b. shape of axes,
c. longevity of the erect thallus.

These variables are not independent, but they provide a basis for comparison. The disposition of axes is determined by the position and number of the filaments of unlimited growth. In many Florideophyceae, the thallus may be simple or exhibit arrangements of branching of all degrees of complexity. In some Ceramiales, formation of filaments of unlimited growth is controlled so that disposition of axes is characteristic of the taxon. Growth of these axes may be coordinated so that elaborate patterns are produced.

The shape of axes is determined by the relative development of filaments of unlimited growth and surrounding filaments of limited growth. Growth of a filament is the result of increase in cell size and number. The length of an axis is determined by divisions of the apical cells of the filament or filaments of unlimited growth and enlargement of the derivative cells. The breadth and depth of an axis are related to divisions of the apical cells of filaments of limited growth and enlargement of their products. There are marked differences between filaments of limited and unlimited growth and the growth of the two types of filament is differently affected by external conditions. The enlargement of derivative cells in both types of filament is usually considerable. Moreover, enlargement is extremely variable and subject to environmental modification. Although there are differences in the rates of division of the two types of apical cell, it is the variation in enlargement of the products which has the greater effect on the shape of the thallus.

The effects of life span must be considered because of changes in the appearance of plants between the first and subsequent years of growth. Studies of perennation have been neglected. In many genera where the texture is comparatively soft the uppermost parts of a plant disintegrate at the end of each growing season. The size of the portion persisting from one season to the next, the number and disposition of new apices, and the extent of growth of new axes all affect the appearance of such thalli. In cartilaginous species, the original apices can function for several years. Increments in length are added to each axis every growing season so that there are differences in appearance in different age groups. Further development of filaments of limited growth in the second and succeeding years is usually negligible so that there is little change in breadth and depth of axes except in certain special cases, such as *Ahnfeltia*. Sometimes the apices are lost by the formation of reproductive structures or animal grazing and a new apical meristem must regenerate from the stump.

General principles of morphogenesis have been analysed in *Grateloupia filicina, Pterocladia capillacea* and species of *Ceramium* and *Gelidium.* It should be appreciated that such analyses are only the first step in investigation of the way by which thallus form comes into being. In order to understand the development of form it is necessary to have information on factors in the environment which influence gross aspects of growth and development. Attempts to analyse material from different situations and relate the variation to specific factors in the environment do not prove that a response is produced by a specific factor, although they do give indications for experimental investigation.

Recently, it has become clear that structure and reproduction are controlled by several factors, acting singly, together, or in sequence. For plants, the most obvious factor to consider is light. It is now 50 years since it was shown that flowering in many angiosperms was controlled by the sequence of light and dark periods. Results obtained with algae have been ambiguous in that three distinct phenomena are involved. The first is a 'photosynthetic effect' where different results are produced according to the quantity of radiation received. The second involves a circadian rhythm, or biological clock mechanism. In this, a cyclic process in the organism is kept in phase by the 24-hour cycle of light and dark. The third is true photoperiodism, defined by Terborgh & Thimann (1964) as characterized by (1) induction, or continuation of the effect following transference to non-inducing conditions after a set time, and (2) sensitivity to short breaks in either the light or dark periods. Most investigations involving algae in which it has been claimed that photoperiodism is involved have shown an effect under one set of conditions but not under another, but the critical experimentation to distinguish between a photosynthetic effect, a circadian rhythm and true photoperiodism has not been undertaken.

The two principal aspects of morphological variation and its environmental control concern (1) the conditions under which growth is initiated and ceases and (2) the conditions responsible for seasonal changes in the growth pattern in terms of cell division and cell enlargement.

Considering first the conditions under which growth is initiated, there are almost no field data. The difficulty with Rhodophyta is that it is impossible to say whether a meristem is active or inactive without destructive analysis. *Pterocladia capillacea* is one species in which it is possible to detect the active apical meristem, at least in British material. Each frond goes through an annual growth cycle and it is easy to detect new growth even with the naked eye, because it is pale in contrast to the purple coloration of old axes. In north Wales, observations of a clonal population each year between 1950 and 1964 showed that reactivation of the meristem could take place as early as February or as late as early June. Such observations indicate that a photoperiodic response is impossible and that a photosynthetic response is more likely. In general, initiation of development can be related to photosynthetic responses although the evidence is far from complete.

Such a conclusion is of interest in relation to problems associated with perennation of basal fragments. Conditions affecting cell division and enlargement are known for some Rhodophyta but there have been few attempts to consider differential effects on principal and lateral filaments such as would explain seasonal and environmental changes in form of marine red algae. Seasonal changes in breadth of red algal thalli relate to the relative duration of light and dark periods and to light intensity. It would appear that such changes are due to a photosynthetic effect, dependent on the total quantity of radiation received.

Considering control of reproduction, it has been appreciated for many years that reproductive structures are formed at particular times of the year in most species. There is some variation at the same locality from year to year and between different locations. Attempts to establish phenological tables suggest that inception and maturation of reproductive structures are responses to the interaction of several environmental factors. As with growth and development, light appears to be the most critical factor, particularly in terms of duration and intensity (Dixon & Richardson, 1970). The first significant information was obtained from Japanese attempts to control reproduction in commercial *Porphyra* cultivation beds. It is now known that in *Bangia* and *Porphyra* the production of spores is under photoperiodic control. Proof was obtained by light-break experiments. In other cases where reproductive behaviour varies with light duration, light breaks are not inhibitory and the responses are due to a photosynthetic effect. The apparent restriction of photoperiodic responses to two entities which occur highest in the intertidal is not surprising. Far-red light penetrates only 1 m in seawater while red light is extinct in 10 m. Elimination of the radiation essential for a phytochrome-mediated system such as photoperiodism raises doubts as to whether it could ever operate in a submerged alga.

In addition to the induction of reproductive bodies, the factors affecting maturation, release and germination of spores are also important. Spores can be released by manipulation but these will not germinate. Quantitative estimates of spore production have considered release only in gross aspects. Tetraspore release in *Nitophyllum punctatum* occurred rhythmically (Sagromsky, 1960) in response to a change between light/dark conditions. Analysis shows that it was not a circadian rhythm. This survey of the responses of red algae with respect to the environment indicates the significance of such observations.

Biological forms

Similar morphological types occur in most algal divisions and environmental modification of algal thalli is a general phenomenon. Can the form of the thallus be used to evaluate the environment, and the occurrence of different morphological types related to the habitat in which they occur? Such a concept was developed for angiosperms by Raunkiaer and proved to be of considerable value. With algae, early systems of classification (Oltmanns, 1905; Funk, 1927) were based on static morphology. The scheme proposed by Gislen (1930) was applicable to all marine organisms. Three categories ('Crustida', 'Corallida' and 'Silvida') were established with several further subdivisions of each. These categories referred to the crustose organisms, the frondose organisms with an incrustation of calcium carbonate, and the frondose organisms without such incrustation. Although used for descriptive purposes none of these systems was of much consequence as a means for the evaluation of the environment. A different approach, analogous to the Raunkiaer scheme, was developed by Feldmann (1937, 1951, 1966), based on longevity of thalli, the means by which persistence was effected and the extent to which perennation occurred. The categories adopted by Feldmann are as follows:

Ephemerophyceae: algae which occur throughout the year, with several generations but no resting states.

Eclipsiophyceae: algae present in a macroscopic state during part of the year but persisting as a microscopic stage for the remainder.

Hypnophyceae: algae present in a macroscopic state for part of the year and persisting in a resting stage for the remainder.

Phanerophyceae: the entire frond perennial, in the erect condition.

Chamaephyceae: the entire frond perennial, in the crustose condition.

Hemiphanerophyceae: only part of the frond perennial, in the erect condition.

Hemicryptophyceae: only part of the frond perennial, in the form of basal prostrate parts.

The Feldmann system was more 'biological' than the preceding schemes and attempts have been made to relate the spectrum of thallus types to particular habitats (Ernst, 1958; Katada, 1963). The Feldmann scheme suffers from the fact that in the algae there are no uniform morphological units. All angiosperms are of similar morphology, with fundamental units of root, shoot and leaf. The form of each is related to growth, longevity and perennation. By comparison, the morphological diversity in the algae is enormous and the Feldmann system is too imprecise for the purpose for which it was intended. No attempt is made in the present Flora to assign entities to the categories of the Feldmann scheme. The extent to which a system of this sort will ever be applicable depends largely upon progress made with the morphogenetic understanding of thalli, not merely of the Rhodophyta, but of all divisions. Progress is slow but it would be unfortunate if the concept of 'life' forms in the algae should be abandoned without further consideration simply because of ignorance of the ways by which algal thalli grow and develop.

Distribution in the British Isles

Summaries of habitat characteristics and distribution are provided for each species in the Flora. There is considerable variation in the completeness of the information on which these summaries are based. Some species are ill-defined, or may be recognisable only when a particular reproductive structure is present, while other species may be present in an area but only in an 'immature' state, such as a basal disc or prostrate axis. Such 'immature' states may not be identifiable even when detectable and they are difficult to find when covered with sand or other debris. Knowledge of different areas in the British Isles varies enormously. Some areas are popular because of floristic luxuriance, and others because phycologists prefer to return to areas which they know rather than to investigate areas about which they are ignorant. Habits are also critical in relation to reported seasonal occurrence and behaviour of species because data are often more indicative of the seasonal behaviour of collectors than of the algae. Studies throughout the year and in areas not normally regarded as worthy of study will often increase considerably the number of species recorded, as in the case of Kent (Price & Tittley, 1972). Counties under-represented will be obvious from the published listings (Dixon, *et al.*, 1966; Price, 1967; Price & Tittley, 1970). It is possible to draw certain general conclusions regarding the distribution of marine Rhodophyta in the British Isles.

1. Some marine Rhodophyta occur all round the British Isles. These include such species as *Bangia atropurpurea, Chondrus crispus, Corallina officinalis, Delesseria sanguinea, Laurencia pinnatifida, Lomentaria articulata, Plocamium cartilagineum, Polysiphonia lanosa* and *Palmaria palmata*. It is not possible to give a simple summary of the distribution of these outside the British Isles other than to say that some occur to the north, some to the south, and some occur both to north and south, to a variable extent.

2. Some marine Rhodophyta are essentially northern in distribution reaching their southern limit in the British Isles. The number of species in this category is not large, but includes *Odonthalia dentata* (to northern Ireland, Isle of Man, Yorkshire) *Ptilota plumosa* to southern, western and northern Ireland, Anglesey and the Isle of Man, Yorkshire), *Rhodomela lycopodioides* (to northern Ireland, Galloway, Northumberland).

3. Numerous marine Rhodophyta are essentially southern in distribution, reaching their northern limits in the British Isles. These include such species as *Crouania attenuata* (southern Ireland, southwest England), *Laurencia obtusa, Pterocladia capillacea* (southern and western Ireland, southern and western England and Wales, Isle of Man). Species of this third group show a marked westerly bias, being restricted to the western shores of the British Isles but absent from the North Sea and, in a few cases, also from the east coast of Ireland. A number of species which would otherwise be placed in this third category show an equivalent westerly bias, but in addition, have a slight extension of range from northwest Scotland to Faeroes or southern Norway, but are absent from the North Sea.

There are many possible sub-groupings but these must be regarded with some doubt because of the incomplete data on which they are based. The most important conclusion is that there are no clearly marked boundaries, such as occur on the Atlantic and Pacific coasts of North America, because of overlaps in range between species. Without experimental investigation of the causes of distribution limits, further speculation is not warranted, other than to suggest that the causes will probably be found in air and water temperature extremes and the interaction of these parameters together with those features of the physical environment, such as aspect and slope, likely to have a modifying effect on temperature conditions.

SYSTEMATICS

Historical introduction

In an historical introduction to the systematics of the Rhodophyta, it is convenient to begin with the *'Species Plantarum'* of Linnaeus (1753). The fundamental criteria used in that work related to the distribution and arrangement of phanerogamic reproductive structures. As a result, the phanerogams were distributed into 23 classes while the algae, together with the fungi, bryophytes and pteridophytes, were relegated to a 24th class, the Cryptogamia, so named because phanerogamic reproductive structures were not displayed. Species of red algae treated by Linnaeus were referred to the three genera *Conferva, Ulva* and *Fucus*. The first received the filamentous species and those with particularly slender, elongate thalli; species with flat membranaceous thalli were referred to *Ulva* and those with fleshy or cartilaginous thalli were placed in *Fucus*. The Linnaean circumscription of these three genera survived for almost half a century although the number of described species increased very considerably during this time. The artificiality of the Linnaean genera was appreciated increasingly so that between 1790 and 1830 considerable numbers of new genera were described. Although Stackhouse (1795–1801, 1809, 1816) was probably the first to suggest breaking down the three Linnaean genera, his work was largely ignored for various reasons whereas the slightly later proposals of Lamouroux (1813) and the elder Agardh (1817, 1820–8) were generally accepted, forming the basis for the currently-accepted genera of all the major groups of algae.

Lamouroux (1813) was the first to segregate on the basis of colour certain algae now placed in the Rhodophyta from others of similar external morphology. For these he established the category Floridées from which the name of the currently-accepted major class of red algae – the Florideophyceae – is derived. Despite this auspicious beginning, the remainder of Lamouroux's system was not based on pigmentation and neither he nor the elder Agardh distinguished clearly between red, green and brown algae. It was Harvey (1836) who divided the algae into four major divisions on the basis of pigmentation, the first use of a biochemical criterion in plant systematics. Harvey's four divisions comprised the Rhodospermae (red algae), Melanospermae (brown algae), Chlorospermae (green algae) and Diatomaceae. The use of colour as a systematic criterion has been extended throughout the algae so that pigmentation is currently of fundamental importance in algal systematics and several major categories in addition to those proposed by Harvey are accepted on this basis at the present time. In general, Harvey's assignment of genera was reasonably accurate by modern standards although *Porphyra* and *Bangia* were placed in the green algae rather than the red because of the greenish colour of thalli in these two genera. The relationship between these two genera (and their relatives) and the other red algae was not appreciated for many years (Berthold, 1882). By comparison, the coralline algae were recognized as red algae at a relatively early date. Although most pre-Linnaean workers had regarded the coralline algae as plants, Linnaeus (1758) was strongly influenced by the work of Ellis (1755) and followed the latter's views that they should be considered as animals. As late as 1816, Lamouroux still regarded the coralline algae and other calcified algae as animal corals. Those coralline algae with erect jointed thalli were treated as plants by Schweigger (1819) and Gray (1821) but the encrusting genera now assigned to the Corallinaceae were not accepted as plants until the somewhat later work of Philippi (1837).

The systematics of the Rhodophyta

The fundamental differences between *Bangia, Porphyra* and their allies on the one hand and the remaining red algae on the other were appreciated by the beginning of the present century, the two groups being considered as subclasses of the class Rhodophyceae, as the Bangioideae and Florideae respectively. Today, major groups of algae are usually regarded as divisions (=phyla) rather than simply as classes, so that the class Rhodophyceae has come to be regarded as a division, Rhodophyta, and the two major subdivisions as classes rather than as subclasses.

Florideophyceae The crucial step in the establishment of present-day systematics came with detailed developmental studies of the carposporophyte. Although the 'cystocarp' of the Florideophyceae had been defined by the younger Agardh as early as 1844 and used by him in diagnoses of new taxa and expanded descriptions of long-known entities, he rarely used little more than the external form of the mature structure. Following the correct interpretation of gamete fusion in Florideophyceae by Bornet and Thuret (1867), various workers examined post-fertilization developments in a range of genera. The resulting confusion was clarified by Schmitz (1883) who showed that in a few Florideophyceae the carposporophyte develops directly from the carpogonium whereas in the majority the carpogonium forms cells or filaments which fuse with another cell and that the carposporophyte develops from the latter. Schmitz termed the structure which participated in the second fusion the 'auxiliary cell'. Although his interpretation that this represented a form of double fertilization was disproved within a few years, the ontogenetic patterns which he demonstrated have served as the basis for red algal systematics.

Schmitz (1889; in Schmitz & Hauptfleisch, 1896) recognized four orders; Gigartinales, Rhodymeniales, Cryptonemiales, Nemalionales. The last-named should be more correctly spelled Nemaliales and this orthography will be used consistently in the remainder of this discussion. Various revisions have been suggested during the past 70 years. The separation by Oltmanns (1904), as the order Ceramiales, of those members of the Rhodymeniales of Schmitz in which the auxiliary cell is formed *after* fertilization has been shown to be well justified.

The separation of the Gelidiales from the Nemaliales has been widely accepted although, as will be shown later, there are serious objections to this proposal.

The monographic treatment by Kylin (1956) accepts six orders, Nemaliales, Gelidiales, Cryptonemiales, Gigartinales, Rhodymeniales, and Ceramiales. Of these, the Ceramiales is probably the most clearly circumscribed. The uniformity of thallus and carposporophyte development in the four families is remarkable. Opinion on the status of the Rhodymeniales is somewhat divided (Papenfuss, 1966; Drew, 1954). Recent studies of the genus *Rhodymenia* have shown that *R. palmata* differs so much from other species of the genus that it is worthy of separation into a distinct genus *(Palmaria),* in a separate family (Palmariaceae) and order (Palmariales). The indefinite nature of the distinction between the Cryptonemiales and certain families of the Gigartinales has been appreciated for many years. When Kylin first separated the Gelidiaceae from the Nemaliales, the remaining families were apparently very homogeneous. Subsequent investigation has disclosed that each family forms a discrete group of genera, but that there is little to hold them together as an order. Proposals elevating various families such as the Acrochaetiaceae, Bonnemaisoniaceae and Chaetangiaceae to ordinal status have been summarized by Dixon (1961, 1973) and Papenfuss (1966). Although there is much evidence for the heterogeneity of the families grouped in the Nemaliales, there is no adequate basis at the present time either for the separation of any of these as independent orders or for the retention of the Gelidiaceae as a distinct order.

In the present treatment of the Florideophyceae, only six orders will be accepted, Nemaliales, Gigartinales, Cryptonemiales, Palmariales, Rhodymeniales and Ceramiales, all of which are represented in the British marine flora.

Bangiophyceae Although certain parts of the generally accepted systematic arrangement of the Florideophyceae are still a matter for controversy, the system is founded on a considerable body of data, much of which has been confirmed and repeated many times. Unsolved problems abound in the Bangiophyceae however, not only with respect to the most elementary details of morphological structure and thallus development but also to such fundamental aspects of reproduction and life history as the nature of a spore and the existence of sexual processes. Consequently, the arrangement of this class must be regarded as very provisional.

The initial incorrect placement of *Porphyra* and *Bangia* with the green algae persisted until the more detailed investigations of Berthold (1882) provided evidence, however doubtful some of this might now appear, of closer affinity with the Rhodophyceae, as they were then termed. Kylin (1937) separated genera with unicellular thalli as the order Porphyridiales, and the multicellular members of the group as the order Bangiales. The major revision by Skuja (1939) divided the class further into four orders – Porphyridiales, Goniotrichales, Bangiales and Compsopogonales with a fifth order, the Rhodochaetales, strongly supported if not formally proposed. The major criterion for the segregation of these orders was the basic morphology of the thallus, as follows:

Porphyridiales: thalli unicellular, occurring either as independent unicells or as irregular colonial masses.

Goniotrichales: thalli filamentous, branched, or unbranched, filaments arising by intercalary cell division.

Bangiales: thalli multicellular, filamentous, tubular or plate-like, with growth through intercalary cell division.

Compsopogonales: thalli filamentous, becoming tubular, branched.

Rhodochaetales: thalli filamentous, uniseriate, composed of elongate cells formed by the transverse division of an apical cell.

This scheme has been widely accepted during the past 35 years although there are objections to certain parts of it. For instance, it is extremely difficult to distinguish between some unicellular representatives referred to the Porphyridiales when these occur in colonial masses and some 'filamentous' members assigned to the Goniotrichales. The proposal by Feldmann (1955) that these two orders should be merged is accepted in the present treatment. As a result, the Bangiophyceae contains four orders, Porphyridiales, Bangiales, Compsopogonales, and Rhodochaetales. Only the first two orders are represented in the marine flora of the British Isles. The one species of the Rhodochaetales is distributed in marine subtropical waters, while the order Compsopogonales is entirely freshwater.

REFERENCES FOR INTRODUCTION

AGARDH, C. A. (1817). *Synopsis Algarum Scandinaviae.* Lund.

—— (1820–8). *Species Algarum* **1**(1), **1**(2), **2**(1). Lund, Griefswald.

BERTHOLD, G. (1882). Die Bangiaceen des Golfes von Neapel und der angrenzenden Meeresabschnitte. *Fauna Flora Golf. Neapel* **8:** 1–28.

BILHAM, E. G. (1938). *The climate of the British Isles.* London.

BONEY, A. D. (1972). *In vitro* growth of the endophyte *Acrochaetium bonnemaisoniae* (Batt.) J. et G. Feldm. *Nova Hedwigia* **23:** 173–186.

BORNET, E. & THURET, G. (1867). Recherches sur la fécondation des Floridées. *Annls Sci. nat.*, sér. 5, Bot. **7:** 137–166.

—— & —— (1876–80). *Notes Algologiques.* **1, 2.** Paris.

CHAPMAN, D. J. & CHAPMAN, V. J. (1961). Life histories in the algae. *Ann. Bot.*, N.S. **25:** 547–561.

CHAPMAN, V. J. (1970). *Seaweeds and their uses.* 2nd Ed. London.

CONWAY, E. (1964). Autecological studies of the genus *Porphyra:* 1. The species found in Britain. *Br. phycol. Bull.* **2:** 342–348.

DANGEARD, P. (1940). Recherches sur les enclaves iridescentes de la cellule des algues. *Botaniste* **31:** 31–63.

DIXON, P. S. (1961). On the classification of the Florideae with particular reference to the position of the Gelidiaceae. *Botanica mar.* **3:** 1–16.

—— (1966). On the form of the thallus in the Florideophyceae. *In:* Cutter, E. (ed.), *Trends in plant morphogenesis.* pp. 45–63. London.

—— (1973). *Biology of the Rhodophyta.* Edinburgh.

—— IRVINE, D. E. G. & PRICE, J. H. (1966). The distribution of benthic marine algae. A bibliography for the British Isles. *Br. phycol. Bull* **3:** 87–142.

—— & RICHARDSON, W. N. (1970). Growth and reproduction in red algae in relation to light and dark cycles. *Ann. N.Y. Acad. Sci.* **175:** 764–777..

DREW, K. M. (1939). An investigation of *Plumaria elegans* (Bonnem.) Schmitz with special reference to triploid plants bearing parasporangia. *Ann. Bot.*, N.S. **3**: 347–367.
—— (1951). Rhodophyta. *In:* Smith, G. M. (ed.), *Manual of Phycology.* pp. 167–191. Waltham.
—— (1954). The organization and inter-relationships of the carposporophytes of living Florideae. *Phytomorphology* **4**: 55–69.
—— (1955). Life histories in the algae with special reference to the Chlorophyta, Phaeophyta and Rhodophyta. *Biol. Rev.* **30**: 343–390.
—— (1956). Reproduction in the Bangiophycidae. *Bot. Rev.* **22**: 553–611.
ELLIS, J. (1755). *An essay towards a natural history of the corallines.* London.
ERNST, J. (1958). The life-forms of some perennial marine algae of Roscoff and their vertical distribution. *Abs. Int. Seaweed Symp.* **3**: 31.
FELDMANN, J. (1937). Recherches sur la végétation marine de la Médi terranée. La Côte des Albères. *Revue algol.* **10**: 1–339.
—— (1951). Ecology of marine algae. *In:* Smith, G. M. (ed.), *Manual of Phycology.* pp. 313–334. Waltham.
—— (1952). Les cycles de reproduction des algues et leurs rapports avec la phylogénie. *Revue Cytol. Cytophysiol. vég.* **13**: 1–49.
—— (1955). Un nouveau genre de Protofloridée: *Colacodictyon,* nov. gen. *Bull. Soc. bot. Fr.* **102**: 23–28.
—— (1966). Les types biologiques d'algues marines benthiques. *Mém. Soc. bot. Fr.* **1966**: 45–60.
FOGG, G. E., STEWART, W. D. P., FAY, P. & WALSBY, A. E. (1973). *The blue-green algae.* New York & London.
FRITSCH, F. E. (1945). *Structure and reproduction of the algae.* **2.** Cambridge.
FUNK, G. (1927). Die Algevegetation des Golfs von Neapel. *Pubbl. Staz. zool. Napoli* **7** (suppl.): 1–507.
GESSNER, F. (1970). Temperature – Plants. *In:* Kinne, O. (ed.), *Marine Ecology,* **1**(1). pp. 363–406. London.
—— & SCHRAMM, W. (1971). Salinity – Plants. *In:* Kinne, O. (ed.), *Marine Ecology,* **1**(2). pp. 705–820. London.
GISLEN, T. (1930). Epibioses of the GullmarFjord. *Skr. svenska Vetensk.-Akad.* **1930** (4): 1–380.
GRAY, S. F. (1821). *Natural arrangement of British plants.* **1, 2.** London.
HARLIN, M. M. (1973). Transfer of products between epiphytic marine algae and host plants. *J. Phycol.* **9**: 243–248.
HARVEY, W. H. (1836). Algae. *In:* Mackay, J. T. (ed.), *Flora Hibernica,* **2.** Dublin.
HELLEBUST, J. A. (1970). Light – Plants. *In:* Kinne, O. (ed.), *Marine Ecology,* **1**(1). pp. 125–158. London.
JANET, C. (1914). *L'alternance sporophyto-gamétophytique de générations chez les algues.* Limoges.
JOHNSON, T. W. & SPARROW, F. K. (1961). *Fungi in oceans and estuaries.* Weinheim.
KATADA, M. (1963). Life forms of seaweeds and succession of their vegetation. *Bull. Jap. Soc. scient. Fish.* **29**: 798–808.
KRAUSS, R. W. (1961). Fundamental characteristics of algal physiology. *In: Algae and metropolitan wastes.* Cincinnati.
KYLIN, H. (1937). Über eine marine *Porphyridium*-Art. *K. fysiogr. Sällsk. Lund Förh.* **7**: 119–123.
—— (1956). *Die Gattungen der Rhodophyceen.* Lund.
LAMOUROUX, J. V. F. (1813). Essai sur les genres de la famille de Thalassiophytes non articulées. *Annls Mus. natn. Hist. nat., Paris* **20**: 115–139, 267–293.
LEE, R. E. (1971). Systemic viral material in the cells of the freshwater red alga *Sirodotia tenuissima* (Holden) Skuja. *J. Cell Sci.* **8**: 623–631.

LEVRING, T., HOPPE, H. A. & SCHMID, O. J. (1969). *Marine algae: a survey of research and utilization.* Hamburg.

LEWIS, J. R. (1964). *The ecology of rocky shores.* London.

LINNAEUS, C. (1753). *Species plantarum* **1, 2.** Stockholm.

—— (1758). *Systema naturae* 10th Ed. **1, 2.** Stockholm.

LINSKENS, H. F. (1963). Beitrag zur Frage der Beziehung zwischen Epiphyt und Basiphyt bei marinen Algen. *Pubbl. Staz. zool. Napoli* **33:** 274–293.

—— (1966). Adhäsion von Fortpflanzungszellen benthontischer Algen. *Planta* **68:** 99–110.

LUMB, F. E. (1961). Seasonal variation of the sea surface temperatures in coastal waters of the British Isles. *Scient. Pap, met. Off., London* **6:** 1–21.

MAGNE, F. (1967). Sur l'existence, chez les *Lemanea* (Rhodophycées, Némalionales), d'une type de cycle de développement encore inconnu chez les algues rouges. *C. r. hebd. Séanc. Acad. Sci., Paris,* sér. D **264:** 2632–2633.

—— (1967a). Sur le déroulement et le lieu de la méiose chez les Lémanéacées (Rhodophycées, Némalionales). *C. r. hebd. Séanc. Acad. Sci., Paris,* sér. D **265:** 670–673.

METEOROLOGICAL OFFICE (1953). *Averages of temperature for Great Britain and Northern Ireland 1921–1950.* M. O. 571. London.

NEWTON, L. (1951). *Seaweed utilization.* London.

NORTH, W. L., STEPHENS, G. C. & NORTH, B. (1972). Marine algae and their relation to pollution problems. *In:* Ruivo, M. (ed.), *Marine pollution and sea life.* pp. 330–340. London.

OLTMANNS, F. (1898). Zur Entwicklungsgeschichte der Florideen. *Bot. Ztg* **56:** 99–140.

—— (1904–5). *Morphologie und Biologie der Algen.* **1, 2.** Jena.

—— (1922–3). *Morphologie und Biologie der Algen,* 2nd Ed. **1, 2, 3.** Jena.

PAPENFUSS, G. F. (1966). A review of the present system of classification of the Florideophyceae. *Phycologia* **5:** 247–255.

PHILIPPI, R. A. (1837). Beweis, dass die Nulliporen Pflanzen sind. *Arch. Naturgesch.* **3**(1): 387–393.

PRICE, J. H. (1967). The distribution of benthic marine algae. A bibliography for the British Isles. Supplement 1. *Br. phycol. Bull.* **3:** 305–315.

—— & TITTLEY, I. (1970). The distribution of benthic marine algae. A bibliography for the British Isles. Supplement 2. *Br. phycol. J.* **5:** 103–112.

—— & —— (1972). The marine flora of the County of Kent, southeast England, and its distribution, 1597–1970. *Proc. 7 Int. Seaweed Symp.* 31–34.

PROVASOLI, L. (1964). Growing marine seaweeds. *Proc. 4 Int. Seaweed Symp.* 9–17.

RAMUS, J. (1969). Pit connection formation in the red alga *Pseudogloiophloea. J. Phycol.* **5:** 57–63.

—— (1969a). The developmental sequence of the marine red alga *Pseudogloiophloea. Univ. Calif. Publs Bot.* **52:** 1–28.

SAGROMSKY, H. (1960). Tagesperiodische Ausschuttung der Tetrasporen bei Rotalgen. *Naturwiss.* **47:** 141.

SCHMITZ, F. (1883). Untersuchungen über die Befruchtung der Florideen. *Sber. Akad. Wiss.* **1883**: 215–258.

—— (1889). Systematische Übersicht der bischer bekannten Gattungen der Florideen. *Flora, Jena* **72:** 435–456.

—— & HAUPTFLEISCH, P. (1896–7). Rhodophyceae. *In:* Engler, A. & Prantl, K. (eds.), *Die natürlichen Pflanzenfamilien* **1**(2): 298–544. Leipzig.

SCHWEIGGER, A. F. (1819). *Beobachtungen auf naturhistorischen Reisen, anatomisch physiologische Untersuchungen über Corallen.* Berlin.

SKUJA, H. (1939). Versuch einer systematischen Einteilung der Bangioideen oder Protoflorideen. *Acta Horti bot. Univ. latv.* **11/12:** 23–40.

SMITH, G. M. (1944). *Marine algae of the Monterey Peninsula.* Stanford.

SPARROW, F. K. (1960). *Aquatic phycomycetes,* 2nd Ed. Ann Arbor.

STACKHOUSE, J. (1795–1801). *Nereis Britannica.* Bathoniae & Londini.

—— (1809). Tentamen marino-cryptogamicum. *Mém. Soc. imp. Nat. Moscou* **2:** 50–97.

—— (1816). *Nereis Britannica,* 2nd Ed. Oxford.

STEPHENSON, T. A. & STEPHENSON, A. (1949). The universal features of zonation between tidemarks on rocky coasts. *J. Ecol.* **37:** 289–305.

—— & —— (1972). *Life between tidemarks on rocky shores.* San Francisco.

STEWART, W. D. P. (ed.). (1974). *Algal physiology and biochemistry.* Oxford.

von STOSCH, H. A. (1965). The sporophyte of *Liagora farinosa* Lamour. *Br. phycol. Bull.* **2:** 486–496.

SUNESON, S. (1950). The cytology of bispore formation in two species of *Lithophyllum* and the significance of the bispores in the Corallinaceae. *Bot. Notiser* **1950**: 429–450.

TERBORGH, J. & THIMANN, K. (1964). Interactions between daylength and light intensity in the growth and chlorophyll content of *Acetabularia crenulata.* *Planta* **63:** 83–98.

TURNER, D. (1802). *A synopsis of the British* Fuci. Yarmouth.

YAMANOUCHI, S. (1906). The life-history of *Polysiphonia violacea.* *Bot. Gaz.* **41:** 425–433; **42:** 401–449.

Taxonomic treatment

ARRANGEMENT OF THE WORK

The following comments are made to help the reader understand the general principles accepted in the preparation of this text, the reasons why certain decisions were made or a particular sequence of information adopted. This seems particularly necessary since users are likely to consult it in the middle for the treatment of a particular taxon, rather than at the beginning as in a normal text.

A starting point for such a discussion is an explanation of the aims of the volume as a whole. It was the authors' intention to produce an up-to-date, critical floristic treatment of the marine Rhodophyta which occur in the British Isles, as defined in its strict geographical sense. On pragmatic grounds, it was decided to include the Channel Islands in the present treatment, despite their closer proximity to and greater floristic affinities with north-western France. An effort has been made to incorporate results of recent studies where these had relevance to taxonomic discrimination or ecological investigation.

The arrangement of taxa follows that given in the most recent check-list of British marine algae (Parke & Dixon, 1976). As stated earlier, divisional status is accepted for the red algae as a whole, as the Rhodophyta, with two classes – Florideophyceae and Bangiophyceae. Six orders (Nemaliales, Gigartinales, Cryptonemiales, Palmariales, Rhodymeniales, Ceramiales) are accepted in the Florideophyceae and two orders (Porphyridiales, Bangiales) accepted in the Bangiophyceae for British marine representatives. Two orders (Gelidiales, Goniotrichales) which are accepted by some workers are rejected in the present treatment and referred to the Nemaliales and Porphyridiales respectively. A full discussion of the evidence on which this arrangement is based ispresented on p. 60. The diagnoses for division, classes and orders are modified from those of Kylin (1956), while the diagnoses for families have been much amplified with particular reference to their representatives which occur in the British Isles. Under each family, some attention is drawn to the criteria by which genera are distinguished when these are doubtful or of critical significance. For each genus, the type species is listed, together with the major synonymy and an artificial key for the species which occur in the British Isles.

For each species, the typification and major synonymy are indicated, together with an outline of the habitat characteristics, geographical distribution, seasonal aspects of growth and reproduction, and form variation. Had space been available, the form range of each species would have been illustrated in the same way as for *Gelidium pusillum*. The information and illustrations are derived from a consideration of living material, collected over the past 20 years, from all parts of the British Isles. This has been supplemented, particularly for species descriptions and information on distribution and phenology, by studies of personal collections made available to us, and of permanent collections in many herbaria, which were also examined in detail in connection with typification. For a few very rare species, newly-collected material from the British Isles was totally unavailable for study but such cases are clearly indicated. The terminology used in the discussion of habitat is modified from that proposed by Lewis (1964) while distribution within the British Isles is indicated in a general manner on the basis of records which the authors have verified. For this reason, the ranges indicated are likely to be conservative and it is unfortunate that space does not permit detailed discussion of the information on distribution which we have rejected. Information on distribution outside the British Isles is, however, based to a great extent on published reports. The final section deals with additional information of use in a taxonomic and ecological study of the taxon under consideration not covered elsewhere. In this are included such significant data as comments on possible specific relationships and other taxonomic problems, economic uses, the occurrence of fungal contaminants, animal infestations, galls, etc.

The treatment of genera and species will be regarded by many as conservative, possible changes or associations being indicated, but accepted only when the evidence is considered to be beyond question. Despite the length of time during which this work has been in preparation, there are still many unresolved problems. It has been the authors' intention to indicate those of which we are aware, even though we could not provide answers in many cases. Because of this approach, we hope those who follow will be better able to take advantage of new data, new materials and new concepts to resolve the outstanding problems.

REFERENCES

KYLIN, H. (1956). *Die Gattungen der Rhodophyceen.* Lund.

LEWIS, J. R. (1964). *The ecology of rocky shores.* London.

PARKE, M. & DIXON, P. S. (1976). Check-list of British Marine Algae – third revision. *J. mar. biol. Ass. U.K.* **56**: 527–593.

RHODOPHYTA Wettstein

RHODOPHYTA Wettstein (1901), p. 46.
Rhodospermae Harvey (1836), p. 160.
Heterocarpeae Kützing (1843), p. 369.
Rhodophyceae Ruprecht (1851), p. 205.
Rhodophycophyta Papenfuss (1946), 218.

Algae of unicellular, filamentous, parenchymatous or pseudoparenchymatous organization; vegetative cells uninucleate or multinucleate; chloroplasts containing chlorophyll and usually both phycoerythrin and phycocyanin so that the colour range of the thalli can be considerable; colour rose-red, crimson, violet, occasionally green, brown or black and even pale-pink or colourless in those plants which have chloroplasts reduced in numbers or lacking completely.

All reproductive bodies without flagella; male gametes (spermatia) formed singly; female gametes formed from an undifferentiated or slightly differentiated vegetative cell or by the conversion of an apical cell into a highly differentiated carpogonium; post-fertilization development involving the formation of sporangia, either directly from the former female gamete or after the intercalation of one or more vegetative cell divisions which produce a multicellular carposporophyte; development of this carposporophyte usually resulting in the induction of cell divisions in adjacent areas of the female gametangial plant to give a 'cystocarp'; reproduction may also occur through one or more types of spore formed in tetrasporangia, bisporangia, monosporangia, polysporangia or parasporangia.

The Rhodophyta contains two classes, the Florideophyceae and the Bangiophyceae. Members of these two classes differ in many respects although precise definition of these differences is not easy. Numerically the Florideophyceae is a much larger class than the Bangiophyceae both on a world-wide scale as well as in terms of their representatives in the British Isles.

REFERENCES

HARVEY, W. H. (1836). Algae. *In:* Mackay, J. T., *Flora Hibernica*. **2.** Dublin.

KÜTZING, F. T. (1843). *Phycologia generalis*. Leipzig.

PAPENFUSS, G. F. (1946). Proposed names for the phyla of algae. *Bull. Torrey bot. Club*, **73:** 217–218.

RUPRECHT, F. J. (1851). *Tange des Ochotskischen Meeres*. *In:* von Middendorff, A. T., *Sibirische Reise*, Botanik, **1**(2), St. Petersburg.

WETTSTEIN, A. (1901). *Handbuch der systematischen Botanik*. Leipzig & Vienna.

FLORIDEOPHYCEAE Cronquist

FLORIDEOPHYCEAE Cronquist (1960), p. 452.
Florideae [as Floridées] Lamouroux (1813), p. 115 (reprint, p. 27).
Eufloridae Johnson (1894), p. 639.
Florideophycidae Newton (1953),p. 407 [as Floridophycideae].

Thalli obviously filamentous or pseudoparenchymatous; pseudoparenchymatous aggregations of filaments either discoid and crustose or erect and frondose; erect thalli terete, compressed or flat, foliose and leaflike, of uniaxial or multiaxial construction; cell division almost entirely restricted to the apical cells although intercalary cell division also present in a few species; cells uninucleate or multinucleate, with many chloroplasts or only a single chloroplast, pyrenoids present only in some cases where there is a single chloroplast in each cell; thalli rose-red, or crimson, violet, brown or black in colour, occasionally pale-pink or colourless in those plants where chloroplasts are reduced in number or lacking completely.

Gametangia well-differentiated; non-motile male gametes (spermatia) produced singly from gametangia (spermatangia); spermatangia produced singly, in clusters, or in large dense masses; female gamete non-motile, attached permanently to the female gametangial thallus, formed by the transformation of an apical cell, and termed a carpogonium; post-fertilization development a one or many-celled tissue, the carposporophyte, which remains firmly attached to the female gametangial plant eventually forming sporangia (carposporangia) from the apical cells and sometimes the intercalary cells also; each carposporangium liberating one carpospore or four carpotetraspores; reproduction also may occur through one or more types of spore formed in tetrasporangia, bisporangia, monosporangia, polysporangia, or parasporangia.

Representatives of the following orders occur in the British Isles:

Nemaliales (formerly known as the Nemalionales, and including in the present treatment those algae referred to the Gelidiaceae and assigned independent ordinal status as the Gelidiales by some workers)

Gigartinales
Cryptonemiales
Rhodymeniales
Palmariales
Ceramiales.

REFERENCES

CRONQUIST, A. (1960). The divisions and classes of plants. *Bot. Rev.* **26:** 425–482.

JOHNSON, T. (1894). The systematic position of the Bangiaceae. *Nuova Notarisia* **5:** 636–647.

LAMOUROUX, J. V. F. (1813). Essai sur les genres de la famille des thalassiophytes non articulées. *Annls Mus. Hist. nat. Paris* **20:** 21–47; 115–139; 267–293. [reprint pp. 84].

NEWTON, L. M. (1953). Marine algae. *Scient. Rep. John Murray Exped.* **9:** 395–420.

Nemaliales

NEMALIALES Schmitz

NEMALIALES Schmitz in Engler (1892), p. 17, *as* Nemalionales, but see Christensen (1967), p. 93.

Thalli erect, filamentous or pseudoparenchymatous, of uniaxial or multiaxial construction; if pseudoparenchymatous, the aggregation of filaments loose or compact. Carposporophyte developing directly from the fertilized carpogonium or following transfer to an auxiliary cell which is a cell of the carpogonial branch; each carposporangium liberating one carpospore or four carpotetraspores; gametangial plant and tetrasporangial plant in each species of similar, slightly dissimilar or totally different organization.

A somewhat artificial grouping containing families which are extremely diverse in structure and reproduction although each is itself remarkably uniform. Various proposals have been made for the subdivision of this order although none is satisfactory and the Nemaliales is retained pending more detailed investigation.

Representatives of the following families referred to the Nemaliales occur in marine situations in the British Isles:

Acrochaetiaceae
Gelidiaceae
Naccariaceae
Bonnemaisoniaceae
Chaetangiaceae
Helminthocladiaceae.

ACROCHAETIACEAE Fritsch

ACROCHAETIACEAE Fritsch (1944), p. 258.

Thallus of simple construction, the form often associated with a particular habitat; endozoic or endophytic species simple, filamentous; epiphytic, epizoic or epilithic species with a unicellular basal cell, a multicellular discoid base, or prostrate creeping filaments and an erect frond of simple or branched filaments. Spermatangia single or clustered, carpogonia simple, carposporophyte simple, without enveloping filaments; carposporangia terminal and also, occasionally, intercalary; reproduction by many types of spore.

Most of the species referable to the Acrochaetiaceae are ill-known but obviously highly variable, while some have been shown to be parts of the life histories of other taxa. The described species form an inter-related reticulum and there appears to be no logical basis at this time for the delineation of genera. The two most widely-used generic names are *Acrochaetium* and *Rhodochorton*, described originally by Nägeli (1861[1862]) and characterized by their reproductive structures. It soon became obvious that this segregation was unsatisfactory but the confusion was made worse, rather than improved, by the description of further genera, particularly by Rosenvinge (1909). The first major attempt to resolve the problems of the Acrochaetiaceae (Drew, 1928) merged all species into a single genus, for which she accepted the name *Rhodochorton* in the belief that this was the oldest available generic name.

Several schemes aimed at resolving the problems of the Acrochaetiaceae have been proposed during the past 35 years (Kylin, 1944; Papenfuss, 1945, 1947; Feldmann, 1962; Woelkerling, 1971) and have involved either the description of additional genera or redefinition of previously-described entities. The various proposals involve different systematic criteria, and markedly different generic circumscription so that the assignment of species is confused at the present time. This confusion is so great that it was argued that the simplest and most suitable course of action is to return to the one-genus concept of Drew. There is no reason for distinguishing between marine and freshwater representatives so that *Audouinella* is accepted as the name for that genus.

AUDOUINELLA Bory

AUDOUINELLA Bory (1823), p. 340.

Type species: *A. miniata* Bory (1823), p. 340 (=*A. hermannii* (Roth) Duby).

Rhodochorton Nägeli (1861 [1862]), p. 355.
Acrochaetium Nägeli (1861 [1862]), p. 402.
Trentepohlia Pringsheim (1862[1863]), p. 29, non *Trentepohlia* Martius (1817).
Thamnidium Thuret in Le Jolis (1863), p. 110.
Balbiania Sirodot (1876), p. 149.
Colaconema Batters (1896), p. 8.
Kylinia Rosenvinge (1909), p. 141.
Grania Kylin (1944), p. 26.
Chromastrum Papenfuss (1945), p. 320.

Plants obviously filamentous, of various forms; some consist of prostrate much-branched filaments embedded in a plant or animal substrate, sometimes loosely aggregated and coherent, such erect filaments as occur being short and simple or sparsely branched, forming only in connection with the development of reproductive structures; some have the thallus differentiated into prostrate and erect filaments, both systems being richly branched, the prostrate filaments either on the surface of the substrate or becoming embedded secondarily in plant or animal substrates of soft texture; some consist of much branched erect filaments attached to the substrate by a single basal cell; erect filaments to 25 mm in length; cells containing one or more chloroplasts with one or more pyrenoids.

Gametangial plants monoecious or dioecious; spermatangia occurring singly, or in clusters of 2 or 3 or more, terminally or laterally on stalks consisting of one or more cells, or sessile upon ordinary vegetative cells; carpogonia simple, sessile or intercalary; carposporophyte developing directly from the carpogonium and consisting of 1, 2, or 3-celled filaments, with carposporangia forming from the terminal cells and occasionally also from intercalary cells; carposporophyte development not resulting in the development of any adventitious, subtending involucral filaments; reproduction also by monospores, tetraspores, bispores, polyspores; when monosporangia occur in any species, they are the most abundant reproductive structures.

The entities grouped here into the genus *Audouinella* are probably the most confused of all the marine Rhodophyta found in the British Isles. In the present account little reduction to synonymy has been carried out although the many instances where this is likely to occur in the future are indicated. The reason for this conservative approach is that it is better to accept entities until the evidence for their abandonment is adequate and to reduce to synonymy only when detailed investigations of morphological and reproductive plasticity have been undertaken, because in this way experimental and critical studies will be encouraged.

Key to Species

1	Plant growing on a plant, animal, or rock substrate, or if endophytic or endozoic with the greater portion emergent from the substrate	2
	Plant entirely endophytic or endozoic, or with the greater portion growing within a plant or animal and only some few-celled filaments emergent from the substrate	10
2	Base of plant merely a single cell, formed from the original spore, with no prostrate system	3
	Base of plant multicellular, either obviously filamentous or pseudoparenchymatous, sometimes with the original spore visible but then with additional filaments of some kind	20
3	Plant with axes containing only 5 cells or fewer before growth is terminated by the conversion of the apical cell to a hair	4
	Plant with axes containing more than 5 cells; if fewer, then not with the apical cell converted to a hair	6
4	Length of cells 4–8 μm	5
	Length of cells 9–12 μm	*A. rosulata*
5	Cells spherical	*A. scapae*
	Cells cylindrical	*A. trifila*
6	Plant of macroscopic size (*c.* 1 mm); epiphytic on *Alaria*	*A. alariae*
	Plant not of macroscopic size (<350 μm); epiphytic on various algae but not *Alaria*	7
7	Epiphytic on *Porphyra* spp.	8
	Epiphytic on various algae but not *Porphyra* spp.	9
8	Cell diameter 4–6 μm	*A. battersiana*
	Cell diameter 7–11 μm	*A. rhipidandra*
9	With a single axis arising from each basal cell	*A. microscopica*
	With more than one axis arising from each basal cell	*A. parvula*
10	Plant entirely within the host	11
	Plant with some few-celled filaments emerging from the host	15
11	Cells squat, 7–10 μm in length, 9–11 μm in diameter	*A. sanctae-mariae*
	Cells elongate, at least 20 μm in length but never more than 10 μm in diameter	12
12	Cell diameter 2–6 μm	13
	Cell diameter 6–10 μm	14
13	Cells of regular cylindrical shape, although of variable size	*A. chylocladiae*
	Cells of highly irregular outline, a few cylindrical, others irregularly enlarged, swollen, furcate or cruciform	*A. bonnemaisoniae*
14	Filaments profusely branched, with a lateral filament originating from almost every cell	*A. asparagopsis*
	Filaments sparsely branched, with 6–10 cells between each lateral filament	*A. brebneri*
15	Plant endophytic	16
	Plant endozoic	17

16 Prostrate filaments profusely branched and interwoven, forming a compact basal disc from which the erect filaments arise . . . . *A. endophytica*
Prostrate filaments branched at right angles, but not aggregated to form a compact disc. . . . . *A. emergens*

17 Prostrate filaments entwined, but never adhering to form a monostromatic plate 18
Prostrate filaments adherent laterally to form a monostromatic plate . 19

18 Cells cylindrical or barrel-shaped, 6–25 μm in length, 6–13 μm in diameter . . . . *A. endozoica*
Cells of highly variable outline, some simple and cylindrical, some curved and contorted, some deeply divided, furcate or even cruciform . *A. infestans*

19 Cells of monostromatic basal plate regularly isodiametric, polygonal, 10–13 μm in diameter; cell walls smooth . . . . . . *A. infestans*
Cells of monostromatic basal plate elongate, at least twice as long as wide; cell walls corrugated, not smooth . . . . *A. membranacea*

20 Chloroplast a distinct spiral band . . . . *A. efflorescens*
Chloroplast not a distinct spiral band 21

21 With a downgrowing prostrate system developing secondarily and penetrating the host 22
With no development of a downgrowing or penetrating prostrate system 23

22 Original spore detectable as a large inflated cell at the base of the erect axes . . . . *A. corymbifera*
Original spore not detectable at the base of the erect axes . *A. nemalionis*

23 Erect axes unbranched or only very sparsely branched 24
Erect axes profusely branched 26

24 Prostrate system consisting of loosely entangled filaments . . *A. sparsa*
Prostrate system of tightly compacted monostromatic sheets of cells formed by the lateral aggregation of filaments 25

25 Cells of upright filaments 15–20 μm in length . . . . . . *A. seiriolana*
Cells of upright filaments 8–13 μm in length . . . . . *A. concrescens*

26 Base a regular monostromatic plate-like structure made up of radiating filaments which are laterally adherent . . . . . . *A. spetsbergensis*
Base made up of filaments which are aggregated and compacted in a highly irregular manner 27

27 Cells containing several very distinct small chloroplasts 28
Cells appearing to contain a single chloroplast although this may be formed by the aggregation of a very large number of small chloroplasts which become partially apparent under some conditions 29

28 Pyrenoids present . . . . *A. floridula*
Pyrenoids absent . . . . *A. purpurea*

29 Chloroplast axile 30
Chloroplast parietal 31

30 Branching of erect and prostrate filaments abundant; prostrate axes aggregating to form a basal disc more than one cell layer deep *A. secundata*
Branching of erect and prostrate filaments sparse; prostrate axes aggregating to form a basal disc only one cell layer deep . . . . . *A. virgatula*

31 Cells in upper parts of erect axes narrow, 4–5 μm in diameter, differing markedly in proportions from cells in other parts . . . . . *A. thuretii*
Cells in upper parts of erect axes not differing in proportions from cells in other parts of plant 32
32 Chloroplast with obvious pyrenoid . . . . *A. daviesii*
Chloroplast not with obvious pyrenoid 33
33 Cells of erect axes 15–20 μm in diameter; cell wall relatively thick, *c.* 2–3 μm . . . . *A. lorrain-smithiae*
Cells of erect axes 10–15 μm in diameter; cell wall not thick, *c.* 1 μm. *A. caespitosa*

Audouinella alariae (Jónsson) Woelkerling (1973), p. 541.

Holotype: C (Jónsson 597). Iceland (Hvammesfjord).

Chantransia alariae Jónsson (1901), p. 132.
Acrochaetium alariae (Jónsson) Collins (1906), p. 192.
Kylinia alariae (Jónsson) Kylin (1944), p. 13.
Chromastrum alariae (Jónsson) Papenfuss (1945), p. 320.

Plants consisting of uniseriate filaments, deep red in colour, with the original spore, from which the erect axes arise, evident throughout the life of the plant; erect axes to 1 mm,

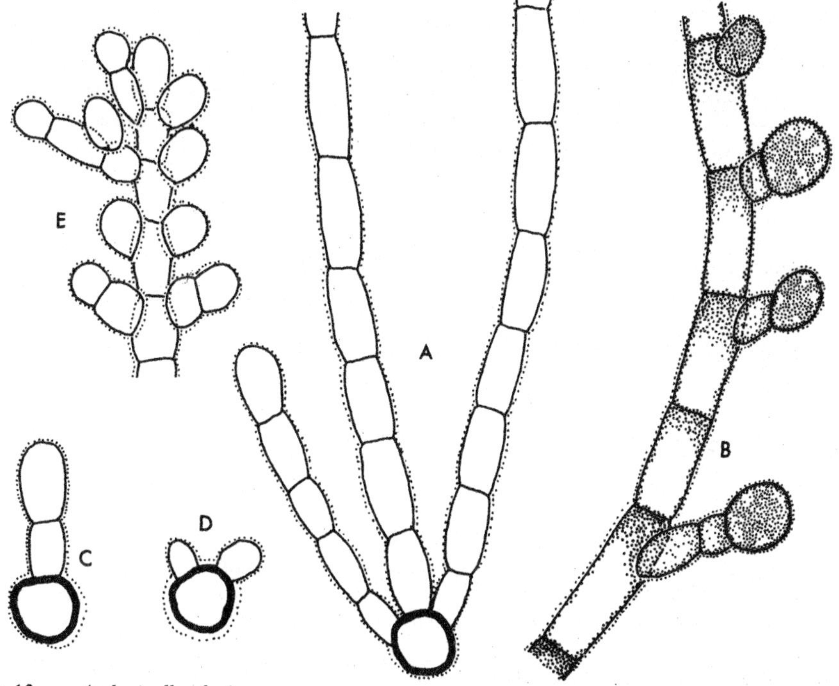

Fig. 13 *Audouinella alariae*
A. Habit ×740; B. Axis with monosporangia ×740; C, D. Monospore germlings ×740; E. Apex ×740.

lower portions unbranched, upper parts much branched, branching opposite, alternate or secund; cells cylindrical, 15–30 μm in length, 8–12 μm in diameter when first formed, increasing in size with age to reach a length of 25–70 μm and diameter of 12–25 μm, each with a single chloroplast; chloroplast lobed with one pyrenoid which is often not conspicuous; apical cells of axes and lateral filaments converted to hairs, particularly in infertile specimens.

Gametangial plants unknown; tetrasporangia unknown, monosporangia 16–25 μm in length, 8–15 μm in diameter, formed both terminally and laterally, often in pairs, on the ultimate branchlets, always sessile.

Epiphytic, principally on *Alaria esculenta* (L.) Grev., in the lower littoral and upper sublittoral.

Widely distributed on northern shores of Scotland, becoming more rare in Ireland, Wales and northern England, as the host species diminishes in frequency.

North Atlantic Ocean; in North America from Massachusetts northwards; Greenland; Iceland; in Europe from Norway to southwest England.

Although the frond of the host species is perennial, any part is of finite longevity so that plants of *A. alariae* cannot be indefinitely perennial. Plants tend to be most conspicuous during July, August and September and fertile at that time although nothing further is known about their behaviour.

Data on form variation too inadequate for comment.

Chloroplast shape is variable, being described by some as stellate and by others as lobed; on the basis of material examined, the latter appears more appropriate.

Woelkerling (1973) refers both *Chantransia unilateralis* Kjellman (1906) and *C. rhipidandra* Rosenvinge (1909) to the synonymy of this species. In the present treatment, the latter is retained, as *Audouinella rhipidandra* (q.v.).

Audouinella asparagopsis (Chemin) Dixon (1976), p. 000.

Lectotype: PC [see Dixon, (1977)]. France (Brignogan).

Colaconema asparagopsis Chemin (1926), p. 902.
Acrochaetium asparagopsis (Chemin) Papenfuss (1945), p. 312.

Plants consisting of uniseriate filaments, entirely endophytic in the outer cell wall of the host species, rose-pink in colour; filaments spreading, profusely branched, branching often alternate but sometimes highly irregular, apparently not anastomosing; cells cylindrical or barrel-shaped, usually with the maximum diameter at the point of origin of a lateral, often curved or bent; 21–30 μm in length, 6–8 μm in diameter, usually with several small discoid or fragmentary chloroplasts devoid of pyrenoid(s).

Gametangial plants unknown; tetrasporangia unknown; monosporangia ovoid or subspherical, 7–10 μm in diameter, formed singly or in clusters of 2–3 sporangia, sometimes with a small subtending cell, usually in a terminal but occasionally a lateral position.

Endophytic in the outer cell wall of the gametangial phase of *Bonnemaisonia hamifera* Hariot, in the lower littoral and sublittoral.

Cornwall, Devon; probably more widely distributed.

Atlantic coast of France.

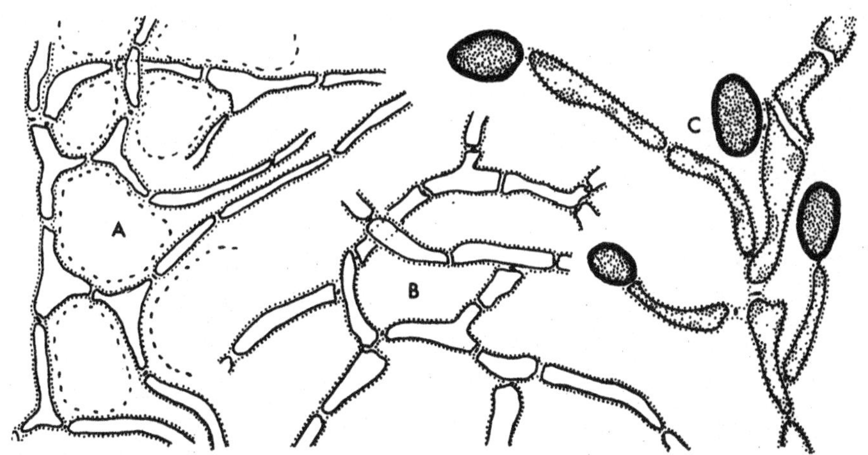

Fig. 14 *Audouinella asparagopsis*
A. Sectional view ×740; B. Surface view ×740; C. Monosporangia ×740.

Collections from localities in the British Isles have all been made during the summer months.

Data on form variation too inadequate for comment.

Audouinella asparagopsis and *A. bonnemaisoniae* are similar in many respects to *Colaconema americana* Jao (1936), growing endophytically in *Bonnemaisonia hamifera* in Massachusetts. More complete studies are needed before the status of these entities can be decided, although the similarities suggest synonymy.

Audouinella battersiana (Hamel) Dixon (1976), p. 590.

Lectotype: BM (Slide 7626). Northumberland (Berwick-on-Tweed).

Acrochaetium battersianum Hamel (1927), p. 83.
Kylinia battersiana (Hamel) Kylin (1944), p. 13.

Plant consisting of a single prominent basal cell, 12–15 µm in diameter, formed from the original spore, and one to three erect axes, to 200 µm in height, formed from branched uniseriate filaments; cells cylindrical, 8–20 µm in length, 4–6 µm in diameter, with a single chloroplast; chloroplast lobed, with a prominent pyrenoid; apical cells frequently converted into elongate hairs, to 40 µm in length.

Gametangial plants dioecious; spermatangial plants often smaller than those with other reproductive structures, rarely exceeding 90 µm, spermatangia small, 3–4 µm in length, without chloroplasts, formed in pairs, occasionally singly, in most cases on a one-celled filament formed laterally on a main axis; carpogonia small, to 3 µm in length, formed singly, either sessile or on a single-celled filament formed laterally on a main axis, gonimoblasts develop directly and give rise to a small cluster of 8–13 ovoid carposporangia, 12–18 µm in length, 8–12 µm in diameter; gametangial plants apparently never bearing monosporangia; monosporangial plants large, to 200 µm, monosporangia

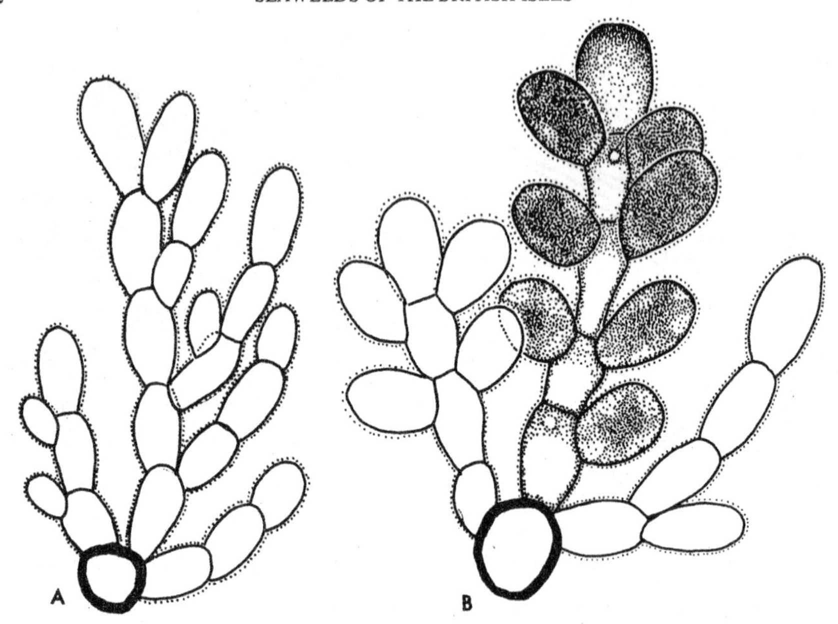

Fig. 15 *Audouinella battersiana*
A. Young plant × 740; B. Plant with monosporangia × 740.

ovoid, 10–12 µm in length, 6–8 µm in diameter, either sessile or occasionally on a one-celled stalk.

Epiphytic on species of *Porphyra*, in the midlittoral.

• Cornwall, Devon, Northumberland, Bute, but probably of more general occurrence.
Not recorded outside the British Isles.

No data on seasonal growth available; records of the occurrence of fertile material in June and July.

Data on form variation too inadequate for comment.

For comments on the status of *A. battersiana*, see *A. microscopica* (p. 102). Hamel (1927) described *Acrochaetium maluinum*, differing from *Audouinella battersiana* only in that the cystocarps were said to develop at a distance from the base rather than close to it. The significance of this claim requires further critical evaluation but it is probable that *Acrochaetium maluinum* should be reduced to the synonymy of the present species.

Audouinella bonnemaisoniae (Batters) Dixon (1976). p. 590.

Lectotype: BM (Slide 7873). Devon (Plymouth).

Colaconema bonnemaisoniae Batters (1896), p. 8.
Chantransia bonnemaisoniae (Batters) Levring (1937), p. 94.
Acrochaetium bonnemaisoniae (Batters) Feldmann & Feldmann (1939), p. 458.

Plants consisting of uniseriate filaments, entirely endophytic in the outer cell wall of the host species, rose-pink in colour; filaments spreading, profusely and irregularly branched, frequently anastomosing; cells of very variable outline, some simple and cylindrical, some barrel-shaped or irregularly enlarged, some deeply divided in various ways to give a furcate, cruciate or highly irregular outline, 15–45 μm in length, 3–6 μm in diameter; chloroplast parietal and simple, apparently fragmenting with age and often not obvious in the larger and more irregular cells, pyrenoid present.

Gametangial plants unknown; tetrasporangia unknown; monosporangia spherical, 8–12 μm in diameter, formed singly or, more often, in clusters of 2–8 sporangia, each with an obvious, cup-shaped subtending cell, usually in a terminal position, occasionally lateral.

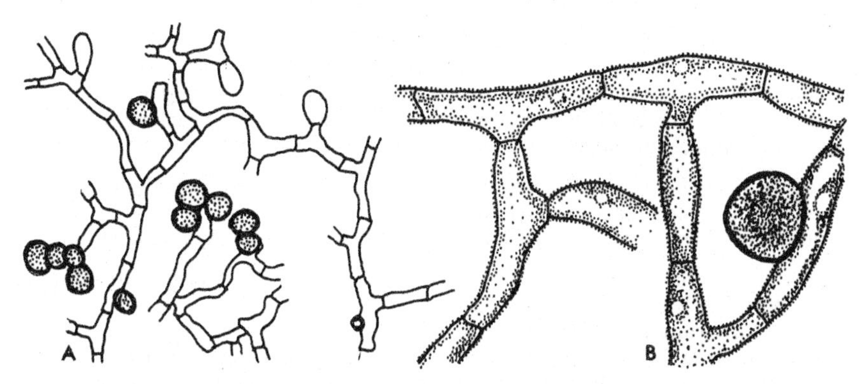

Fig. 16 *Audouinella bonnemaisoniae*
A. Habit ×350; B. Cell detail and monosporangium ×650.

Endophytic in the outer cell wall of the gametangial phase of *Bonnemaisonia asparagoides* (Woodw.) C. Ag., in the lower littoral and sublittoral.

Cornwall, Devon, Northumberland, Galway, Mayo, Dunbarton, Channel Islands; probably of more general occurrence.

Distributed from Norway to Portugal, although sparsely reported; Mediterranean.

Collections have all been obtained during the summer months.

Data on form variation too inadequate for comment.

For comments on the status of *A. bonnemaisoniae, A. asparagopsis* and *Colaconema americana* Jao, see p. 81.

Audouinella brebneri (Batters) Dixon (1976), p. 590.

Lectotype: BM (Slide 10066). Devon (Plymouth).

Rhodochorton brebneri Batters (1897), p. 437.
Chantransia brebneri (Batters) Rosenvinge (1909), p. 82.

Plants consisting of uniseriate filaments, endophytic, penetrating deeply into the host species, usually also with some emergent filaments, pale rose-pink in colour; filaments spreading, widely separated, sparsely branched; cells cylindrical or irregularly inflated, 18–30 μm in length, 8–10 μm in breadth, chloroplast parietal, apparently devoid of pyrenoid; hairs sometimes forming on the emergent filaments.

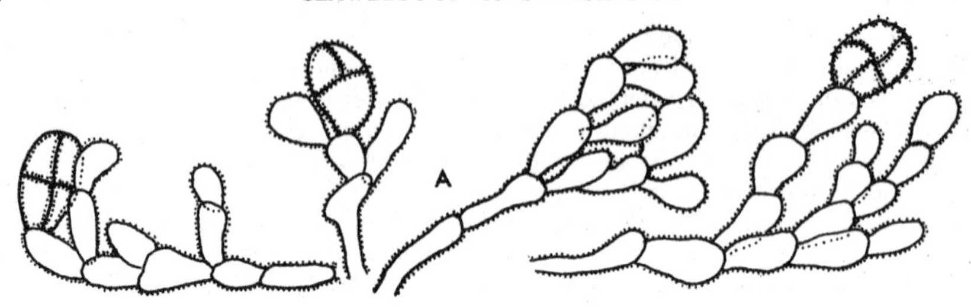

Fig. 17 *Audouinella brebneri*
A. Plants with tetrasporangia ×350.

Gametangial plants unknown; tetrasporangia relatively large, ovoid, 18–30 μm in length, 20–30 μm in breadth, formed singly or in small clusters on the emergent filaments, tetraspores cruciately arranged.

Endophytic in *Gloiosiphonia capillaris* (Huds.) Carm. ex Berk., penetrating deeply into the host, in the lower littoral and sublittoral, to 5 m.

Known only from the type material, collected at Rennie Rocks, Plymouth, Devon.

Data on seasonal aspects of growth too inadequate for comment. Tetrasporangial plants reported in September.

Data on form variation insufficient for comment.

The tetrasporangia of *A. brebneri* are much larger than in any other species of *Audouinella,* and it is possible that this entity represents the product of internal development of the carpospores of the host species, *Gloiosiphonia capillaris.*

Audouinella caespitosa (J. Agardh) Dixon (1976), p. 590.

Lectotype: LD (Herb. Alg. Agardh. 18011). France (Brest).

Callithamnion caespitosum J. Agardh (1851), p. 18.
Acrochaetium caespitosum (J. Agardh) Nägeli (1861 [1862]), p. 407.
Chantransia caespitosa (J. Agardh) Batters (1896), p. 9.

Plants consisting of uniseriate filaments, differentiated into erect and prostrate axes, red or blackish-red but sometimes becoming bleached and greenish in colour; prostrate axes poorly developed, branched but not greatly, not compacted into a solid discoid base; cells cylindrical, 45–60 μm in length, 15–20 μm in diameter; erect axes to 10 mm, tufted, much entangled at the base and not branched in the lower parts, branched sparsely to densely in the upper parts, increasing towards the apices, branching highly variable, alternate or sometimes unilateral, occasionally opposite; cells cylindrical, 30–50 μm in length, 10–13 μm in diameter when first formed, increasing in size with age to reach a length of 45–60 μm and diameter of 12–15 μm, each cell containing a single chloroplast; chloroplasts parietal, relatively large, apparently devoid of pyrenoid.

Gametangial plants unknown; tetrasporangia of doubtful occurrence; monosporangia ovoid, 20–30 μm in length, 9–15 μm in diameter, occurring singly or in groups of two or three, borne on a single-celled stalk which arises from the lowermost cells of the ultimate lateral filaments, on the adaxial or uppermost surface, usually releasing contents undivided but sometimes with transverse septa so that interpretation is uncertain; they have been

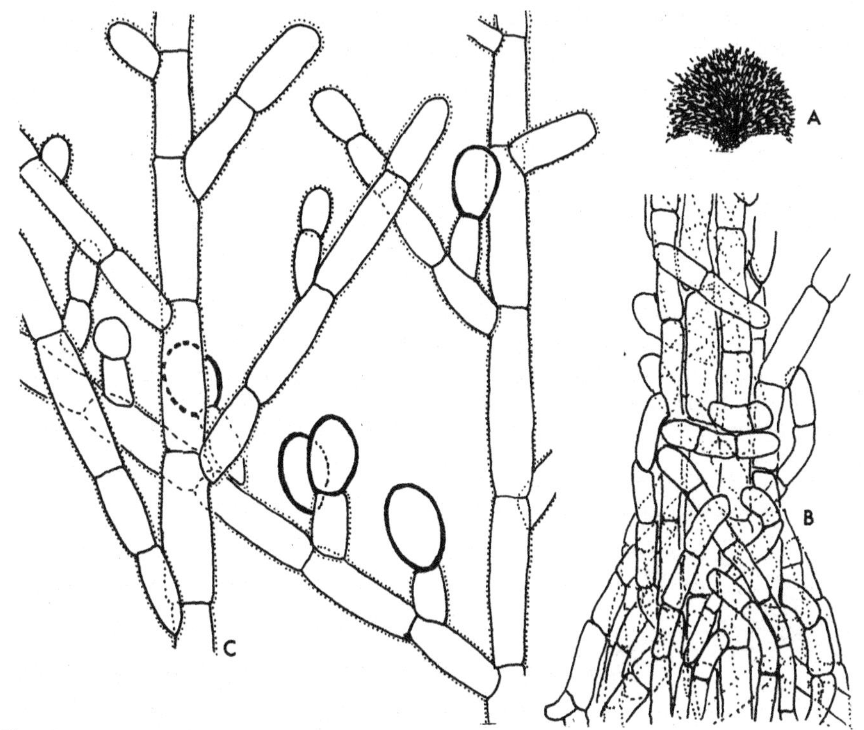

Fig. 18 *Audouinella caespitosa*
A. Habit×3; B. Base of plant×400; C. Axes with monosporangia×400.

considered by some to be tetrasporangia or bisporangia although the possibility of monospore germination *in situ* should not be excluded.

Epiphytic on various algae in the sublittoral, to a depth of 8 m. Endophytic penetration by the prostrate system may occur with host species of soft texture such as *Codium* spp. Epizoic growth on shells of live *Patella* reported extensively in France and detected occasionally in southwest England.

Cornwall, Devon, Dorset, Channel Islands.

Atlantic and Channel coasts of France.

The prostrate system is relatively long-lived in plants growing on *Patella*, with the upright system best-developed between April and September, although it is not known whether epiphytic plants can be perennial. Monosporangia have been recorded between April and August.

The size of erect axes varies considerably but this variation does not appear to be related to any specific factor. The depth of penetration of the prostrate system varies according to the compactness of the surface of the host species.

Numerous taxa have been described from all parts of the world in what can be considered a single entity. In northwest Europe, the entity most closely related to *A.*

caespitosa has been termed '*Acrochaetium codii*', although there is considerable confusion surrounding this binomial. There does not appear to be a formal diagnosis published and for this reason the binomial is not cited in the synonymy of *A. caespitosa.*

Hamel (1927) suggested that *A. lorrain-smithiae* should be reduced to the synonymy of *Audouinella caespitosa. A. lorrain-smithiae* has been retained in the present treatment, pending further investigation, but its eventual reduction to synonymy is highly probable.

Audouinella chylocladiae (Batters) Dixon (1976), p. 590.

Lectotype: BM (Slide 7877). Devon (Bovisand Bay).

Colaconema chylocladiae Batters (1896), p. 8.
Acrochaetium chylocladiae (Batters) Batters (1902), p. 58.

Plants consisting of uniseriate filaments, entirely endophytic in the outer cell wall of the host species, pale rose-pink in colour; filaments spreading, sparsely branched, occasionally anastomosing; cells cylindrical, elongate, 20–34 μm in length, 2–3 μm in diameter; chloroplast apparently parietal and simple, occupying only a small part of the cell, devoid of pyrenoid.

Gametangial plants unknown, tetrasporangia unknown; monosporangia ovoid, 6–8 μm long, 4–6 μm in diameter, formed both laterally and terminally.

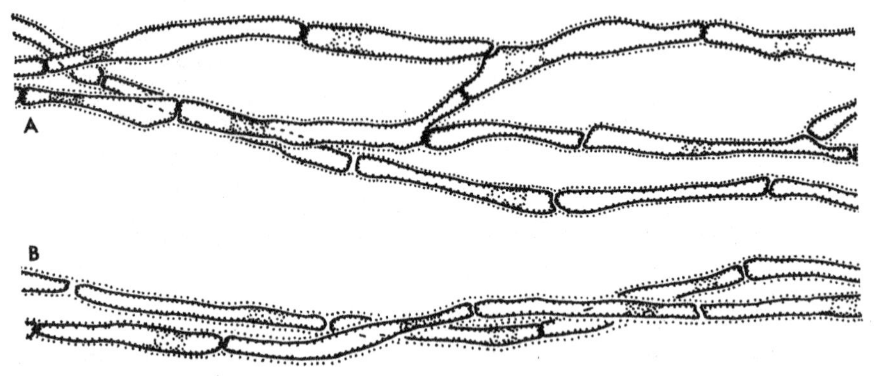

Fig. 19 *Audouinella chylocladiae*
A. Habit of plant in *Gastroclonium* ×740; B. Habit of plant in *Sertularia* ×740.

Endophytic in the outer cell wall of *Gastroclonium ovatum* (Huds.) Papenf., or endozoic in *Sertularia,* sublittoral, to 10 m.

Devon, Bute, Dublin, Cork, Wexford; probably of more general occurrence.

Atlantic coasts of France, Spain, Portugal; Mediterranean.

No data on seasonal growth. Monosporangia have been observed in summer and autumn collections.

Data on form variation too inadequate for comment.

Audouinella chylocladiae differs from all other species of this genus in its very elongate and very narrow cells.

The material from *Sertularia* was distinguished as f. *pulchra,* but that name is a *nomen nudum.*

Audouinella concrescens (Drew) Dixon (1976), p. 590.

Holotype: UC (294561 = Gardner 4828) U.S.A. (Carmel Bay, California).

Rhodochorton concrescens Drew (1928), p. 167.

Plants consisting of uniseriate filaments, differentiated into erect and prostrate axes; prostrate axes epizoic, growing parallel to the surface, much branched, aggregated laterally to form a compact monostromatic plate of cells; cells of monostromatic plate elongate, 10–24 μm in length, 6–12 μm in diameter, often with fusions between cells of the same or adjacent filaments; erect axes short, composed of 2–12 cells, usually unbranched, cells cylindrical, 8–24 μm in length, 9–11 μm in diameter, with thick walls and numerous chloroplasts; chloroplasts discoid or ellipsoid, devoid of pyrenoids, closely packed and often difficult to distinguish, giving the appearance in herbarium specimens or fixed material of a single lobed parietal structure.

Gametangia unknown; tetrasporangia ovoid, 22–26 μm in length, 18–22 μm in diameter, borne terminally on the short erect axes, tetraspores cruciately arranged.

Epizoic on animals with a chitinous surface (hydroids, ectoprocts, crustaceans), particularly from the sublittoral, to 30 m.

Argyll, Devon, Pembroke; probably of more general occurrence.

Atlantic coast of France; Pacific coast of North America; New Zealand.

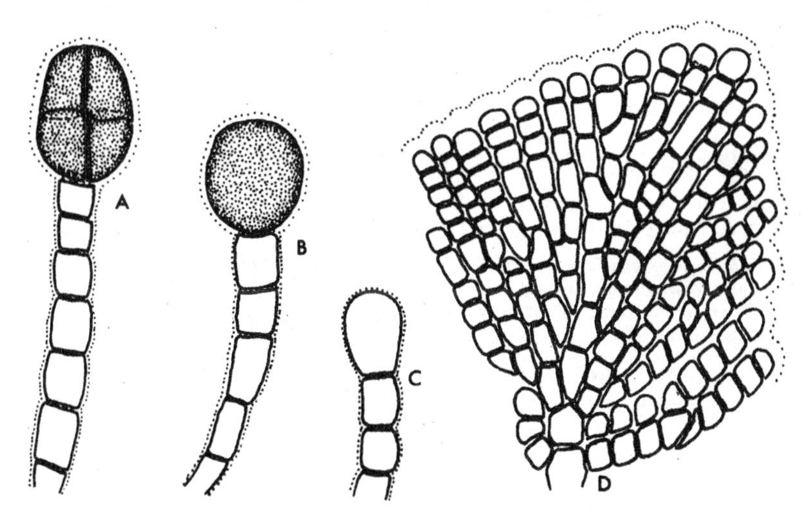

Fig. 20 *Audouinella concrescens*
A. Erect axis with tetrasporangium ×450; B. Erect axis with undivided tetrasporangial initial ×450; C. Sterile erect axis ×450; D. Prostrate base in surface view ×450.

The prostrate system is capable of perennation. Erect axes seem to be more ephemeral although insufficient information is available to indicate precisely the time of development. Tetrasporangia have been reported only in November and December in the British Isles.

The greatest variation is in the development of erect axes. These may contain as few as three cells before the apical cell is converted into a tetrasporangium, although the erect axes may contain 12 or more cells and even an occasional lateral branch. In culture, erect axes to 4 mm developed in 4 months; such plants are impossible to distinguish from *Audouinella spetsbergensis*.

The original description, based upon a herbarium specimen, described the chloroplast as single, lobed or reticulate, and parietal. Examination of fresh material shows that each cell contains numerous discoid or ellipsoid chloroplasts which are closely packed giving the appearance of a single parietal structure in herbarium specimens.

The relationship between *A. concrescens* and *A. spetsbergensis* is a critical problem requiring resolution. The organizations are identical apart from the degree of development of the upright system and, in culture, axes up to 4 mm have been obtained in *A. concrescens*. In both species, each cell contains numerous chloroplasts although the degree of packing is less dense in *A. spetsbergensis* than in *A. concrescens*. The two entities are both accepted in the present treatment pending further studies.

As *A. concrescens* and *A. membranacea* occur on chitinous animals and may occur together, there has been some confusion particularly if identification has been done without microscopic examination. Until recently, all pinkish patches occurring on hydroids were referred to *A. membranacea* and records of this type need to be re-examined critically. The two algae do differ in various respects. *A. concrescens* occurs only on the outer surface of the wall whereas *A. membranacea* occurs *within* the wall; the prostrate system of *A. concrescens* is always a compact plate with laterally adherent filaments, whereas in *A. membranacea* it is very variable, the filaments either loose or distant, twisted together or forming an irregular plate, even in a single specimen.

Audouinella corymbifera (Thuret in Le Jolis) Dixon (1976), p. 590.

Holotype: cannot be located; it would seem best to typify by the illustration given by Bornet & Thuret (1876), Pl. V (see Dixon, 1977). France (Belle-Ile-en-Mer).

Chantransia corymbifera Thuret in Le Jolis (1863), p. 107, pro parte.
Acrochaetum corymbiferum (Thuret in Le Jolis) Batters (1902), p. 59.
Rhodochorton corymbiferum (Thuret in Le Jolis) Drew (1928), p. 183.
Acrochaetium bornetii Papenfuss (1945), p. 313.

Plants consisting of uniseriate filaments, differentiated into erect and prostrate axes, red and brownish-red in colour; prostrate axes developing from the original spore which remains throughout as a globular cell, penetrating deeply into the host thallus and spreading widely, much branched, undulate, cells elongate, 25–40 μm in length, 8 μm in diameter; erect axes to 2 mm, arising from the original spore as well as by conversion from the endophytic system, much branched, branching largely unilateral, cells cylindrical, 22–35 μm in length, 7–9 μm in diameter, when first formed, increasing in size with age to reach a length of 40–55 μm and breadth of 8–10 μm, each with a single chloroplast; chloroplast parietal, lobed, containing a pyrenoid.

Gametangial plants dioecious; spermatangia 4–5 μm, without chloroplasts, developing in corymbose clusters on a short filament of one or two cells which is formed laterally on a principal axis; carpogonia 10 μm in length, with an elongate (10 μm) trichogyne, formed singly on a short filament of 1 or 2 cells developing laterally on a principal axis; gonimoblasts developing directly and giving rise to dense corymbose clusters of short filaments, the terminal cells of which form carposporangia; carposporangia ovoid, 14 μm in

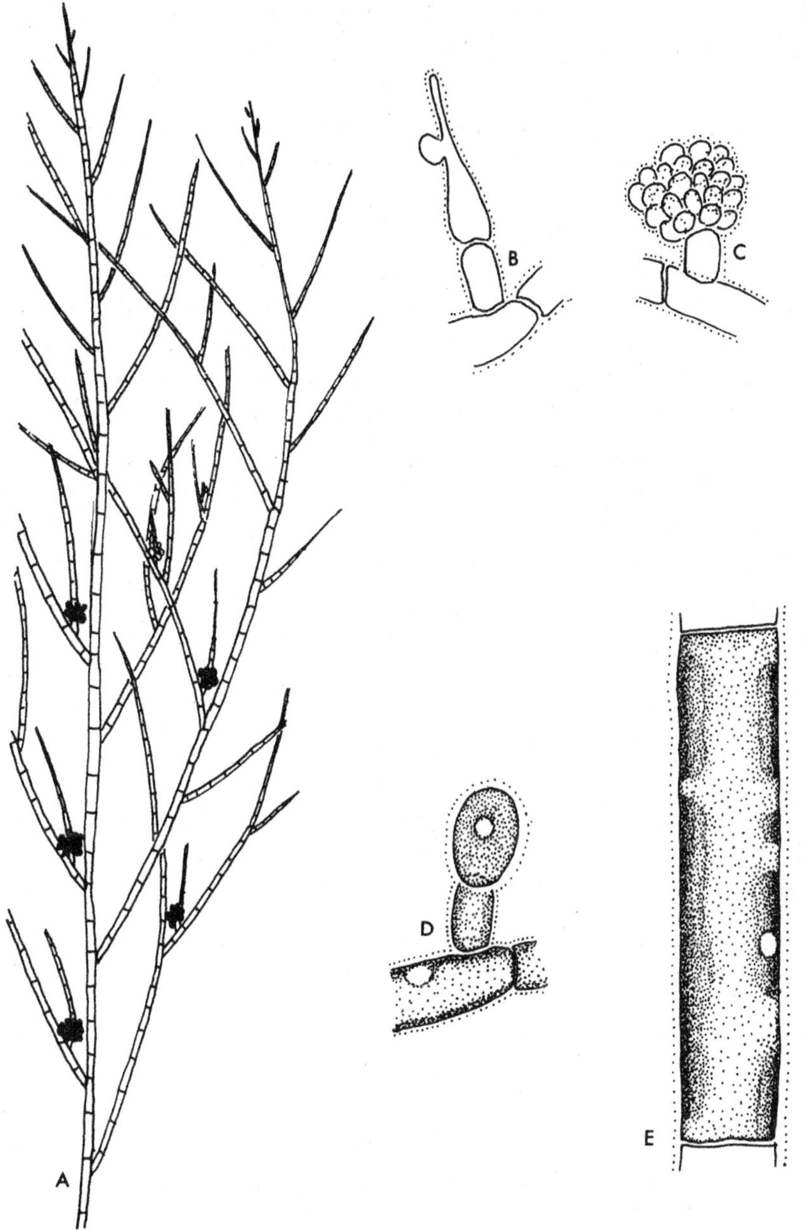

Fig. 21 *Audouinella corymbifera*
A. Habit ×160; B. Carpogonium ×800; C. Spermatangial cluster ×800; D. Monosporangium ×800; E. Cell structure ×1650.

length, 9 μm in diameter; tetrasporangia unknown; monosporangia ovoid, 15–18 μm in length, 8–10 μm in diameter, borne on the gametangial plant.

Epiphytic, occurring in the sublittoral to 5 m.

Devon, Sussex, Clare; probably of more general occurrence.

Widely reported from the warmer seas, Mediterranean, Bermuda, North Carolina and California.

Data on seasonal behaviour and form variation too inadequate for comment.

The original description of *Chantransia corymbifera* by Thuret (in Le Jolis, 1863) was based on a mixture of two elements, one growing on *Ceramium* and the other on *Helminthocladia*. A subsequent, more detailed treatment was restricted to the plant growing on *Helminthocladia* (Bornet & Thuret, 1876) and it is this plant for which the epithet *corymbifera* must be retained. Papenfuss (1945) argued that the epithet *corymbifera* must be rejected as a name based on two discordant elements and replaced this epithet by *bornetii*. This argument is not correct and the original epithet has been retained in the present treatment.

The material from *Ceramium* originally included in the treatment of *Chantransia corymbifera* was subsequently described (Bornet, 1904) as *Chantransia efflorescens* var. *thuretii* and later assigned specific status as *Acrochaetium thuretii* (see *Audouinella thuretii*).

Audouinella daviesii (Dillwyn) Woelkerling (1971), p. 28.

Lectotype: BM. Probably north Wales (see Dixon, 1977).

Conferva daviesii Dillwyn (1809), p. 73.
Acrochaetium daviesii (Dillwyn) Nägeli (1861 [1862]), p. 405.
Chantransia daviesii (Dillwyn) Thuret in Le Jolis (1863), p. 106.
Rhodochorton daviesii (Dillwyn) Drew (1928), p. 172.
Trentepohlia mirabilis Suhr (1839), p. 73.
Acrochaetium mirabile (Suhr) Nägeli (1861 [1862]), p. 405.
Chantransia mirabilis (Suhr) Batters (1896), p. 9, nom illeg, non *C. mirabilis* Heydrich (1892), p. 475.

Plants consisting of uniseriate filaments, differentiated into erect and prostrate axes, red or brownish-red in colour; prostrate axes epizoic, epiphytic or, rarely, with slight endophytic penetration, much branched, branching highly irregular, forming an entangled mass, often tightly compacted, largely superficial, with the original spore not obvious beyond the earliest stages of development; erect axes to 10 mm, tufted, much branched, branching highly irregular but usually in one plane; cells cylindrical, 18–30 μm in length, 8–12 μm in diameter when first formed, increasing in size with age to reach a length of 30–55 μm and diameter of 9–15 μm, each with a single chloroplast; chloroplast parietal, lobed, containing one pyrenoid.

Gametangial plants unknown in Europe; tetrasporangia ovoid, 16–24 μm in length, 12–20 μm in diameter, arising singly or in clusters on a single stalk cell or a branched cluster of cells which develop first on the lowermost cell or cells of lateral axes and rarely then develop elsewhere; monosporangia ovoid, 12–24 μm in length, 8–16 μm in diameter, arising singly or in clusters on a single stalk cell or a branched cluster of cells which develop first on the lowermost cell or cells of lateral axes and subsequently elsewhere, tetraspores cruciately arranged.

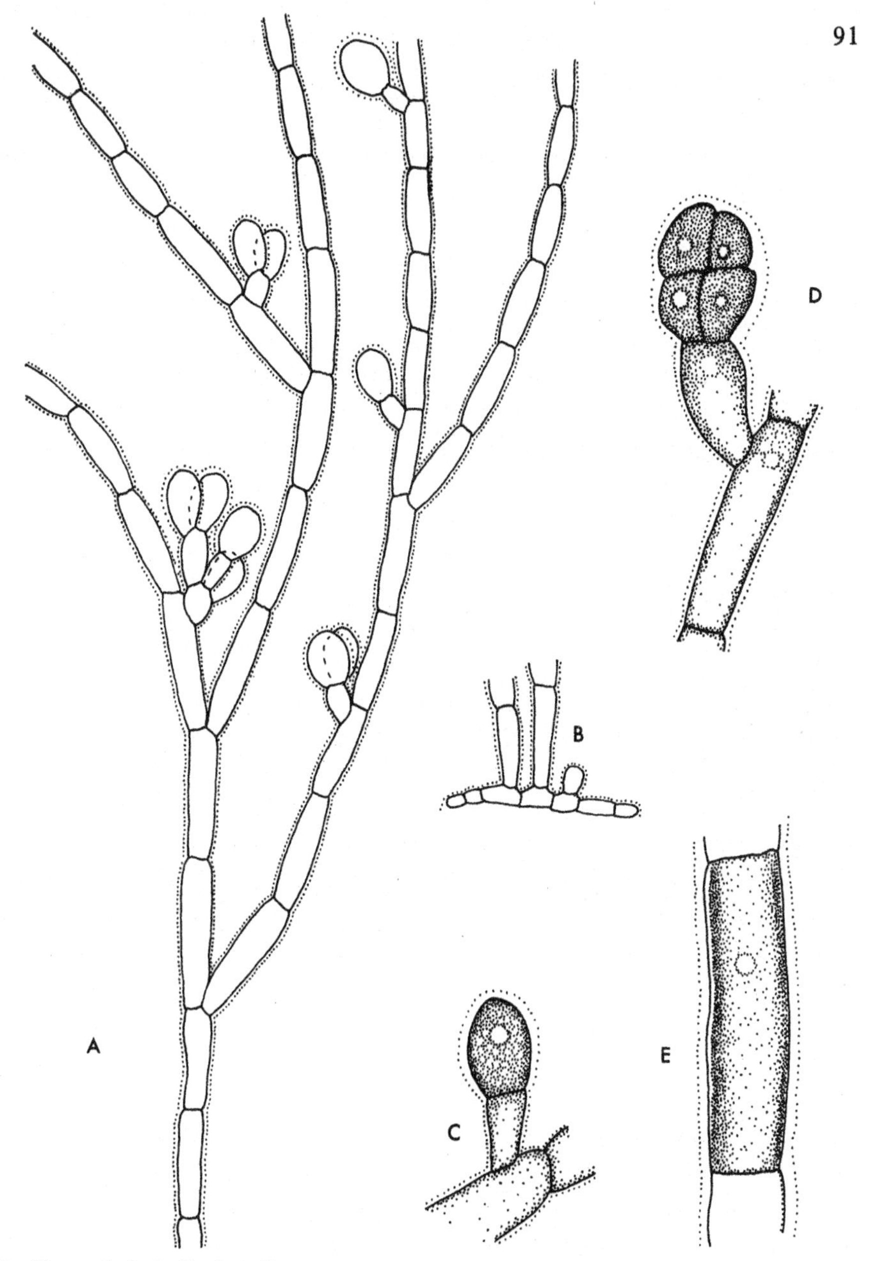

Fig. 22 *Audouinella daviesii*
A. Habit ×450; B. Base of plant ×450; C. Monosporangium ×450; D. Tetrasporangium ×740; E. Cell detail ×1350.

Epiphytic or (more rarely) epizoic on a wide variety of substrate species in the lower littoral and sublittoral, to a depth of 5 m. *Palmaria palmata* (L.) O. Kuntze, is the species on which it is found most frequently and on which its development is most luxuriant.

Generally distributed throughout the British Isles.

Norway to Morocco; Canary Islands; U.S.A. (New Jersey to Maine); virtually cosmopolitan.

The erect axes are most conspicuous between March and October; the prostrate system possibly persists for long periods. Both monosporangia and tetrasporangia occur between May and October and may even persist longer.

The variation of *A. daviesii* is not very great other than in overall size.

Gametangial plants have been reported only rarely and never in Europe. Woelkerling (1971) lists three reports of their occurrence, two from Australia and one from North America.

There are many non-European entities possibly referable to *A. daviesii* but in need of further investigation before final attribution is possible.

Audouinella efflorescens (J. Agardh) Papenfuss (1945), p. 326.

Holotype: LD (Herb. Alg. Agardh 35129). Sweden (Kullaberg) (see Dixon, 1977).

Callithamnion efflorescens J. Agardh (1851), p. 15.
Acrochaetium efflorescens (J. Agardh) Nägeli (1861 [1862]), p. 405.
Chantransia efflorescens (J. Agardh) Kjellman (1875), p. 14.
Rhodochorton efflorescens (J. Agardh) Drew (1928), p. 151.

Plants consisting of uniseriate filaments, differentiated into erect and prostrate axes, pale red in colour but often completely colourless towards the apices; prostrate axes branched irregularly and compacted to a varying degree, with the original spore often visible even in old plants except where the base becomes particularly solid and compact, sometimes partially endophytic when growing on substrates of soft texture; erect axes either to 6 mm, tufted (gametangial plants), or forming a felt-like covering, to 2 mm (tetrasporangial plants), sparingly branched in the lower parts, increasingly so towards the apices, downgrowing filaments appressed to the erect axes in the lower parts; cells cylindrical, very elongate, 45–90 μm in length, 4–7 μm in diameter when first formed and undergoing little change in size or shape with age, each with one or more chloroplasts; chloroplasts parietal, spiral-shaped, possibly fragmenting or fusing with age.

Gametangial plants monoecious; spermatangia small, 4–5 μm, without chloroplast, formed singly or in pairs on short, one-celled lateral filaments which are often clustered; carpogonia small, 5 μm, with a short trichogyne, to 4 μm, formed singly in a terminal or intercalary position, gonimoblast developing directly and giving rise to a cluster of short, branched filaments, each of 2–3 cells, the ultimate cells of which develop into carposporangia, 8–12 μm in length, 7–9 μm in diameter; tetrasporangia ovoid, 14–25 μm in length, 8–12 μm in diameter, formed singly or in pairs on short, usually one-celled stalk, tetraspores cruciately arranged; monosporangia ovoid, 10–18 μm in length, 5–9 μm in diameter, formed singly or in pairs, on short one-celled lateral filament, occurring mixed with tetrasporangia or on a thallus alone.

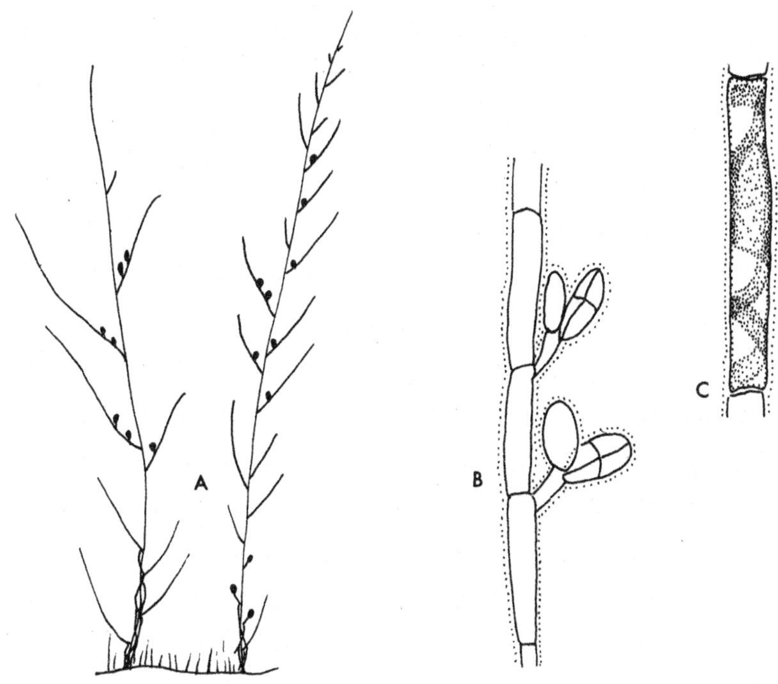

Fig. 23 *Audouinella efflorescens*
A. Habit ×50; B. Tetrasporangia ×600; C. Cell detail ×1200.

Occurs on a wide variety of substrates, but principally epiphytic on various algae and *Zostera*, although also reported from hydroids, ascidians, stones and shells, from the sublittoral to a depth of 15 m.

Apparently restricted to northern and eastern shores, from Caithness to Yorkshire.

Denmark, Sweden; widely distributed in the Arctic Ocean.

Tetrasporangial plants appear during March and persist until July/August while gametangial plants appear later and persist until August/September. Neither plant is apparent during the winter months. Tetrasporangia occur from May to July with gametangia from April and the resulting cystocarps from July/August. Monosporangia occur sparsely from April to August.

There are marked morphological differences between gametangial and tetrasporangial plants. The thallus in the former is erect and tufted while in the latter it is a low, felt-like growth, so different that it was regarded as an independent species (*Rhodochorton chantransioides* Reinke).

Although there has been no objection to the association of *Audouinella efflorescens* and *Rhodochorton chantransioides*, there is no direct proof of the relationship either by cytological or cultural investigation and no appreciation of the role of monosporangia.

Audouinella emergens (Rosenvinge) Dixon (1976), p. 590.

Holotype: C (Rosenvinge 6574). Denmark (Møllegrund, off Hirshals).

Chantransia emergens Rosenvinge (1909), p. 128.
Achrochaetium emergens (Rosenvinge) Weber-van Bosse (1921), p. 194.

Plants consisting of uniseriate filaments, entirely endophytic in the outer cell wall of the host species or occasionally with a few one- or two-celled emergent filaments, rose-pink in colour; filaments spreading, branched, branching highly variable in extent but usually arising at right angles; cells cylindrical or barrel-shaped, 8–14 μm in length, 2–4 μm in diameter; chloroplast parietal, large and simple, forming an almost complete cylinder adjacent to the cell wall, apparently devoid of pyrenoid.

Gametangial plants unknown; tetrasporangia unknown; monosporangia ovoid, 4–6 μm long, 3–4 μm in diameter, either sessile and endophytic or developing above the surface of the host species on the short emergent filaments.

Endophytic in the outer cell wall and between the cells of species of *Polysiphonia,* from the upper sublittoral.

Galway, Isle of Man; probably of more general occurrence.

Denmark, Norway.

Collections appear to be limited to the spring and summer, with monosporangia reported in May.

Data on form variation too inadequate for comment.

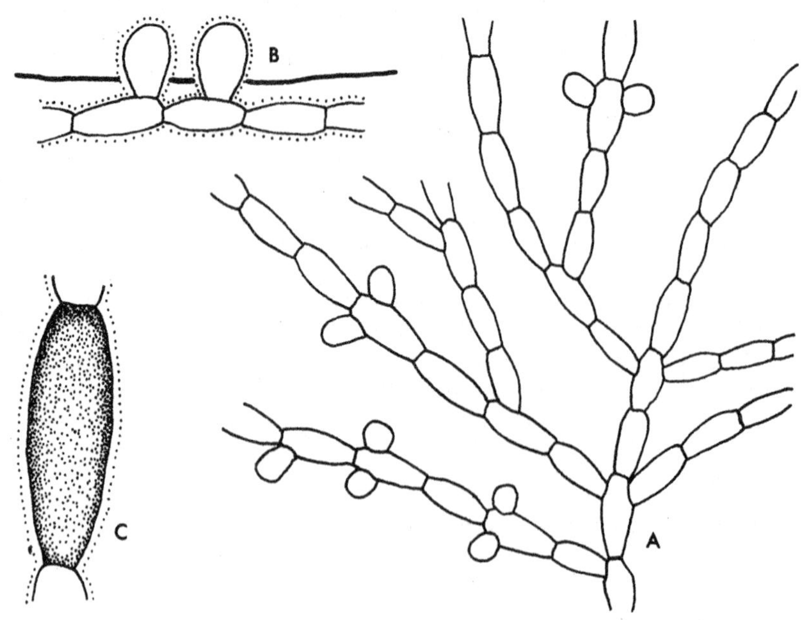

Fig. 24 *Audouinella emergens*
A. Habit ×1150; B. Side view with monosporangia ×1600; C. Cell detail ×3000.

Audouinella emergens is similar to *A. endophytica* and to the plant described by Rosenvinge (1909) as *Chantransia immersa,* growing endophytically in species of *Polysiphonia* and *Rhodomela* in Denmark. More information is needed before the status of these entities can be evaluated although the points of similarity are such as to suggest synonymy. Woelkerling (1973) reduces this entity to the synonymy of *Colaconema minima (Acrochaetium minimum)*; although possible, this suggestion is not adopted here until more detailed trans-oceanic comparisons can be made between the European and North American entities.

Audouinella endophytica (Batters) Dixon (1976), p. 590.

Lectotype: BM (Slide 7020). Devon (Plymouth).

Acrochaetium endophyticum Batters (1896a), p. 386.

Plant consisting of uniseriate filaments, endophytic in the outer cell wall and between the outermost cells of the host species, usually with short emergent filaments of one to four cells, pale to rose pink in colour; filaments spreading, profusely and irregularly branched to form a complex network; cells elongate, cylindrical, 5–20 μm in length, 2–4 μm in diameter, those of the short emergent filaments usually more cuboid, 4–6 μm in length, 2–4 μm in diameter; chloroplast parietal, apparently devoid of pyrenoid.

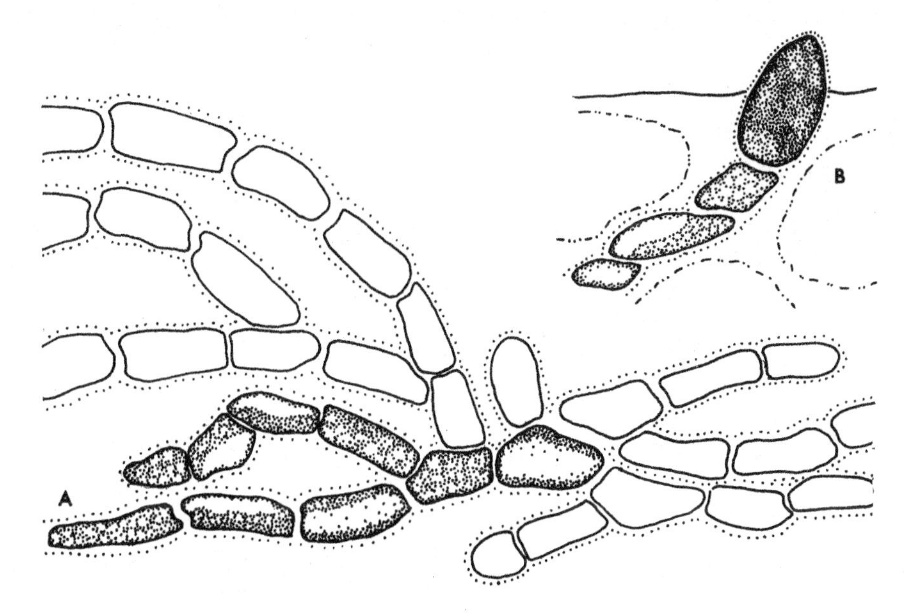

Fig. 25 *Audouinella endophytica*
A. Habit in surface view ×2000; B. Emergent axis with monosporangium in side view ×2000.

Gametangial plants unknown; tetrasporangial plants unknown; monosporangia spherical or slightly ovoid, 6–7 μm in length and diameter, formed singly and terminally on the emergent filaments.

Endophytic in *Heterosiphonia plumosa* (Ellis) Batt. from the sublittoral, to 5 m.
Generally distributed throughout the British Isles.
Apparently not otherwise recorded.

No data on seasonal growth. Monosporangia have been detected at all times of the year.

Data on form variation too inadequate for comment.

For comments on the relationship of *Audouinella endophytica, A. emergens* and the plant described as *Chantransia immersa* see p. 95.

Audouinella endozoica (Darbishire) Dixon (1976), p. 590.

Lectotype: original illustration (Darbishire, 1899, pl. 1) in the absence of material. Kerry (Valencia).

Chantransia endozoica Darbishire (1899), p. 15.
Acrochaetium endozoicum (Darbishire) Batters (1902), p. 58.

Plants consisting of uniseriate filaments, endozoic in the outer membranes of Bryozoa with some emergent filaments which may be branched and which may reach a length of 85 μm, rose-pink in colour; filaments spreading, richly branched; cells elongate or barrel-

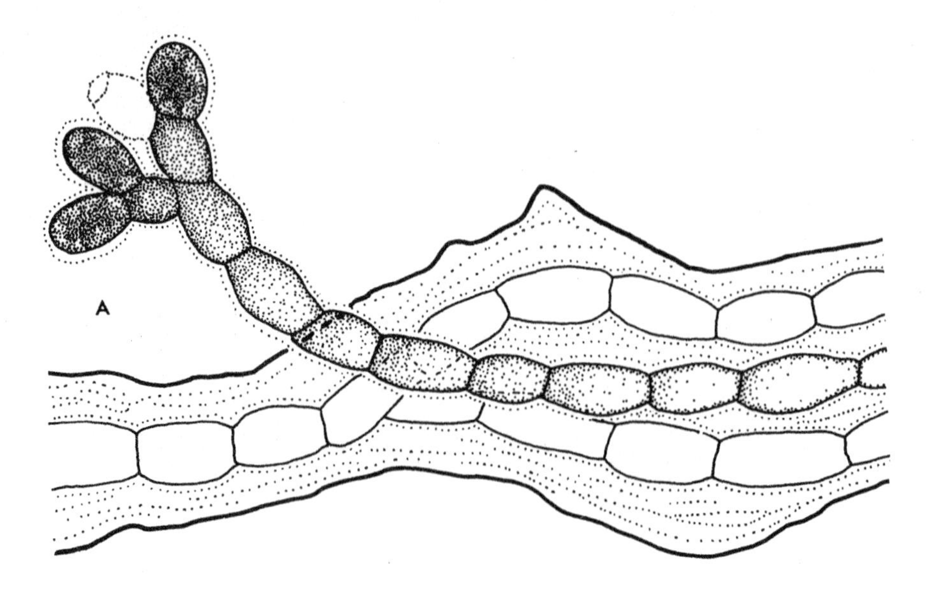

Fig. 26 *Audouinella endozoica*
A. Habit of plant in *Alcyonidium* with emergent axis bearing monosporangia.

shaped, 6–25 μm in length, 6–13 μm in diameter; chloroplasts parietal and simple, apparently devoid of pyrenoid.

Gametangial plants unknown; tetrasporangia unknown; monosporangia ovoid, 10–12 μm in length, 8–10 μm in diameter, formed both laterally and terminally.

Endozoic in the outer membranes of *Alcyonidium* and other Bryozoa, in the sublittoral, to 15 m.

Widely distributed throughout the British Isles.

Southern Norway to French Channel coast.

A. endozoica appears to persist throughout the year although most obvious between March and September. Monosporangia found between March and May.

Data on form variation too inadequate for comment.

Audouinella floridula (Dillwyn) Woelkerling (1971), p. 30.

Lectotype: BM. Galway coast (see Dixon, 1977).

Conferva floridula Dillwyn (1809), p. 73.
Rhodochorton floridulum (Dillwyn) Nägeli (1861 [1862]), p. 358.
Chromastrum floridulum (Dillwyn) Papenfuss (1945), p. 323,
Kylinia floridula (Dillwyn) Papenfuss (1947), p. 437.

Plants consisting of uniseriate filaments, differentiated into erect and prostrate axes, brownish-red in colour, the basal parts often colourless or greenish in appearance; prostrate axes branched irregularly and compacted to a varying degree, with the original spore not persistent; erect axes to 30 mm, tufted or matted, sparingly branched in the lower parts, increasingly branched towards the apices but highly irregular and variable; cells cylindrical, 25–45 μm in length, 15–20 μm in diameter when first formed, increasing in size with age to reach a length of 60–120 μm and diameter of 20–30 μm, with 3–8 chloroplasts; chloroplasts sinuate, parietal, each with one pyrenoid; herbarium specimens not adhering well to paper.

Gametangial plants unknown in the field in Europe; tetrasporangia ovoid, 20–40 μm in length, 18–33 μm in diameter, arranged in secund series on the upper parts of the erect axes, occurring singly or in clusters of 2–5 sporangia, either sessile or on a single-celled stalk, occasionally the stalk may contain two or even three cells, tetraspores cruciately arranged.

Lower littoral and sublittoral, to 5 m, associated with sandbinding and producing a spongy carpet.

Generally distributed throughout the British Isles.

France, Spain, Portugal.

Perennial, occurring in quantity at all times of the year. Growth is most luxuriant during the winter while during the summer the compacted sandy carpet may become bleached or disrupted. Tetrasporangia most frequent between November and March.

The sandbinding capacity of this species results from the development of several types of specialized lateral filaments which interweave with the erect filaments to produce a complex network to which sand particles adhere.

The product of germination of tetraspores has been interpreted as being gametangial although details are not fully known. The germlings are of similar size to *Audouinella rosulata* although different in many details.

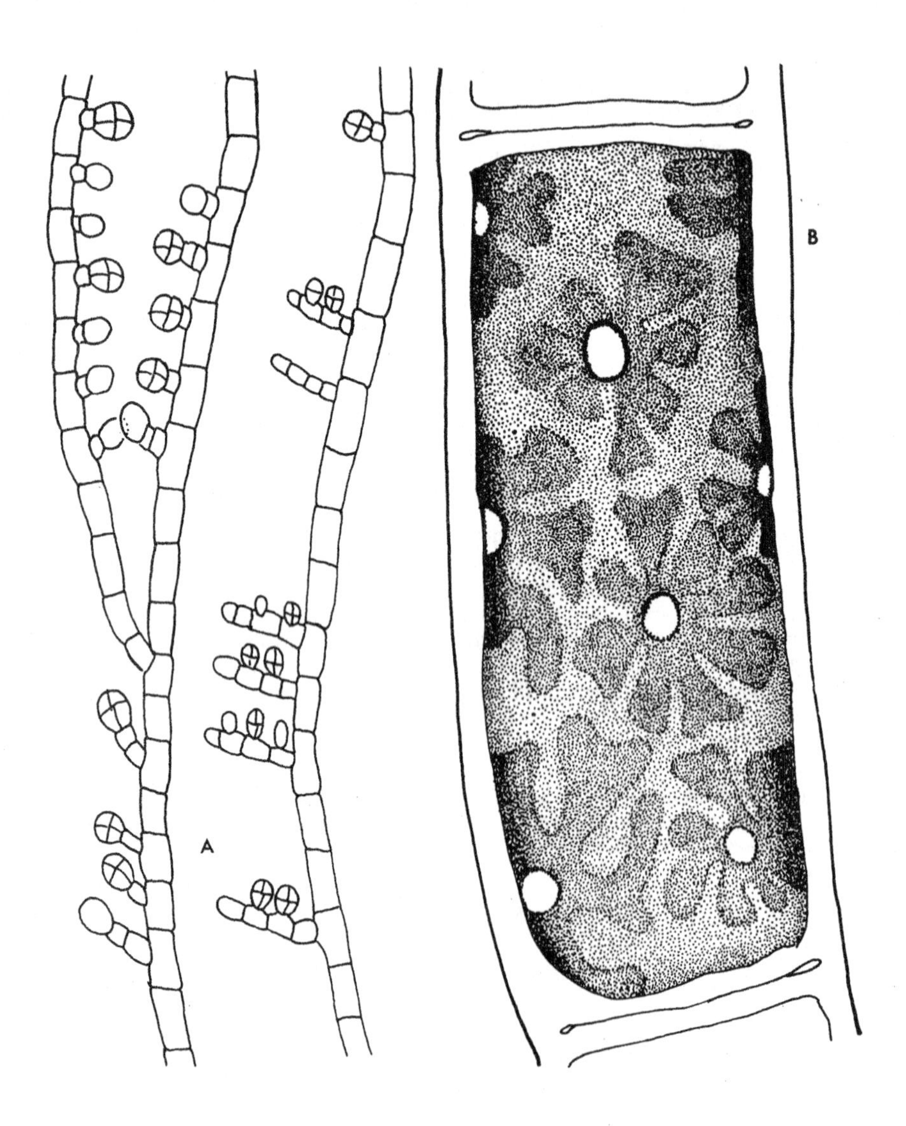

Fig. 27 *Audouinella floridula*
A. Habit ×120; B. Cell structure ×1500.

Audouinella infestans (Howe & Hoyt) Dixon (1976), p. 590.

Lectotype: NY. U.S.A. (Beaufort, North Carolina).

Acrochaetium infestans Howe & Hoyt (1916), p. 116.
Chromastrum infestans (Howe & Hoyt) Papenfuss (1945), p. 324.
Kylinia infestans (Howe & Hoyt) Papenfuss (1947), p. 438.

Plant consisting of uniseriate filaments, endozoic in the inner layers of the perisarc of hydroids, with some emergent filaments which are often branched and which may attain a length of up to 90 μm, pale rose-pink in colour; endozoic filaments spreading, profusely and irregularly branched, tortuous and intricate, sometimes closely packed; cells of very variable outline, some simple and cylindrical, some curved and contorted, some deeply divided in various ways to give a furcate or even cruciate outline, 12–60 μm in length, 2–6 μm in diameter, when tightly packed into pseudoparenchyma, becoming quadrate or polygonal, 10–13 μm in diameter; emergent filaments consisting of one or more cells, the larger containing 10 or more cells, often with secund branching; cells 6–12 μm in length, 4–7 μm in diameter, usually with very elongate hairs which may attain a length of 120–170 μm; chloroplasts small, discoid, usually with a pyrenoid.

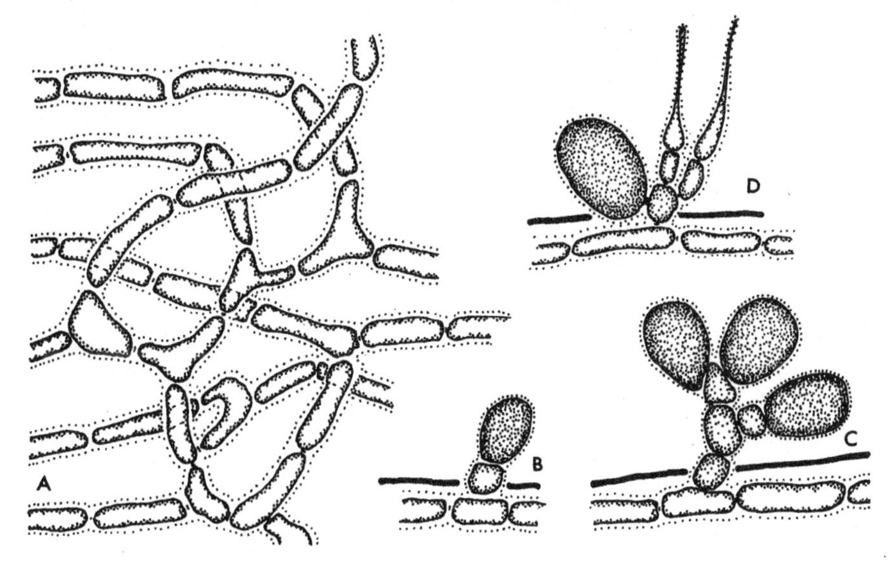

Fig. 28 *Audouinella infestans*
A. Axes in surface view ×650; B. Emergent axis, with monosporangium ×650; C. Emergent axis with several monosporangia; D. Emergent axes, with monosporangium and hairs ×650.

Gametangial plants unknown; tetrasporangia unknown; monosporangia ovoid, 10–14 μm in length, 6–9 μm in diameter, borne singly in a terminal position on an emergent filament or in a cluster of 1–3 sporangia or terminally on a short secund lateral filament of the emergent system.

Endozoic in the inner layers of the perisarc of hydroids, lower littoral.

Widely distributed throughout the British Isles, except for eastern and southeastern coasts of England.

U.S.A. (North Carolina); Mediterranean.

Monosporangia present at most times of the year.

Data on form variation too inadequate for comment.

Audouinella lorrain-smithiae (Lyle) Dixon (1976), p. 590.

Provisional lectotype: BM (Slide 3679). Channel Islands (Guernsey).

Chantransia lorrain-smithiae Lyle (1920), p. 13.
Acrochaetium lorrain-smithiae (Lyle) Newton (1931), p. 256.

Plants consisting of uniseriate filaments, differentiated into erect and prostrate axes, rose-carmine or greenish in colour; prostrate axes well developed and usually compacted into a solid discoid base; erect axes to 9 mm, branching sparse in the lower parts other than for the production of abundant down-growing filaments, increasing towards the apices and

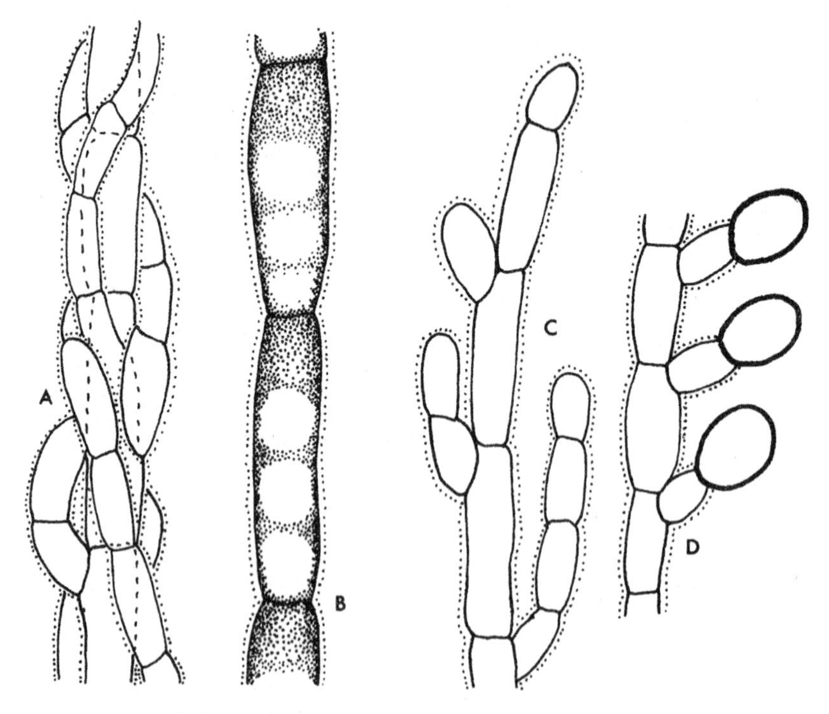

Fig. 29 *Audouinella lorrain-smithiae*
A. Base of plant ×450; B. Cell structure ×450; C. Apex ×450; D. Monosporangia ×450.

highly variable; cells cylindrical, 45–60 μm in length, 15–20 μm in diameter when mature, with thick walls.

Gametangial plants unknown; monosporangia ovoid, 30–35 μm in length, 15–20 μm in diameter, occurring singly or rarely in groups of two or three, borne on single-celled stalk which arises from the lowermost cells of the ultimate lateral filaments on the adaxial surface.

Epiphytic on *Saccorhiza polyschides* (Lightf.) Batt. at an unstated level, presumably upper sublittoral.

Devon, Channel Islands.

Not recorded outside the British Isles.

The only known material was collected during the autumn months (September–November); this bore sporangia interpreted as monosporangia.

Data on form variation too inadequate for comment.

It has been suggested (Hamel, 1927) that the present entity should be reduced to the synonymy of *A. caespitosa*. This suggestion was based only on the illustration given by Lyle (1920, fig. 1a–f). Hamel's suggestion is likely to be correct although *A. lorrain-smithiae* is retained pending examination of further material.

Audouinella membranacea (Magnus) Papenfuss (1945), p. 326.

Lectotype: original illustration (Magnus, 1875, pl. 2, fig. 8), in the absence of material. Denmark (off Korsör).

Callithamnion membranaceum Magnus (1875), p. 67.
Rhodochorton membranaceum (Magnus) Hauck (1885), p. 69.
Colaconema membranacea (Magnus) Woelkerling (1973), p. 566.

Plants consisting of uniseriate filaments, differentiated into erect and prostrate axes, bright scarlet in colour; prostrate axes endozoic, growing parallel to the surface, branched, branching opposite, alternate or irregular, sometimes loose and distant, sometimes twisted together in an irregular manner, sometimes adhering laterally to form a compact sheet of cells, the form being determined by the substrate species, cells of varying shape, 14–25 μm in length, 7–8 μm in diameter, cylindrical or polygonal, often with corrugations of the wall; erect filaments short, composed of 3 to 6 cells, usually unbranched but occasionally with a cluster of branches at the apical end, emerging above the surface of the host species, cells cylindrical, 10–16 μm in length, 7–8 μm in diameter, with chloroplasts; chloroplasts several in each cell, ribbon-shaped, more readily observable in the prostrate than the erect filaments where they frequently pack to form a single parietal sheet; hairs reported as of rare occurrence and not observed in any British specimens.

Gametangia unknown; tetrasporangia ovoid, 30–60 μm in length, 10–20 μm in diameter, cruciately divided, occurring in a terminal position on the erect filaments or on the lateral branches of the latter; monosporangia unknown.

Endozoic in *Sertularia*, in the sublittoral, to 30 m. There are also some reports of its occurrence in sponges.

Widely distributed throughout the British Isles.

Norway to the Atlantic coast of France; U.S.A. (Connecticut to Maine); California.

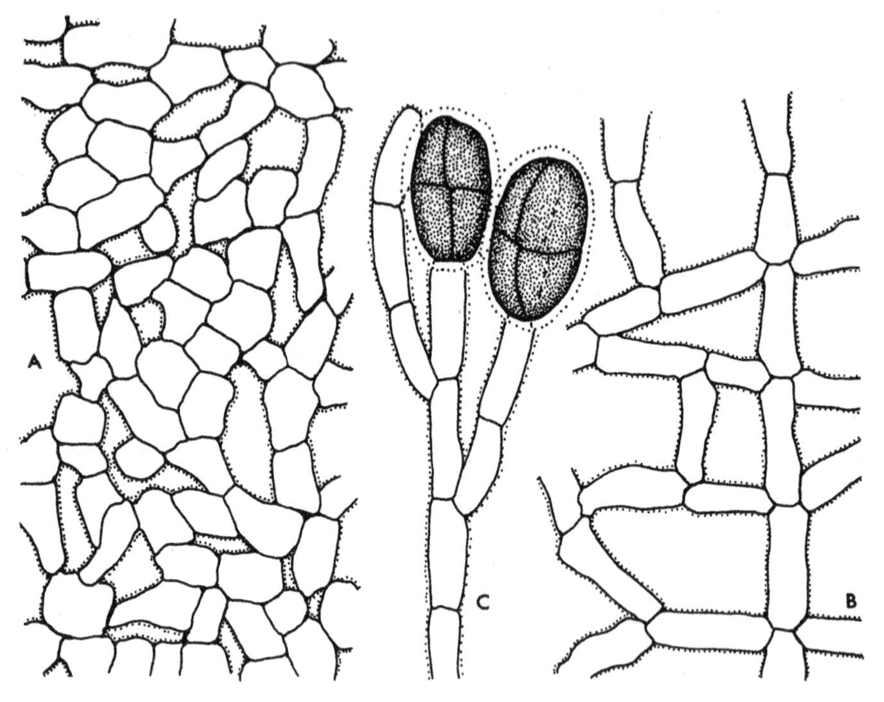

Fig. 30 *Audouinella membranacea*
A. Compact prostrate axes in face view ×350; B. Diffuse prostrate axes in face view ×350; C. Erect axes with tetrasporangia ×350.

Present at all times of the year, clearly long-lived and possibly perennial. Little indication of any seasonal occurrence of tetrasporangia although they are least abundant during the autumn.

There is considerable variation related to the host species in which the plant is growing. Rosenvinge (1924) gives a detailed account of the differing aspects of plants in terms of host species and cell size, cell shape, the degree of wall corrugation, the pattern of branching and the degrees of twisting and adhesion of the filaments.

Audouinella microscopica (Nägeli in Kützing) Woelkerling (1971), p. 33.

Holotype: L (Herb. Lugd. Bat. 940.285.306). Possible syntypes: Hauck & Richter, Phyk. Univ., no. 454. Devon (Torquay).

Callithamnion microscopicum Nägeli in Kützing (1849), p. 640.
Acrochaetium microscopicum (Nägeli in Kützing) Nägeli (1861 [1862]), p. 407.
Rhodochorton microscopicum (Nägeli in Kützing) Drew (1928), p. 163.
Kylinia microscopica (Nägeli in Kützing) Kylin (1944), p. 13.
Chromastrum microscopicum (Nägeli in Kützing) Papenfuss (1945), p. 322.

Plant consisting of a single basal cell, formed from the original spore, and usually a single erect axis to 220 µm in height, formed from a branched uniseriate filament, branching abundant, usually alternate; cells initially cylindrical, 5–7 µm in length, 3–5 µm in diameter when first formed, but increasing in size and changing in shape with age so that they can achieve a length of 15 µm and their apical diameter (10–13 µm) is greater than the basal (8–12 µm), with a single chloroplast; chloroplast lobed or stellate, with an obvious pyrenoid; apical cells frequently converted into elongate hairs, to 50 µm in length.

Gametangial plants dioecious; spermatangial plants usually smaller than those with other reproductive structures, rarely exceeding 60 µm, spermatangia small, 2–3 µm, without chloroplast, formed singly or in pairs on a one-celled filament formed laterally on a main axis; carpogonia to 5 µm in length with a short trichogyne to 5 µm, formed singly, either sessile or on a single-celled filament formed laterally on the main axis, gonimoblasts developing directly and giving rise to an irregular cluster of 6 to 14, almost spherical, carposporangia, each 5 µm in diameter; tetrasporangia not recorded in the British Isles (and apparently only once elsewhere), sessile, dimensions uncertain, about 30 µm in length and 20 µm in diameter, tetraspores said to be tetrahedrally arranged; monosporangial plants large, to 220 µm, monosporangia ovoid, 15–22 µm in length, 10–14 µm in diameter, formed laterally on the major axis, usually sessile.

Epiphytic on various algae in the lower littoral and sublittoral to 5 m.

Widely distributed throughout the British Isles.

Norway to Atlantic coast of France; Mediterranean; U.S.A. (Massachusetts to Maine); Atlantic Canada; Greenland; West Indies; Pacific Ocean.

The taxonomic problems must be solved before comments on seasonal aspects of growth and reproduction, and form variation have any validity. If the prevailing views as

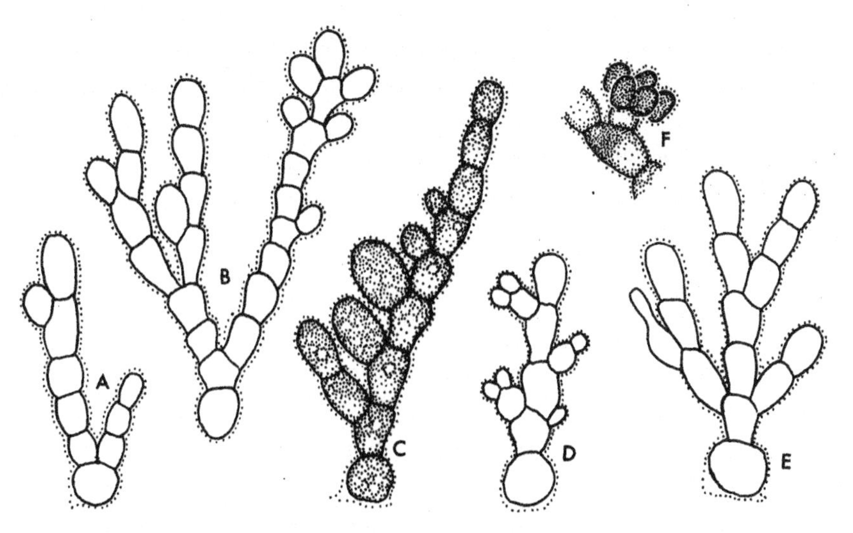

Fig. 31 *Audouinella microscopica*
A – C. Habit ×750; D. Spermatangial plant ×750; E. Carpogonial plant ×750; F. Carposporophyte ×750.

to the large number of possible synonyms are proved correct, the extent of form variation will be of considerable magnitude.

Species of *Audouinella* where the thallus arises from a prominent single basal cell are extremely confused. Material referred to *Acrochaetium microscopicum* from various parts of Europe was regarded by Hamel (1927) as representative of several distinct taxa. The points of difference between these are not great and a full investigation of their significance is needed so that the taxa can be accepted or rejected; *Audouinella battersiana* (q.v.) is one such segregate.

Kylin (1906) and Rosenvinge (1909) both described a number of new species from northern Atlantic waters, some said to occur in the British Isles. Lund (1959) dismisses most of the northern records of *A. microscopica* as *A. parvula* (q.v.) and refers *A. rhipidandra* (q.v.) also to this entity. Woelkerling (1971) has recently indicated various taxa, from all parts of the world, which are either possibly conspecific with *A. microscopica* or very similar to it. Of these, two have been reported from Britain, *A. trifila* (q.v.) and *A. parvula* (q.v.). These two species are retained in the present treatment pending the studies needed to determine their status more precisely although their continued independence is highly unlikely.

There is only one report of the occurrence of tetrasporangia in this species (Schiffner, 1931). This report must be treated with caution in view of the tetrahedral spore arrangement figured, the absence of critical description, measurements or even magnification of the figure. The dimensions quoted in the description of the species were estimated by comparison of cell size and sporangial size in Schiffner's illustration and the assumption that cell size in his material was similar to that in British specimens.

Audouinella nemalionis (De Notaris ex Dufour) Dixon (1976), p. 590.

Lectotype: BM Erbar. Critt. Ital., no. 952 (see Dixon, 1977). Italy (Cornegliano, near Genova).

Callithamnion nemalionis De Notaris ex Dufour (1863), no. 952.
Chantransia nemalionis (De Notaris ex Dufour) Ardissone & Strafforello (1877), p. 167.
Acrochaetium nemalionis (De Notaris ex Dufour) Bornet (1904), p. XX.

Plants consisting of uniseriate filaments, differentiated into erect and prostrate endophytic axes, red or brownish-red in colour; prostrate axes developing after initial spore germination, penetrating deeply into the host thallus and spreading widely, much branched, undulate or sinuate, cells 24–48 μm in length, 8–11 μm in diameter, usually containing a small, platelike chloroplast with pyrenoid; erect axes to 5 mm, tufted, arising both from the original spore and from the widespread portions of the endophytic system, much branched, branching increasingly unilateral in the outermost portions, cells cylindrical, 28–50 μm in length, 7–11 μm in breadth when first formed, increasing in size with age to reach a length of 36–60 μm and breadth of 9–12 μm, each with a single chloroplast; chloroplast parietal, lobed, containing a pyrenoid.

Gametangial plants unknown; tetrasporangia unknown; monosporangia ovoid, 18–20 μm in length, 9–11 μm in diameter, arising singly or in clusters of two or three, occasionally sessile but usually on a one- or two-celled lateral filament.

Occurs on *Nemalion helminthoides* (Vell. in With.) Batt. in the midlittoral, both as a

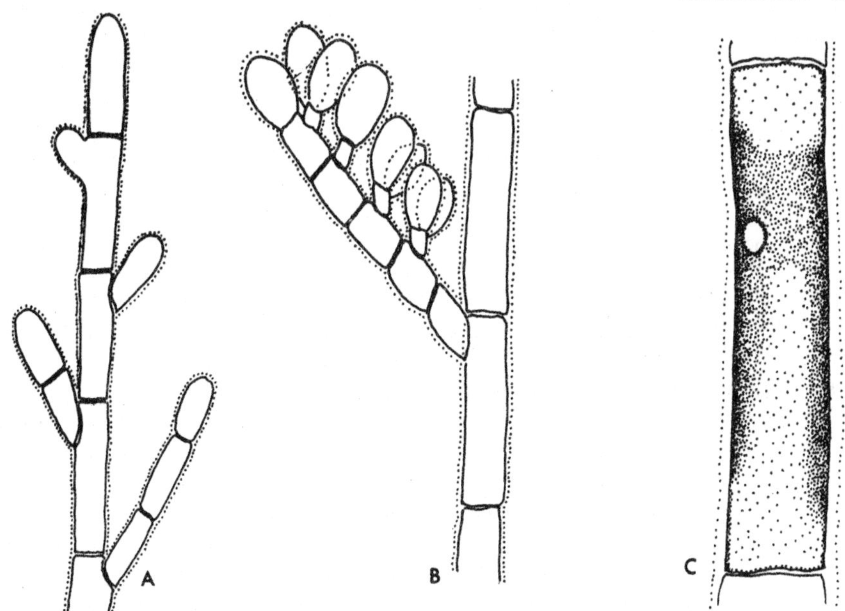

Fig. 32 *Audouinella nemalionis*
A. Apex ×750; B. Lateral axis with monosporangia ×750; C. Cell detail ×2000.

deeply-penetrating endophyte and subsequently with the development of a considerable system of emergent axes. Also reported on other substrate species outside Europe.

Isle of Man.

Denmark; Mediterranean; Canary Islands; Bermuda.

Data on seasonal behaviour and form variation too inadequate for comment.

Very probably records for Denmark and the British Isles are referable to *A. corymbifera* devoid of gametangia.

Audouinella parvula (Kylin) Dixon (1976), p. 590.

Lectotype: original illustration (Kylin, 1906, fig. 9), in the absence of material. Sweden (source of material used for lectotype illustration not known).

Chantransia parvula Kylin (1906), p. 124.
Chantransia hallandica γ parvula (Kylin) Rosenvinge (1909), p. 97.
Acrochaetium parvulum (Kylin) Hoyt (1920), p. 470.
Kylinia parvula (Kylin) Kylin (1944), p. 13.
Chromastrum parvulum (Kylin) Papenfuss (1945), p. 322.

Plants consisting of a single enlarged basal cell with a thick cell wall formed from the original spore and up to six uniseriate branched axes, to 120 μm in length, arising both horizontally and in contact with the substrate and upright and away from it, composed of 10–20 cells, profusely branched, most branches consisting only of a single cell; cells initially cylindrical 14–18 μm in length, changing shape with age so that the apical diameter (6–8 μm) is greater than the basal (5–7 μm), with chloroplast; chloroplast axile, stellate, in the apical end of the cell; apical cells occasionally converting into hairs.

Gametangial plants said to be either dioecious or monoecious; spermatangial plants often smaller than those with other reproductive structures, rarely exceeding 40 μm, spermatangia small, 2–3 μm, without chloroplasts, formed singly or in pairs, on a one-celled filament formed laterally on a major axis; carpogonia small, to 3 μm in length, with a short trichogyne, to 3 μm, formed singly, usually sessile; gonimoblasts develop directly and give rise to an irregular radiating cluster of 4–8 carposporangia; tetrasporangia not known; monosporangia ovoid, 10–14 μm in length, 6–9 μm in diameter, sessile, secund or opposite in pairs; gametangia and monosporangia often formed on the same thallus.

Epiphytic on various algae in the sublittoral, to 7 m.

Orkney, Galway; probably of wider occurrence.

Norway to Atlantic coast of France; Canary Islands; Mediterranean; U.S.A. (North Carolina).

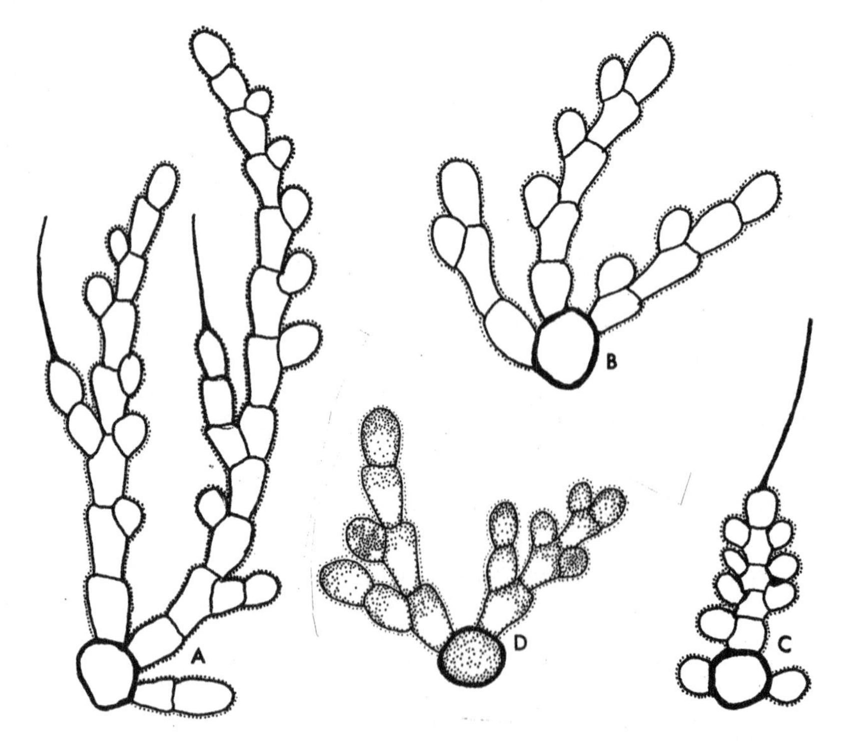

Fig. 33 *Audouinella parvula*
A – C. Habit ×740; D. Cell detail ×740.

The only collections in the British Isles have been made between August and October.

Little variation other than in overall size.

Some authors merge this species with *Acrochaetium hallandicum* (Kylin) Hamel; others keep the two entities distinct. The differences appear to involve little more than size but both species are poorly known and it would seem better at this time to retain *Audouinella parvula* as a distinct species pending further investigation of all taxa where the thallus arises from a single cell.

Borsje (1973) has compared *A. parvula* with gametangial plants formed in culture by the germination of the tetraspores of *A. virgatula*.

Audouinella purpurea (Lightfoot) Woelkerling (1973), p. 536.

Holotype: cannot be located (see Dixon, 1977). Argyll (Iona).

Byssus purpurea Lightfoot (1777), p. 1000.
Rhodochorton purpureum (Lightfoot) Rosenvinge (1900), p. 75.
Conferva violacea Roth (1797), p. 190, non *C. violacea* Hudson (1778), p. 592.
Rhodochorton rothii (Turton) Nägeli (1861 [1862]), p.356, pro parte, excl. typ., non C. *rothii* Turton (1806), p. 1809.
Thamnidium intermedium Kjellman (1875), p. 28.
Rhodochorton intermedium (Kjellman) Kjellman (1883), p. 184.
Rhodochorton parasiticum Batters (1896a), p. 389.
Rhodochorton islandicum Rosenvinge (1900), p. 75.

Plants consisting of uniseriate filaments, differentiated into erect and prostrate axes, dark red or purplish-red in colour; prostrate axes branched irregularly and compacted to varying degree, with the original spore not persisting beyond the earliest stages of development; erect axes to 30 mm, densely tufted to form a compact, soft, smooth turf, sparingly branched in lower parts, branching increasing towards the apices but never dense; cells cylindrical, 20–30 μm in length, 10–16 μm in diameter when first formed, increasing in size with age to reach a length of 35–60 μm and diameter of 14–20 μm, each with a single chloroplast; chloroplast reticulate, deeply lobed and becoming fragmented with age into several parts, apparently with no pyrenoid, adhering well to paper in herbarium specimens.

Gametangial plants unknown in the field in Europe; tetrasporangia ovoid, 25–35 μm in length, 15–24 μm in diameter, arranged in dense corymbose clusters, on branched stalks of 2–3 cells, usually occurring only in a terminal position but also occurring in lateral positions in the upper portions of the erect axes, tetraspores cruciately arranged.

Occurs in a wide range of habitats, from rock exposed to sea spray, throughout the littoral particularly where light intensity is much reduced, as in caves or under dense fucoid cover, where *A. purpurea* tends to occur on rock, to the sublittoral epiphytic populations on the stipes of *Laminaria hyperborea* (Gunn.) Fosl.

Generally distributed throughout the British Isles.

Norway to Mauretania; Mediterranean; U.S.A. (New York to Maine); Atlantic Canada and Greenland; Pacific coast (Alaska to Mexico).

Plants perennial, producing tetrasporangia between October and March; gametangial plants known only in culture.

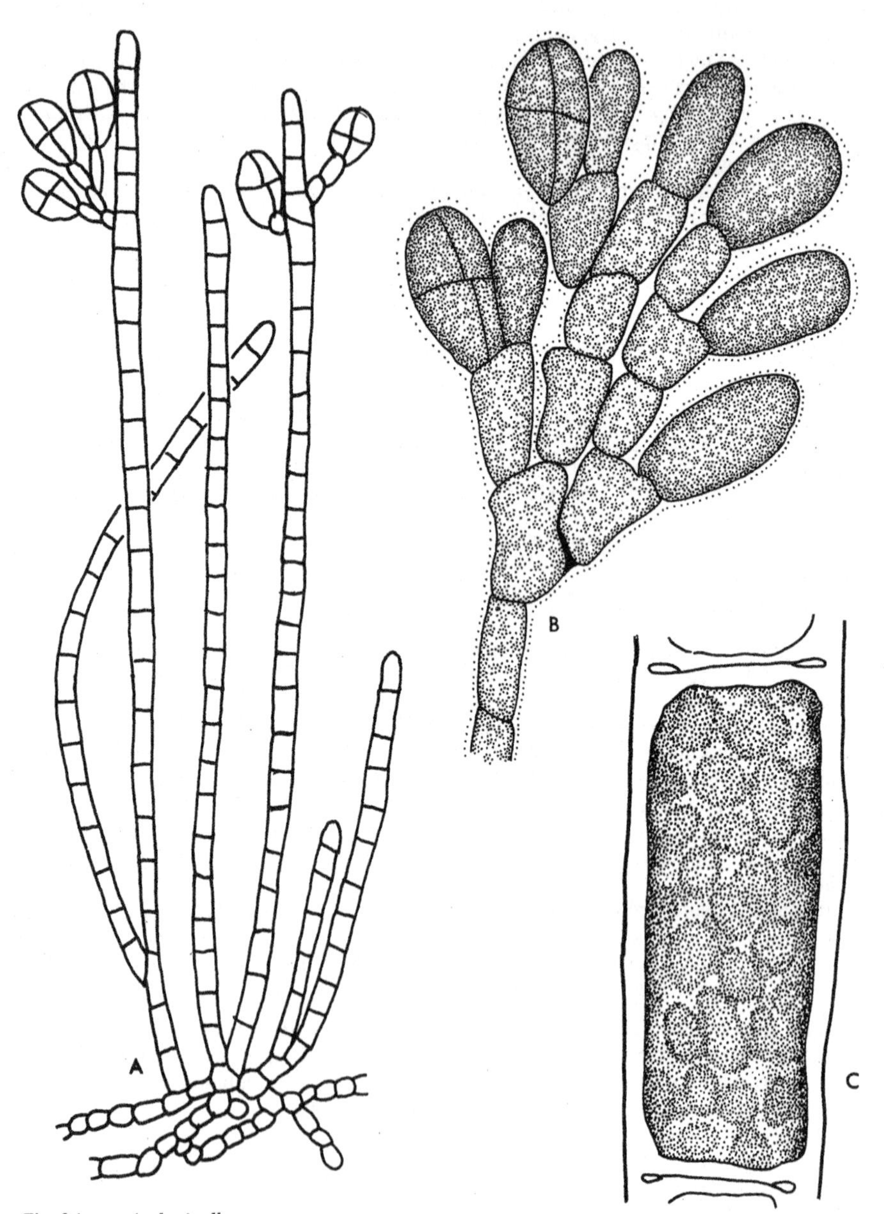

Fig. 34 *Audouinella purpurea*
A. Habit ×120; B. Terminal tetrasporangia ×750; C. Cell structure ×1500.

Audouinella purpurea is a very polymorphic species. Several formae of distinct morphology have been described although intergrades occur frequently and the morphological variation appears to be closely related to environmental conditions. For these reasons, these formae have not been accepted in the present treatment other than for purposes of convenience.

The entity known as forma *purpurea* tends to occur on maritime rock and on walls, roofs, soil or rock on clifftops and it is a low-growing stunted form. In such locations, as well as in the upper littoral, plants reach maximum length where they are subjected to the effects of freshwater and may be up to 30 mm (=f. *intermedium*). Under exposure to wave action of severity, the plants in the littoral take the form of small hemispherical cushions (=f. *globosa*), with specialized lateral filaments forming a compact mass of interwoven filaments. With diminished wave action, the erect filaments form a dense turf but do not become laterally united as in the preceding form; this has been referred to as 'f. *typica*', although this is nomenclaturally incorrect.

Sporangia containing fewer than four spores have been reported widely. West (1969), working with North Pacific material indicated various patterns of development and spore germination and much more investigation is needed of these in other locations before any definitive explanation of the life history is possible. Tetraspores may germinate to give gametangial plants but not in all cases; the gametangial plants produced may be monoecious or dioecious. Carposporophytes are formed and these transform directly into tetrasporangial plants.

Audouinella rhipidandra (Rosenvinge) Dixon (1976), p. 590.

Lectotype: C (Rosenvinge 1412). Denmark (Frederikshavn).

Chantransia rhipidandra Rosenvinge (1909), p. 91.
Acrochaetium rhipidandrum (Rosenvinge) Hamel (1927), p. 25.
Rhodochorton rhipidandrum (Rosenvinge) Drew (1928), p. 151.
Kylinia rhipidandra (Rosenvinge) Kylin (1944), p. 13.
Chromastrum rhipidandrum (Rosenvinge) Papenfuss (1945), p. 322.

Plants consisting of a single enlarged basal cell, formed from the original spore, and two or three uniseriate branched axes, to 350 μm in length, composed of 10–20 cells, profusely branched, most branches consisting of only a single cell although some may contain 3–4 cells and form lateral axes; cells initially cylindrical, 14–28 μm in length, changing shape with age so that the apical diameter (9–11 μm) is greater than the basal (7–9 μm), with chloroplast; chloroplast axile, stellate, with pyrenoid, in the apical end of the cell; apical cells frequently converting into hairs.

Gametangial and tetrasporangial plants apparently not recorded for the British Isles; elsewhere, gametangial plants dioecious; spermatangial plants often smaller than those with other reproductive structures, rarely exceeding 100 μm, spermatangia small, 4–6 μm, without chloroplast, formed singly on a one-celled filament produced laterally on a major axis; carpogonia small, to 5 μm in length, with a short trichogyne, to 3 μm, formed singly, usually sessile; gonimoblasts developing directly and giving rise to a subglobose cluster of carposporangia; tetrasporangia apparently unknown completely; monosporangia ovoid, 14–18 μm in length, 9–10 μm in diameter, sessile, usually opposite, in pairs, formed either on a gametangial plant or on a distinct plant.

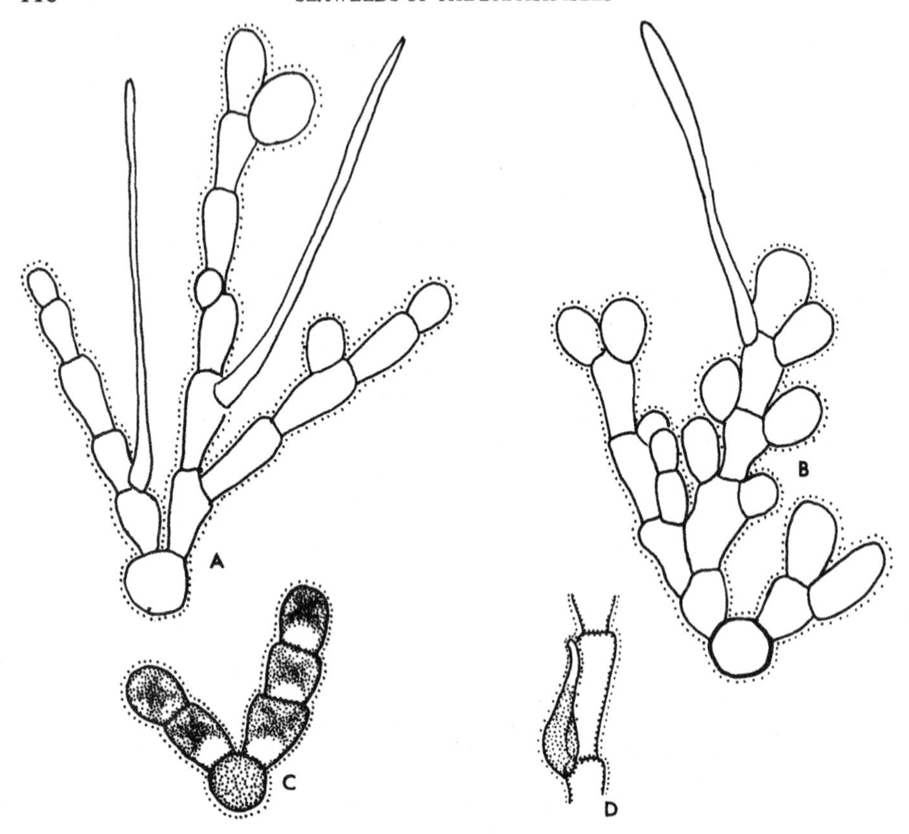

Fig. 35 *Audouinella rhipidandra*
A, B. Adult plants ×750; C. Young plant showing cell structure ×750; D. Carpogonium ×750. All drawn from non-British material.

Epiphytic on *Porphyra* spp. throughout the midlittoral.
Reported only from the west coast of Ireland (Galway).
Widely distributed between southern Norway and Portugal.

Data on seasonal behaviour and form variation too inadequate for comment.

Audouinella rhipidandra has been retained pending investigation of the relationships between all species of *Audouinella* where the thallus arises from a single basal cell. As stated under *A. microscopica* (q.v.), these relationships are extremely confused. Woelkerling (1973) refers *A. rhipidandra* to the synonymy of *A. alariae*, although the two entities are retained in the present treatment.

Borsje (1973) has compared *A. rhipidandra* with gametangial plants formed in culture by the germination of the tetraspores of *A. virgatula*.

Audouinella rosulata (Rosenvinge) Dixon (1976), p. 590.

Holotype: C (Rosenvinge 5374). Denmark (Tønneberg Banke).

Kylinia rosulata Rosenvinge (1909), p. 141.
Acrochaetium rosulatum (Rosenvinge) Papenfuss (1945), p. 307.

Plant consisting of a single basal cell, formed from the original spore, and one or more short uniseriate filaments appressed to or inclined away from the substrate, each filament composed of 2–5 cells, sometimes simple, occasionally branched, the branches composed of 1 or 2, at most 3, cells; cells cylindrical, barrel-shaped, 9–12 μm in length, 4–6 μm in diameter, or sometimes asymmetric, with the apical diameter (5–6 μm) greater than the basal (4–5 μm), containing a lobed chloroplast; apical cells frequently converting into hairs.

Gametangial plants monoecious; spermatangia ovoid, minute, 2 μm in length, 1 μm in diameter, borne on an elongate stalk; carpogonia small, 5 μm in length, with an elongate trichogyne, to 7 μm; gonimoblast develops directly from the carpogonium and apparently gives rise to a carposporophyte consisting of 2–3 carposporangia; tetrasporangia unknown; monosporangia ovoid, 6–9 μm in length, 5–6 μm in diameter, produced terminally or laterally.

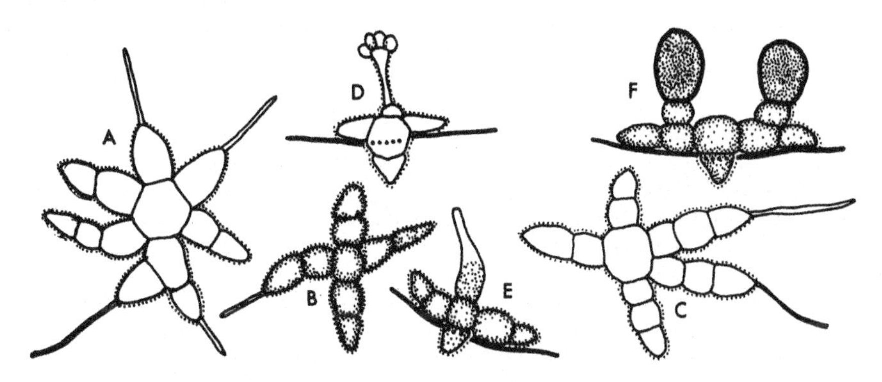

Fig. 36 *Audouinella rosulata*
A – C Habit ×740; D. Spermatangial plant ×740; E. Carpogonial plant ×740; F. Monosporangial plant ×740.

Epiphytic in the sublittoral, to 30 m, apparently restricted to *Sporochnus* and *Carpomitra*.

Southwest England (Devon, Cornwall).

Southern Norway to Portugal; California.

Plants occur between May and September, bearing both gametangia and monosporangia.

Data on form variation too inadequate for comment.

The statement by Kylin (1944) regarding the occurrence of stellate chloroplasts was the result of misidentification; the material used by Kylin has been described by Feldmann (1958) as *Acrochaetium kylinioides*.

Gametangial plants produced in culture by the germination of tetraspores of *Audouinella floridula* bear a slight resemblance to *A. rosulata* although differing in spermatangial arrangement, the form of the chloroplast and the presence or absence of pyrenoids.

Boillot & Magne (1973) obtained a filamentous growth in culture of carpospores of *A. rosulata,* which they refer to *Acrochaetium pectinatum* (Kylin) Hamel. The latter has not been recognized in the field in the British Isles.

Audouinella sanctae-mariae (Darbishire) Dixon (1976), p. 590.

Lectotype: original illustration (Darbishire, 1910, fig. 1, 2) in the absence of material. Northumberland (St Mary's Island, Whitley Bay).

Chantransia sanctae-mariae Darbishire (1910), p. 41.
Acrochaetium sanctae-mariae (Darbishire) Hamel (1927), p. 93.

Plants consisting of uniseriate filaments, entirely endophytic and penetrating deeply (0·4 mm) into the host species, rose-pink in colour; filaments largely unbranched except in the innermost and outermost parts where sparse alternate branching occurs; cells squat, cylindrical, 7–10 μm in length, 9–11 μm in breadth, apparently with a single, parietal, plate-like chloroplast, (?) devoid of pyrenoid; elongate hairs of rare occurrence.

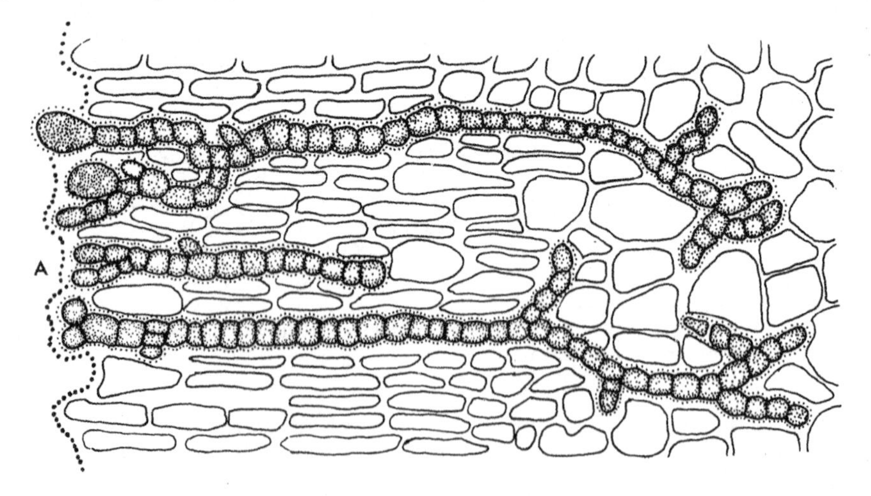

Fig. 37 *Audouinella sanctae-mariae*
A. Habit ×300 (after Darbishire).

Gametangial plants unknown; tetrasporangia unknown; monosporangia ovoid, 15 μm long, 11 μm in diameter, formed terminally, immediately below the surface of the host species and penetrating it when mature; regeneration of monosporangia within the old sporangial wall apparently of common occurrence.

Occurs as an endophyte in the 'lower portion of a reproductive frond' of *Himanthalia elongata* (L.) S. F. Gray.

Known only from the type locality.

Impossible to comment on seasonal behaviour and form variation. The original treatment does not even indicate date of collection.

Audouinella sanctae-mariae differs in many respects from other species of this genus. For these reasons, it is retained despite obvious suspicions as to the status of a taxon known only from a single collection, particularly when the present location of this is unknown.

Audouinella scapae (Lyle) Dixon (1976), p. 590.

Lectotype: BM (Slide 3713). Orkney (Scapa Flow).

Kylinia scapae Lyle (1929), p. 245.
Acrochaetium scapae (Lyle) Papenfuss (1945), p. 307.

Plants consisting of a single basal cell, formed from the original spore and several, usually four, prostrate uniseriate axes in contact with the substrate, each composed of 3–5 cells, sometimes simple, occasionally branched, the branches consisting only of a single cell; cells somewhat spherical, 4·5–7 μm in length and diameter, with chloroplast; apical cells frequently converting into hairs.

Reports of gametangial plants doubtful; tetrasporangia not known; monosporangia ovoid, 11 μm in length, 9 μm in diameter, sessile, terminal or lateral.

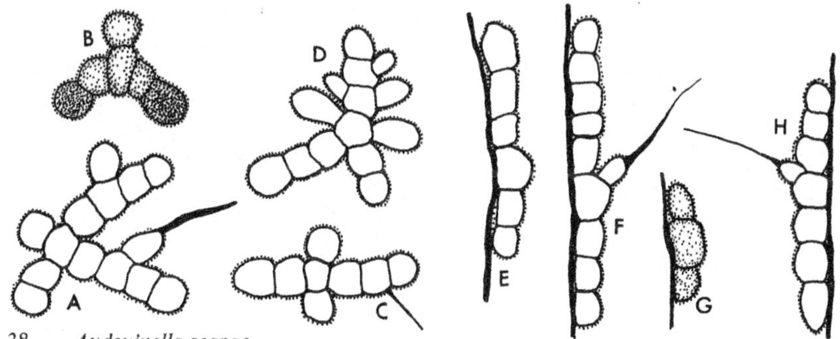

Fig. 38 *Audouinella scapae*
A – D. Habit, in face view ×740; E – H. Habit, in side view ×740.

Epiphytic on *Cladophora* spp., in the sublittoral to 3·5 m.

Known only from the type material, collected at Scapa Flow, Orkney.

Data on seasonal behaviour and form variation too inadequate for comment.

The initial treatment (Lyle, 1929) indicated that the material occurred intermixed with the alga here referred to *A. parvula* (q.v.). Examination of the type material of *A. scapae* indicates that there is little difference between these two entities. Although the status of *A. scapae* is highly suspect, complete reduction to synonymy is withheld pending examination of further collections of all species of this complex where a small thallus arises from a single cell.

Audouinella secundata (Lyngbye) Dixon (1976), p. 590.

Lectotype: C. Faeroes (Quivig).

Callithamnion dawiesii β secundatum Lyngbye (1819), p. 129.
Acrochaetium secundatum (Lyngbye) Nägeli (1861 [1862]), p. 405.
Chantransia secundata (Lyngbye) Thuret in Le Jolis (1863), p. 106.
Chantransia virgatula γ secundata (Lyngbye) Rosenvinge (1909), p. 112.
Chromastrum secundatum (Lynbgye) Papenfuss (1945), p. 323.
Kylinia secundata (Lyngbye) Papenfuss (1947), p. 437.
Colaconema secundata (Lyngbye) Woelkerling (1973), p. 575.

Plants consisting of uniseriate filaments, differentiated into erect and prostrate axes, red or brownish-red in colour; prostrate axes abundantly branched and aggregated into a compact disc, often more than one cell layer in thickness, with the original spore evident only in the very earliest stages of development; erect axes to 2 mm, often loosely entangled in the lower parts, much branched, branching highly variable, with the lower portions of the erect axes usually unbranched but relatively abundant branching in the upper portions, branching usually unilateral or secund, with a lateral branch arising from every cell of the erect axes for considerable distances; cells cylindrical, 12–24 μm in length, 6–8 μm in diameter, when first formed, increasing in size with age to reach a length of 25–45 μm and diameter of 8–15 μm, each with a single chloroplast; chloroplast axial, stellate or lobed, containing one pyrenoid.

Gametangial plants unknown; tetrasporangia ovoid, 16–24 μm in length, 10–18 μm in diameter, cruciately divided, arising singly in a terminal position on short filaments formed on the upper parts of the erect axes; monosporangia of similar dimensions to tetrasporangia and arising in similar positions; tetrasporangia and monosporangia frequently, but not always, occurring on the same specimen and often intermixed.

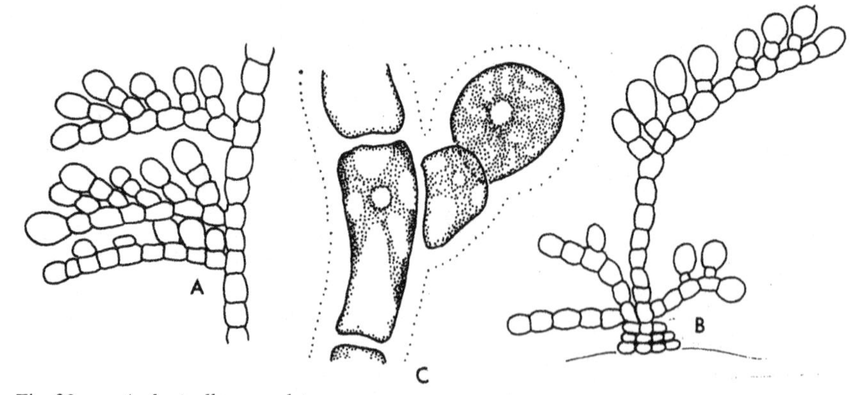

Fig. 39 *Audouinella secundata*
A. Habit, showing secund arrangement of branches ×200; B. Habit, showing base and monosporangium ×200; C. Cell and monosporangial structure ×740.

Epiphytic on larger perennial algae and *Zostera*, throughout the lower littoral and sublittoral to 5 m. Occurs in all conditions of exposure to wave action, from the most sheltered to the most exposed.

Generally distributed throughout the British Isles.

Northern Norway to Mauretania; Canary Islands; Mediterranean; U.S.A. (New Jersey to Maine).

The erect axes are most conspicuous between March and September; the prostrate axes probably perennate. Monosporangia and tetrasporangia occur between April and July.

There is little variation other than in overall size.

The relationship between *A. secundata* and *A. virgatula* has been questioned many times. The differences between the two entities are essentially of branching pattern and size (size of plant, cell dimensions, number of cell layers in the basal disc) and it is often impossible to distinguish between some small plants of *A. virgatula* and those referred to *A. secundata*. Despite its doubtful status, *A. secundata* has been retained as an independent taxon in the present treatment pending detailed investigation of the relationship.

Audouinella seiriolana (Gibson) Dixon (1976), p. 590.

Possible lectotype: BM-K, with a preparation (BM Slide 11013) made from this. Anglesey (Puffin Island).

Rhodochorton seiriolanum Gibson (1891), p. 204.
Acrochaetium seiriolanum (Gibson) Hamel (1927), p. 102.

Plants consisting of uniseriate filaments, differentiated into prostrate and erect axes, rose-red in colour; prostrate axes well developed, much branched but tightly compacted into a monostromatic sheet of cells so that the filamentous organization of the base is obvious only in young specimens or at the edge; erect axes to 1 mm but usually much less, virtually unbranched except occasionally during the formation of reproductive structures; cells cylindrical, 15–20 μm in length, 8–11 μm in diameter, with chloroplast; chloroplast lobed, with a pyrenoid.

Gametangial plants unknown; tetrasporangia 21–25 μm in length, 18–25 μm in diameter, occurring singly, usually in a terminal position but occasionally laterally also, sessile although rarely subtended by a single-celled stalk; tetraspores cruciately arranged.

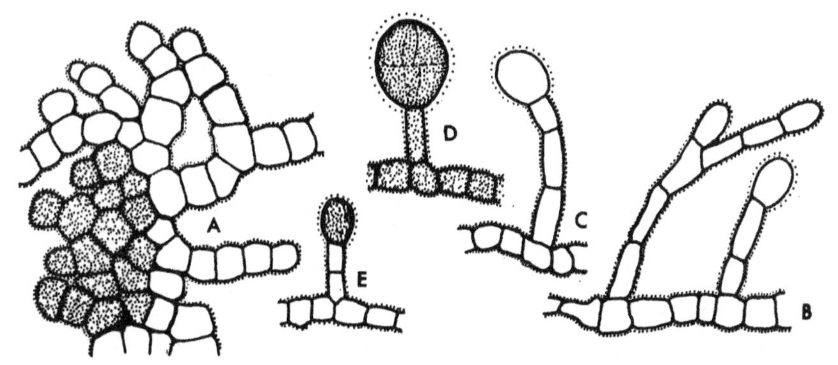

Fig. 40 *Audouinella seiriolana*
A. Prostrate axes in surface view ×740; B, C. Prostrate and erect axes in side view ×740; D. Tetrasporangium ×740; E. Tetrasporangial initial ×740.

Epiphytic on various algae in the lower littoral and upper sublittoral but, apparently, not reported from deeper water.

Known only from the type locality.

Basal disc perennial, but little other data on growth, development or reproduction.

Sporangial regeneration occurs extensively in this species. The regenerant may develop into a vegetative filament rather than a new sporangium but the remnant of the old sporangial wall forms a 'collar' about it.

Audouinella sparsa (Harvey) Dixon (1976), p. 590.

Lectotype: TCD. Argyll (Appin).

Callithamnion sparsum Harvey (1833), p. 348.
Acrochaetium sparsum (Harvey) Nägeli (1861 [1862]), p. 405.

Plant consisting of uniseriate filaments, differentiated into erect and prostrate axes, red or blackish-red in colour; prostrate axes branched irregularly but not extensively, anastomosing but not forming a solid discoid base, the original spore not persistent; erect axes to 0·5 mm, tufted, simple throughout or sparsely branched in the upper parts and unbranched below; cells cylindrical, 12–15 μm in length, 9–12 μm in diameter when first formed, increasing only slightly in size with age, each with a single chloroplast.

Reports of gametangial plants not substantiated; reports of bisporangia and tetrasporangia doubtful; monosporangia sessile, 15–25 μm in length, 15–20 μm in diameter, borne on the upper parts of the erect fronds, particularly in an axillary position, sometimes with transverse septa suggestive of germination *in situ,* which might explain the occasional reports of bisporangia or tetrasporangia.

Epiphytic on various algae, but most frequently on species of *Laminaria,* in the sublittoral to a depth of 5 m. The prostrate base might be perennial and with time this becomes increasingly endophytic.

Widely distributed throughout the British Isles.

Atlantic coast of France.

The prostrate axes and parts of the erect system persist for considerable periods. Growth is erratic and there are no sharply defined periods for growth and development. Reproductive stages are extremely scarce in the British Isles and it is not possible to give any precise data on their periodicity.

Data on form variation insufficient for comment.

The relationships between *Audouinella sparsa* and other members of the complex are in need of examination. In particular, epiphytic specimens referred to *A. purpurea* are similar to *A. sparsa* and occur in similar habitats. These two entities differ in cell size in the erect axes, although not in the prostrate systems.

The reports of reproductive stages in *A. sparsa* are very confused. Carmichael, in his unpublished account of this entity, mentioned the occurrence of 'capsules obovate sessile mostly axillary' and this became changed by Harvey so that the plant was said to possess both tetraspores and 'berry-like receptacles (favellae)'. The latter would appear to be the 'standard' reproductive structure for the genus *Callithamnion,* to which *Audouinella sparsa* was referred at that time. The tetraspores refer to the structures previously mentioned by Carmichael which Harvey admits not having seen. In a later examination of

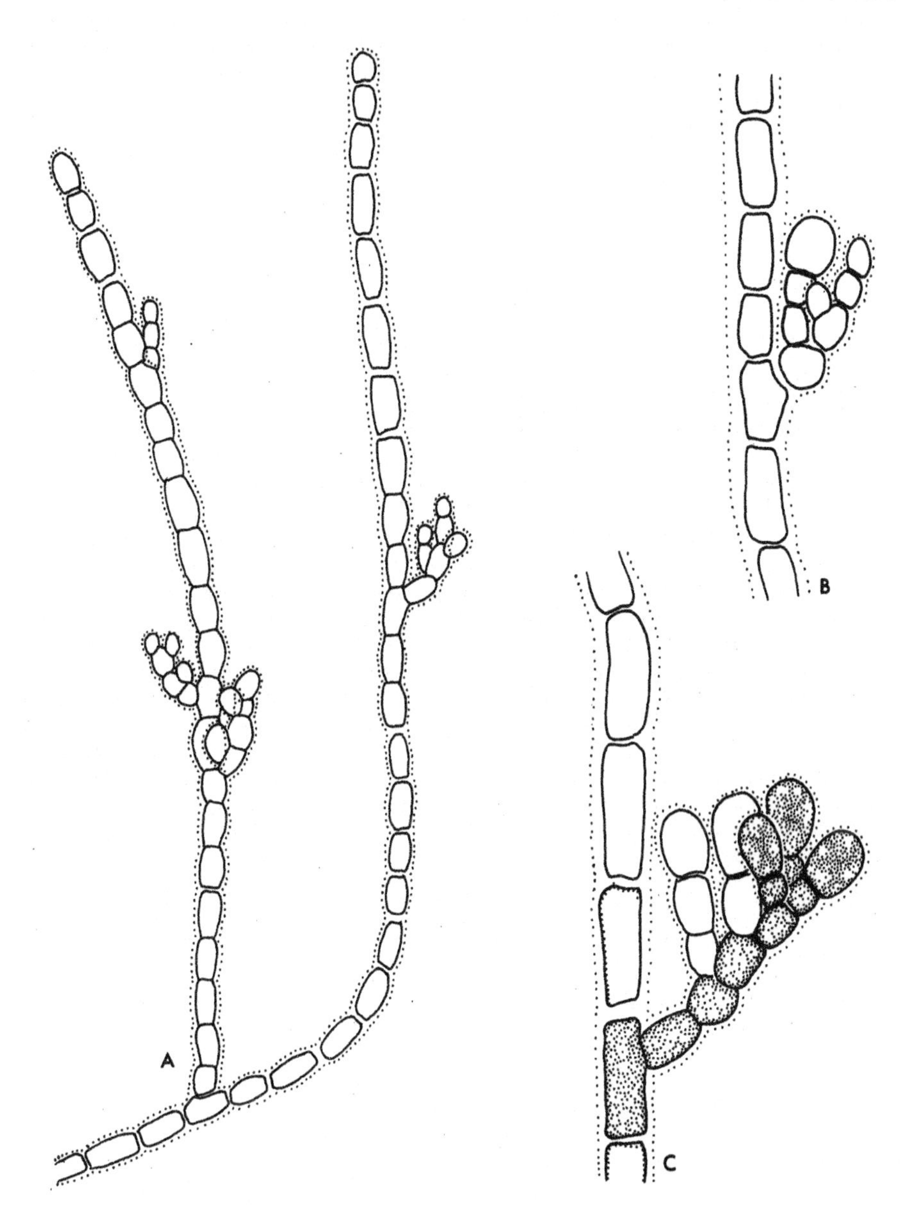

Fig. 41 *Audouinella sparsa*
A. Habit ×450; B. Monosporangial cluster ×740; C. Monosporangial cluster showing cell and sporangial detail ×740.

Carmichael's original material, Batters (1896a, p. 390) indicates that he was unable to find any sign of tetrasporangia although sparse monosporangia were observed. Re-examination of the same material confirms Batters's statement and it would appear that the only reproductive structure which can be accepted for *Audouinella sparsa* with certainty is the monosporangium.

There are many entities in the Mediterranean which might possibly be referable to this species.

Audouinella spetsbergensis (Kjellman) Woelkerling (1973), p. 585.

Holotype: UPS. Spitzbergen (Fairhavn).

Thamnidium spetsbergense Kjellman (1875), p. 31.
Rhodochorton spetsbergense (Kjellman) Kjellman (1883), p. 187.
Thamnidium mesocarpum f. *penicilliformis* Kjellman (1875), p. 30.
Rhodochorton mesocarpum f. *penicilliformis* (Kjellman) Kjellman (1883), p. 187.
Rhodochorton mesocarpum var. *penicilliforme* (Kjellman) Rosenvinge (1893), p. 792.
Rhodochorton penicilliforme (Kjellman) Rosenvinge (1894), p. 66.

Plants consisting of uniseriate filaments, differentiated into erect and prostrate axes, deep red in colour; prostrate axes adhering laterally to form a compact monostromatic plate of cells; cells of monostromatic plate elongate, 6–30 μm in length, 8–12 μm in diameter, often with fusions between cells; erect axes to 5 mm, arising from the monostromatic plate, singly or in clusters, unbranched or with a few short, simple lateral filaments near the apical end; cells of erect axes cylindrical, elongate, 10–35 μm in length, 9–14 μm in diameter, each with several chloroplasts; chloroplasts discoid or ribbon-shaped; hairs not known.

Gametangia not reported in British material, elsewhere spermatangia described but carpogonia unknown; spermatangia ovoid, 4–6 μm in length, 4–5 μm in diameter, formed in clusters, either laterally or terminally on erect axes or short apical lateral filaments; tetrasporangia ovoid, 25–30 × 18–22 μm; formed singly, in a terminal position on the short apical lateral filaments although at least one sporangium is formed subsequently in a lateral position, tetraspores arranged in a cruciate manner, sporangial regeneration occurring after spore release; tetrasporangia and spermatangia frequently reported on the same thallus.

On hydroids, worm tubes and epiphytic on algae, in the sublittoral, to 10 m.

Cornwall, Donegal, Northumberland, Shetland; probably distributed generally throughout the British Isles.

From arctic Russia and Iceland to the Atlantic coast of Spain; Greenland to U.S.A. (Rhode Island); Pacific coast of North America from Alaska to Oregon.

The basal monostromatic sheet of cells is perennial, while erect axes arise in March–April and survive until September or November. Sporangia formed on erect axes have been observed from May onwards.

The major variation in appearance is related to seasonal changes in the occurrence of the erect axes, and it is possible that specimens in which development of erect axes is less than normal are being referred to other entities.

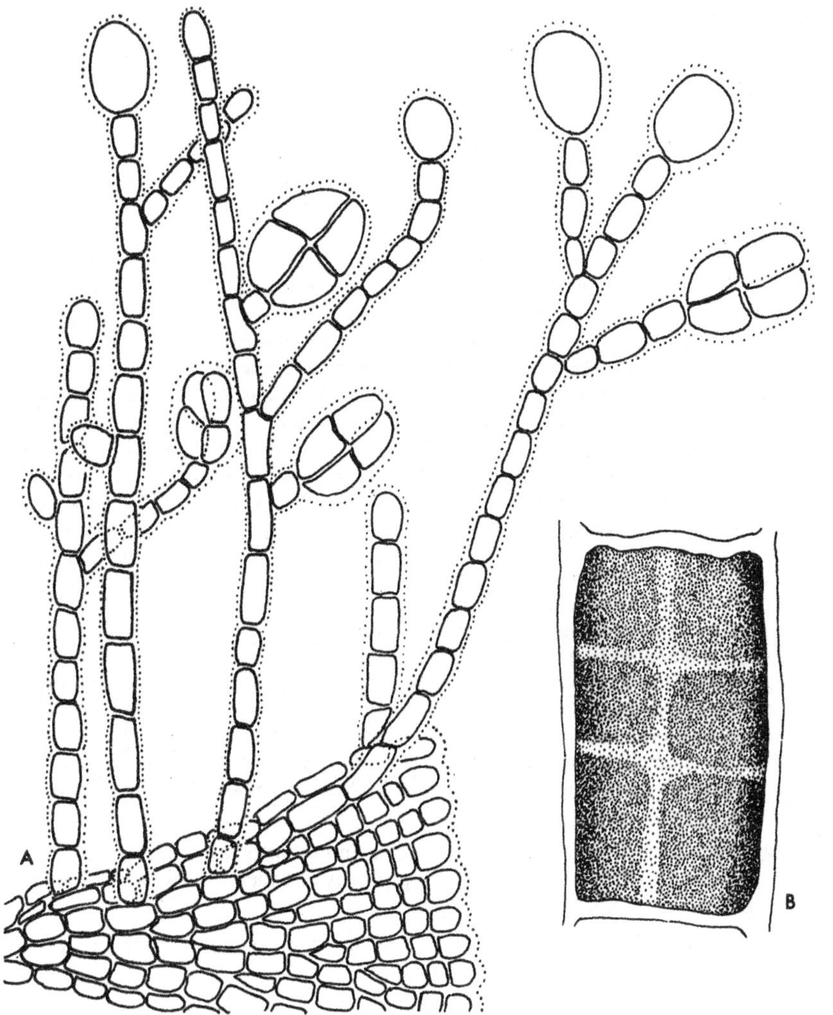

Fig. 42 *Audouinella spetsbergensis*
A. Habit, showing prostrate base, erect axes, and tetrasporangia ×300; B. Cell detail ×2400.

The relationships of *Audouinella spetsbergensis* and those species which possess a monostromatic basal plate, but shorter erect axes, are urgently in need of investigation. Comments have been made (Rosenvinge, 1924) regarding the relationship with *A. seiriolana* but there are many differences, in particular the single parietal chloroplast of that entity. A more probable relationship is that with *A. concrescens* (q.v.) but more collections are needed before any opinion of value can be made.

Audouinella thuretii (Bornet) Woelkerling (1971), p. 9.

Lectotype: In the absence of original material of an appropriate date (Dixon, 1977), the entity must be typified by the illustration given by Bornet (1904, pl. I). France (Cherbourg).

Chantransia efflorescens var. *thuretii* Bornet (1904), p. xvi.
Chantransia thuretii (Bornet) Kylin (1907), p. 119.
Acrochaetium thuretii (Bornet) Collins & Hervey (1917), p. 97.
Rhodochorton thuretii (Bornet) Drew (1928), p. 152.

Plants consisting of uniseriate filaments, differentiated into erect and prostrate axes, red or reddish-brown in colour, although often less deeply coloured towards the apices; prostrate axes much branched, branching highly irregular, compacted to form an irregular disc, with the original spore not obvious beyond the earliest stages of development; erect axes to 3 mm, arising both from the position of the original spore as well as secondarily from other parts of the disc, branched, branching sparse, irregular, but usually unilateral; cells cylindrical, of somewhat variable dimensions, 15–30 μm in length, 8–10 μm in diameter towards the base of the erect filaments, 25–50 μm in length, 4–5 μm in breadth towards the apices to form hair-like structures, differing in appearance from the hairs normally found in this genus, which are here lacking, each with chloroplast; chloroplast parietal, lobed, containing a pyrenoid.

Gametangial plants monoecious; spermatangia 4–5 μm, without chloroplasts, formed in small corymbose clusters on short, 1 or 2-celled filaments; carpogonia to 5 μm in length, with a short (3 μm) trichogyne, formed terminally on a 1 or 2-celled filament at the base of a major lateral axis; gonimoblasts develop directly, and give rise to a dense cluster of 10–14 short filaments, the end cells of which form carposporangia, carposporangia ovoid, 15–20 μm in length, 11–14 μm in diameter; tetrasporangial plants not known in Britain, elsewhere tetrasporangia spherical or subspherical, 24–26 μm in length, 20–23 μm in diameter; tetraspores cruciately arranged; monosporangia ovoid, 16–24 μm in length, 10–12 μm in diameter, formed in dense clusters at the base of major lateral axes, occurring on both gametangial and, elsewhere, tetrasporangial plants as well as on their own in some cases.

Epiphytic on various algae in the lower littoral and sublittoral to 5 m.
Channel Islands, Devon, Galway, Mayo; probably of wider occurrence.
Central Norway to Atlantic coast of France.

The species occurs only from May to September, with monosporangia throughout, gametangia in June and cystocarps from June to September.

Data on form variation too inadequate for comment.

The original description of *Chantransia corymbifera* by Thuret (*in* Le Jolis, 1863) was based on a mixture of two elements, one growing on *Ceramium* and the other on *Helminthocladia.* A later, more detailed treatment of *Chantransia corymbifera* was restricted to a consideration of the plant from *Helminthocladia.* The material from *Ceramium* was subsequently described as *Chantransia efflorescens* var. *thuretii* (Bornet, 1904) and later assigned specific status as *Acrochaetium thuretii.*

The distribution and occurrence of reproductive structures exhibit certain peculiarities. These may be due to confusion with *A. corymbifera* although this cannot be the complete explanation. Tetrasporangia, for instance, have been reported only in Denmark and California and it is possible that the failure to detect these reproductive structures is caused by the dismissal of tetrasporangial specimens as *A. daviesii*. *A. thuretii* and *A. daviesii* are very similar in appearance. Rosenvinge (1909) claimed that *A. daviesii* had 'thicker filaments, shorter cells and smaller sporangia', although these comments could not be verified during the present studies.

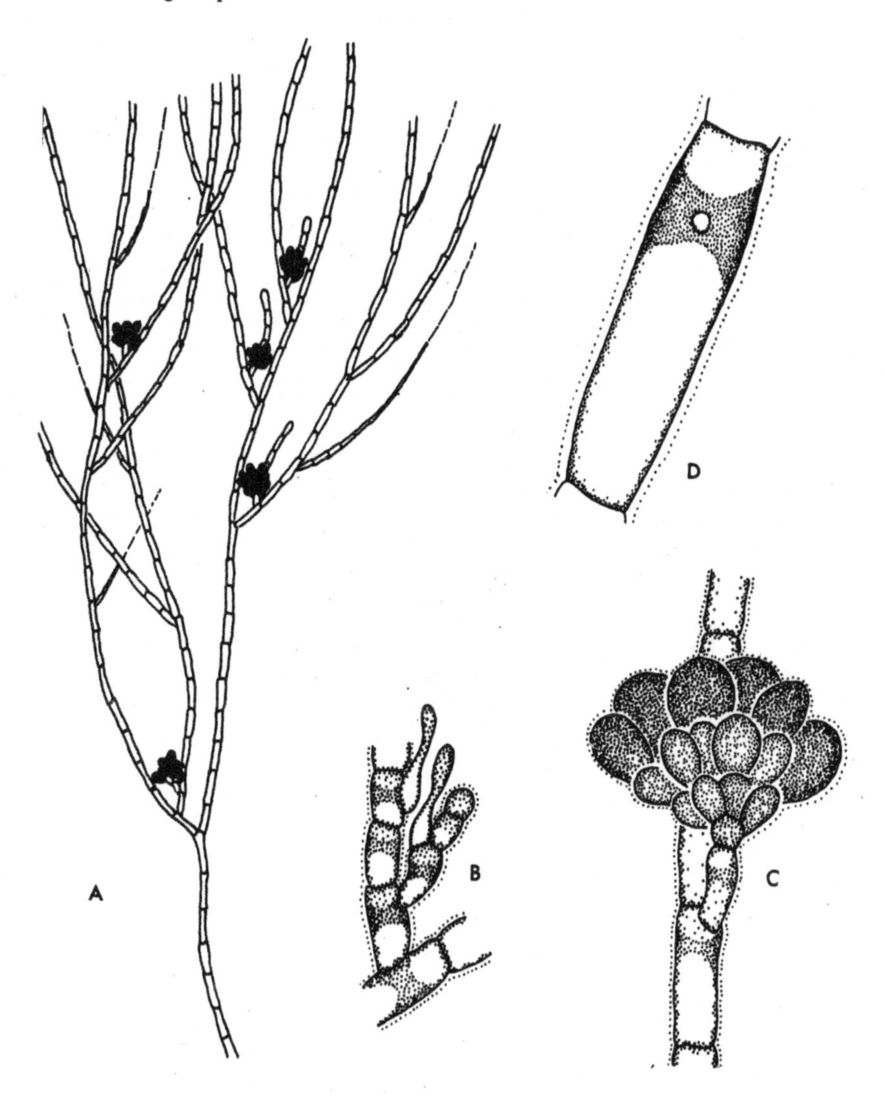

Fig. 43 *Audouinella thuretii*
A. Habit ×10; B. Lateral axis with carpogonia ×450; C. Carposporophyte ×450; D. Cell detail ×1200.

Woelkerling (1973) regards *A. thuretii* as synonymous with the plant from the Mediterranean which has been described many times under the name *Acrochaetium savianum* (Menegh.) Näg. Further studies of the relationships between these two taxa are required before this can be accepted, particularly as the epithet *saviana* is the older name. Feldmann (1939) regarded *A. saviana* as the Mediterranean form of *A. daviesii*.

Audouinella trifila (Buffham) Dixon (1976), p. 590.

Holotype: BM. Dorset (Swanage).

Chantransia trifila Buffham (1892), p. 25.
Acrochaetium trifilum (Buffham) Batters (1902), p. 58.

Plant consisting of a single basal cell, formed from the original spore, and three, sometimes two, uniseriate filaments inclined away from the substrate, each composed of two, three or four cells, sometimes simple, occasionally branched, the branches usually consisting only of a single cell; cells cylindrical, 5–8 μm in length, 5–6 μm in diameter, with a simple chloroplast containing a pyrenoid; apical cells frequently converting into hairs.

Gametangial plants unknown; tetrasporangia unknown; monosporangia spherical, 7–8 μm in diameter, formed either laterally or terminally from an apical cell, usually only a single monosporangium formed on each plant, although two sometimes present.

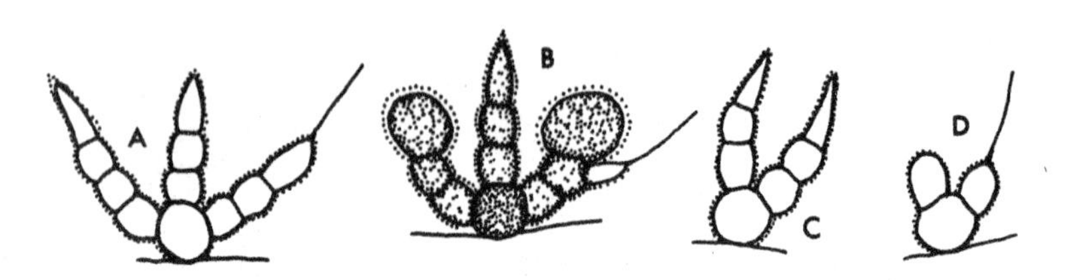

Fig. 44 *Audouinella trifila*
A – D. Habit ×740.

Epiphytic on various algae in the upper sublittoral and possibly elsewhere but overlooked.

Devon, Dorset.

Atlantic coast of France; Mediterranean.

Plants collected in August and September with monosporangia; reported from Algeria in February.

Data on form variation inadequate for comment.

As stated under *A. microscopica* (q.v.), species of *Audouinella* where a small thallus arises from a single cell are extremely confused. *A. trifila* is accepted in the present treatment, although its continued status as an independent entity is doubtful.

Audouinella virgatula (Harvey) Dixon (1976), p. 590.

Lectotype: TCD. Devon.

Callithamnion virgatulum Harvey (1833), p. 349.
Chantransia virgatula (Harvey) Thuret in Le Jolis (1863), p. 106.
Acrochaetium virgatulum (Harvey) Bornet (1904), p. XXII.
Chromastrum virgatulum (Harvey) Papenfuss (1945), p. 323.
Acrochaetium griffithsianum Nägeli (1861 [1862]), p. 406.

Plants consisting of uniseriate filaments, differentiated into erect and prostrate axes, red or

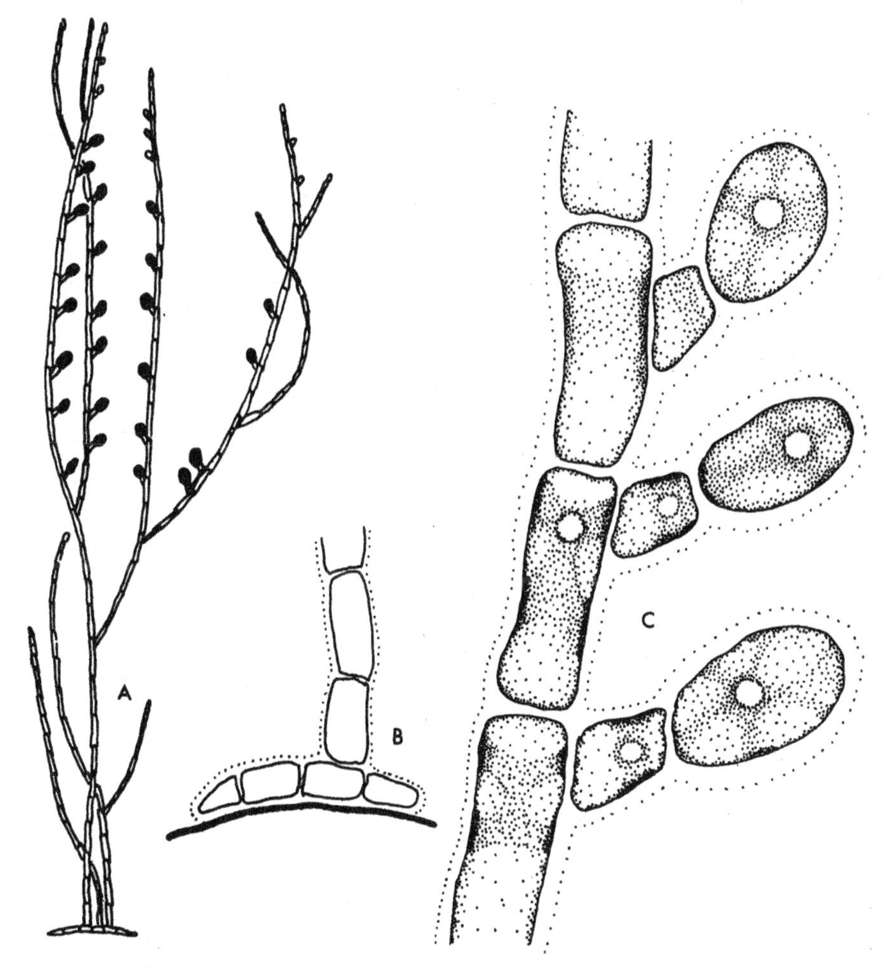

Fig. 45 *Audouinella virgatula*
A. Habit ×120; B. Base of plant ×450; C. Axes showing cell detail and monosporangia ×740.

brownish-red in colour; prostrate axes frequently branched and aggregated into a compact disc, with the original spore evident only in the youngest stages of development; erect axes to 7 mm, often loosely aggregated at base, much branched, branching sparse and highly irregular, main and lateral axes often terminating in a hair; cells cylindrical, 15–20 μm in length, 6–12 μm in diameter when first formed, increasing in size with age to reach a length of 30–70 μm and diameter of 10–18 μm, each with a single chloroplast; chloroplast axial, deeply lobed, containing one pyrenoid.

Gametangial plants unknown; tetrasporangia ovoid, 18–22 μm in length, 14–18 μm in diameter, arising singly on short filaments formed on the upper parts of the erect axes, either terminal or lateral, but always sessile, tetraspores cruciately arranged; monosporangia of similar dimensions to tetrasporangia and arising in similar positions; tetrasporangia and monosporangia often but not always occurring on the same specimen and then frequently intermixed.

Epiphytic on many algae throughout the lower littoral, in pools, and the sublittoral to a depth of 10 m.

Generally distributed throughout the British Isles.

Northern Norway to Mauretania; Canary Islands; U.S.A. (North Carolina to Maine).

Erect axes most conspicuous between March and November, prostrate axes probably perennate. Monosporangia and tetrasporangia occur throughout the year although most abundant between April and October.

There is some variation with age in terms of overall size.

The relationship of *A. secundata* to *A. virgatula* has been questioned many times; *A. secundata* (q.v.) is retained in the present treatment pending further investigation.

Borsje (1973) has grown tetraspores of *A. virgatula* in culture and obtained gametangial plants similar to *A. parvula* (q.v.) and *A. rhipidandra* (q.v.).

Excluded species

Acrochaetium minutum (Suhr) Hamel

This species has been included in the algal flora of the British Isles on the basis of material collected by Mrs K. Holmes at Weymouth (Dorset) in April, 1892. The taxon is confused taxonomically and nomenclaturally and there seems little reason to include it in the present treatment on the basis of current knowledge.

Acrochaetium pallens (Zanardini) Nägeli.

This species occurs most frequently in the Adriatic, with few records elsewhere, even in the Mediterranean. Its occurrence in the British Isles has been accepted on the basis of a single report (Batters, 1902) from Seaton, Devon. There is a slide preparation at BM (10089) which is probably the basis for this; it is of *Audouinella daviesii*.

GELIDIACEAE Kützing

GELIDIACEAE Kützing (1843), p. xxiv.

Thallus erect, usually with prostrate axes, terete or strap-shaped, branched, rarely simple, soft or cartilaginous, uniaxial; gonimoblasts developing internally to give swollen terminal or subterminal cystocarps; tetrasporangia scattered or in spathulate tips, cruciate, tetrahedral or irregular.

Two genera, *Gelidium* and *Pterocladia,* occur in the British Isles. The basic criterion for separation relates to the structure of the cystocarp, which in *Gelidium* is bilocular with openings on both faces, whereas in *Pterocladia* it is unilocular with an opening or openings on one face only. In terms of British material, cystocarps are rare in *Gelidium* and have never once been found in *Pterocladia* so that to use this feature as the practical basis for separation, as in many keys, is meaningless. There are also serious disadvantages to the use of the distribution of internal rhizoids, or rhizines, as proposed by Okamura (1934). However, faced with a bewildering assemblage of sterile or tetrasporangial specimens, one can do little more than compare external form and internal structure and agree with Bornet's comment on *Gelidium* – 'genre diabolique'.

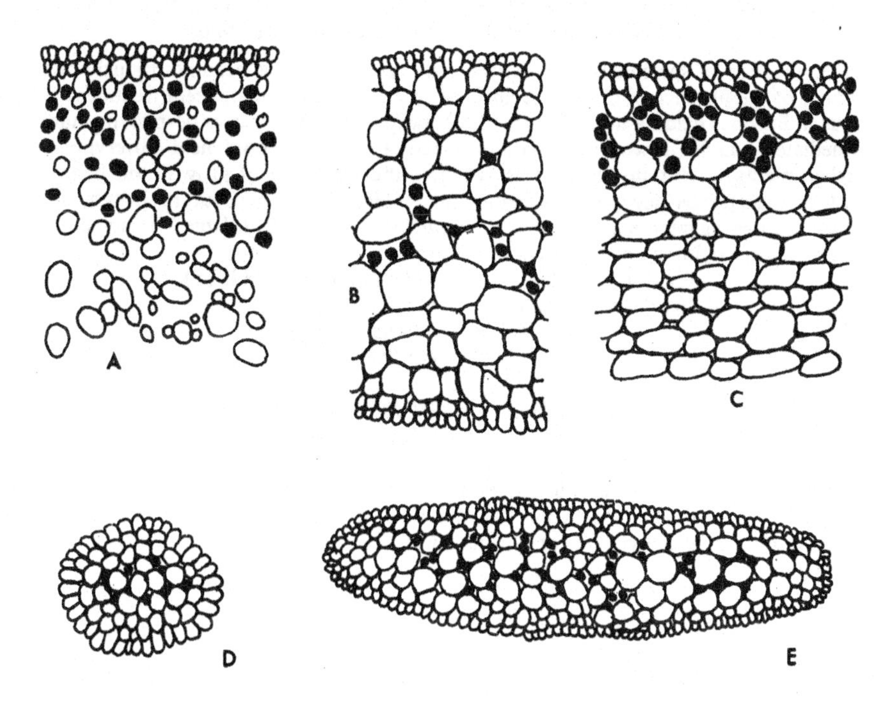

Fig. 46 Transverse sections of axes of *Gelidium* and *Pterocladia*
A. *Gelidium latifolium* ×300; B. *Pterocladia capillacea* ×300; C. *Gelidium sesquipedale* ×300; D, E. *Gelidium pusillum* ×300.

GELIDIUM Lamouroux **nom. cons.**

GELIDIUM. Lamouroux (1813), p. 128.

Type species: *G. corneum* (Hudson) Lamouroux (1813), p. 128 (reprint p. 41)

Cornea Stackhouse (1809), p. 57.
Acrocarpus Kützing (1843), p. 405.

Thalli simple or branched, differentiated into erect fronds and prostrate axes; axes cylindrical, compressed or markedly flattened, of uniaxial construction; filaments of limited growth closely compacted to produce an inner region of large colourless cells ('medulla') and an outer region of smaller cells, with chloroplasts ('cortex'); internal rhizoids (rhizines) distributed immediately inside the cortex so that in most specimens they lie at the outer edge of the medulla, although in smaller specimens the medulla is so slight that the internal rhizoids are more central in distribution. Gametangial thalli dioecious or monoecious; spermatangia formed in superficial colourless patches over the surface of the ultimate divisions of the frond; carpogonia sessile, arising in groups in the ultimate divisions of the frond, the carposporophytes developing internally, the outer cortex lifting on both faces of the frond to form a bilocular cystocarp which develops at the extremity of an axis; tetrasporangia cruciate, tetrahedral or irregularly divided, distributed in groups or singly in the ultimate divisions and penultimate parts of the erect fronds.

Identification of specimens in the genus *Gelidium* has always been notoriously difficult because of the absence of clearly circumscribed entities. The most recent treatment of the genus in Europe (Feldmann & Hamel, 1936) accepted various species despite the many intermediates which, it was admitted, occurred. From sequential field observations of marked plants in Britain and a more detailed understanding of growth and behaviour, most of the species accepted by Feldmann & Hamel which occur in Britain can be reduced to two aggregate groups, *Gelidium pusillum* and *G. latifolium*. The attribution of species, *sensu* Feldmann & Hamel, is as follows:

Present Treatment	Feldmann & Hamel (1936)
Gelidium pusillum	*Gelidium pusillum*
	Gelidium crinale
	Gelidium pulchellum
Gelidium latifolium	*Gelidium latifolium*
	Gelidium attenuatum

The second aggregate also includes much of the material which has been referred previously to *'Gelidium corneum'*. In addition to these two aggregates, the present treatment also recognizes, unchanged from the previous treatment by Feldmann & Hamel, *Gelidium sesquipedale*, which is one of the few species of this genus with little variation of external form.

KEY TO SPECIES

The extent of morphological variation makes artificial keys of minimal significance in genera such as *Gelidium* and *Pterocladia*. The following key to the accepted species of both genera should be used with caution.

1 Axes predominantly cylindrical, tips spathulate in winter collections *Gelidium pusillum*
Axes predominantly flattened, tips sometimes terete in winter collections . 2
2 Outline of frond highly irregular; internal rhizoids distributed in the sub-surface layer . . . . *Gelidium latifolium*
Outline of frond regularly triangular or parallel-sided 3
3 Outline of frond triangular; internal rhizoids distributed in the centre of the frond . . . . *Pterocladia capillacea*
Outline of frond parallel-sided; internal rhizoids distributed in a narrow sub-surface layer . . . . *Gelidium sesquipedale*

Gelidium latifolium (Greville) Bornet & Thuret (1876), p. 58.

Lectotype: E. Devon (Sidmouth).

Gelidium corneum var. *latifolium* Greville (1830), p. 143.

No attempt is made to cite the synonymy for this species. Complete citation is impossible because of the many basionyms, the large number of different combinations resulting from changing ideas on the status and inter-relationships of the entities involved, and the innumerable misapplications of names. Selection of a few synonyms from this assemblage is meaningless. The names generally applied to the British entities of this species have been indicated previously (p. 126).

Thallus of varying form but consisting usually of erect fronds and creeping prostrate axes; prostrate axes of variable extent but, when present, attached to the substrate by lateral protuberances; erect fronds to 200 mm, red, brownish-red, pinkish-red, greenish-red in colour, much branched in most specimens, sometimes dendroid with age, axes markedly compressed or flattened, 1–5 mm in breadth, branched once, twice or three times, opposite or alternate, usually distichously but occasionally radially, producing a frond of very variable outline; medulla of large colourless cells with an outer cortex of small cells containing chloroplasts, with internal thick-walled rhizoids occurring immediately inside the cortex.

Gametangial thalli dioecious or monoecious; spermatangia formed in superficial colourless patches over the surface of the ultimate divisions of the frond; carpogonia sessile, formed in small areas immediately behind the apices of the ultimate divisions of the frond which are usually colourless or with reduced pigmentation; carposporophytes developing internally, the outer cortex lifting on both sides of the frond to form a bilocular cystocarp; cystocarp 2–3 mm in diameter, with a pore said to be formed in the wall of each locule although apparently often wanting in British material, usually with a short, spike-like projection beyond the cystocarp; tetrasporangia ovoid or spherical, 30–40 μm in length, 30–35 μm in diameter, distributed usually in groups in spathulate tips to the ultimate parts of the erect fronds but in isolated examples scattered over all parts, tetraspores with cruciate, tetrahedral or irregular arrangement.

Epilithic in the lower littoral, although usually restricted to pools, and in the sublittoral to 15 m; occasionally on coralline algae or epiphytic on *Laminaria hyperborea* (Gunn.) Fosl.

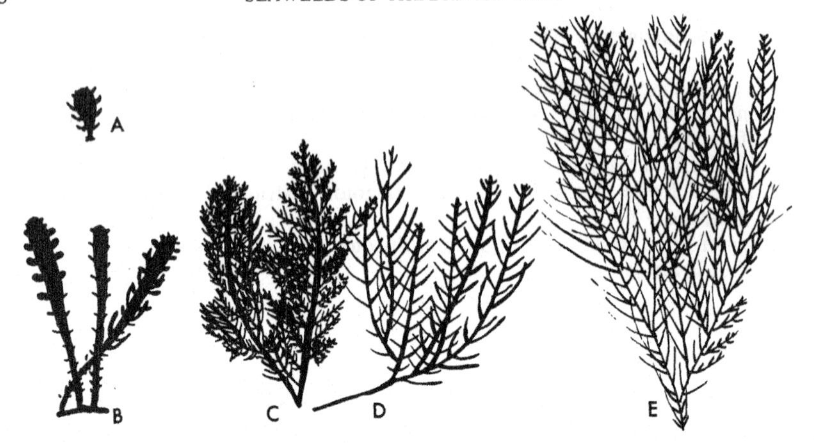

Fig. 47 Range of form of *Gelidium latifolium* complex in the British Isles ×1. Specimens A, B, were collected in midlittoral shallow pools; specimens C, D, were collected on the edges of pools in the midlittoral; specimen E, was collected in deep pools at low water of spring tides.

Generally distributed throughout the British Isles although somewhat sporadic on the eastern and northern coasts of England and Scotland.

Southern Norway to Rio de Oro.

Erect fronds perennial, persisting for up to three years, although many are annual in lower littoral pools. Fronds arise from prostrate axes between February and June in southwest England and the west coast of Ireland, between March and June in Anglesey and the Isle of Man, and May/June in Scottish localities.

Tetrasporangia produced in fronds of all sizes between May and October, persisting until December when decay eliminates the tetrasporangial spathulate tips. Gametangia and cystocarps not common in the smaller, broader plants *('G. latifolium')* and completely unknown in the larger plants of the upper sublittoral *('G. attenuatum')*. Gametangia formed between June and August; cystocarps persisting until December or even April in sublittoral specimens.

Erect fronds of very variable appearance. Specimens are always much broader than those of the *G. pusillum* aggregate, although in winter the tips of principal axes may be elongate and terete. The most luxuriant specimens occur in deep pools and upper sublittoral.

The life history of this species is unknown. In the British Isles, the rarity of gametangial material and the total absence of gametangia in the larger plants *('G. attenuatum')* cannot be explained. Attempts to locate germlings in the field proved impossible; all 'young' axes arose from prostrate axes, so that extensive perennation is the most significant aspect of the life history.

The larger specimens of the *G. latifolium* aggregate *('G. attenuatum')* have often been confused with specimens of *Pterocladia capillacea* (S. G. Gmel.) Born. & Thur. These two entities can always be distinguished by internal anatomy and often by external form. The internal thick-walled rhizoids are restricted to the centre of the frond in *P. capillacea* but occur immediately beneath the cortex in *G. latifolium*. The frond outline in the latter is irregular, but characteristically triangular in *P. capillacea*.

Gelidium latifolium has been used as a commercial source of agar in various places, either alone or mixed with *Pterocladia capillacea*. Attempts to develop an industry on this material in Ireland failed after successful harvests in the first year, due to ignorance of the life history and growth rates.

Gelidium pusillum (Stackhouse) Le Jolis (1863), p. 139.

Provisional lectotype: B M (see Dixon & Irvine (1977). Devon (Sidmouth).

Fucus pusillus Stackhouse (1795), p. 16.

No attempt is made to cite the synonymy for this species. Complete citation is impossible because of the many basionyms, the large number of different combinations resulting from changing ideas on the status and inter-relationships of the entities involved, and the innumerable misapplications of names. Selection of a few names from this assemblage would be meaningless. The names generally applied to the British entities within this species have been indicated previously (p. 126).

Thallus of varying form but consisting usually of both erect fronds and creeping prostrate axes; prostrate axes of variable extent but, when present, attached to the substrate by lateral protuberances; erect fronds to 150 mm, black, purple, red, brown or yellow-green in colour, simple in small specimens but much branched in larger, sometimes dendroid with age, axes cylindrical or slightly compressed, 0·3–1 mm in diameter although the tips may be spathulate, *c.* 2 mm broad, particularly during the winter, some axes unbranched or branched once, twice or three times, opposite or alternate, usually in a distichous arrangement but sometimes distributed radially, producing a frond of very variable outline; medulla made up of larger colourless cells with a cortex of smaller cells containing chloroplasts; internal thick-walled rhizoids occurring immediately inside the cortex in larger specimens, occurring occasionally also in the centre of the frond in smaller examples.

Gametangial thalli dioecious or monoecious; spermatangia forming superficial patches in ultimate divisions of the frond; carpogonia sessile, formed in small areas immediately behind the apices of the ultimate divisions of the frond which are usually colourless or with reduced pigmentation; carposporophytes developing internally, the outer cortex lifting on both faces of the frond to form a bilocular cystocarp; cystocarp 2 mm in diameter, with a pore said to be formed in the wall of each locule although apparently often wanting in British material, usually in a terminal position but often with a small spike-like extension and occasionally with a prominent axis developing during the maturation of the cystocarp; tetrasporangia ovoid or spherical, 20–40×20–30 μm, distributed in groups or singly in the ultimate or penultimate parts of the erect fronds, tetraspores with cruciate, tetrahedral or irregular arrangement.

Epilithic in the midlittoral, lower littoral and occasionally in the upper sublittoral.

Generally distributed throughout the British Isles.

Central Norway to Cape Verde; Mediterranean; U.S.A. (Florida to Maine); widely distributed in the Pacific Ocean.

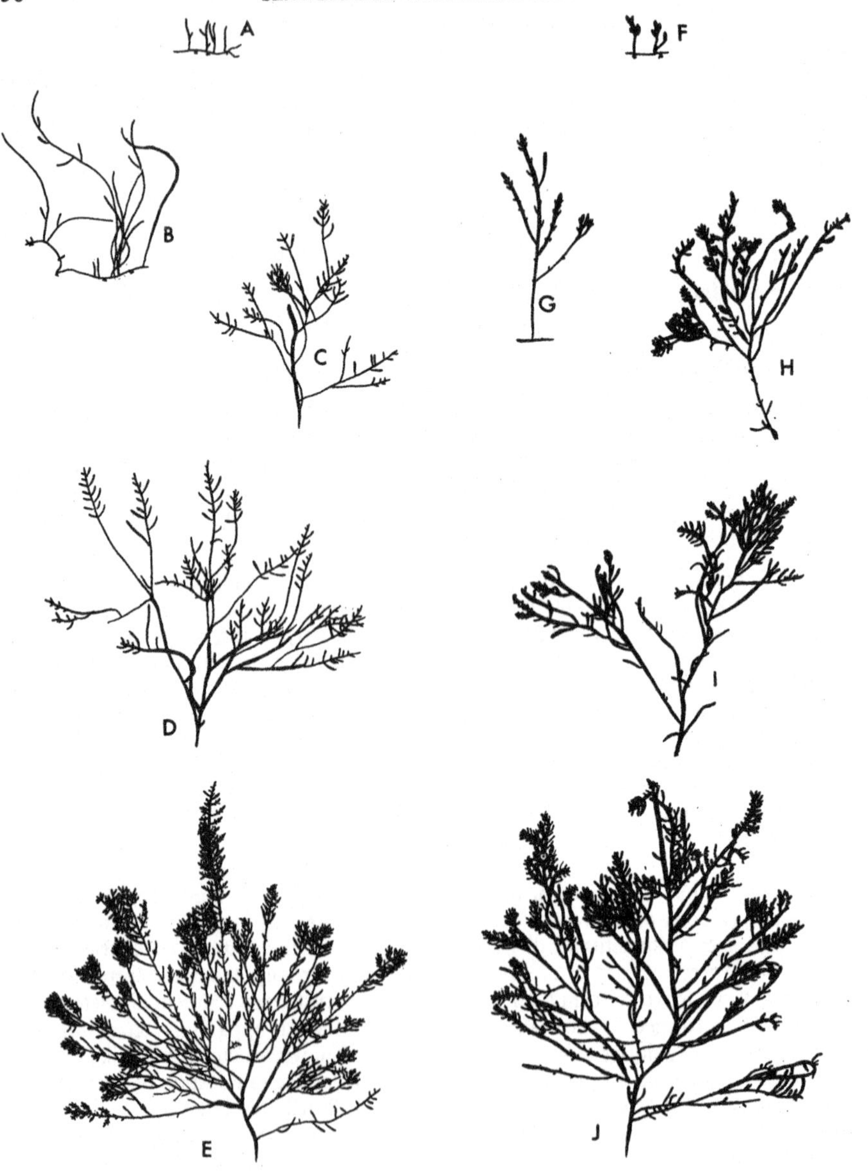

Fig. 48 Range of form of the *Gelidium pusillum* complex in the British Isles ×1. Specimens A – E were collected along a transect from high water of neap tides (A) to low water of spring tides (E) in August. Specimens F – J were collected at the same positions in February.

Erect fronds perennial, persisting for up to three years. They arise from prostrate axes throughout the period of active growth, which may commence in February and continue until December in the west of Ireland and southwest England. In Scotland, it is shorter, usually between May and September. Tetrasporangia arise on plants of all sizes from June to November. Gametangia and cystocarps occur between July and September. They are uncommon in the largest plants *('G. pulchellum')*, rare in those of intermediate size *('G. crinale')* and virtually unknown in the smallest *('G. pusillum')* even in southwest England and Ireland; gametangial plants are completely unknown on eastern and northern shores.

The erect fronds are of very variable appearance. The length of axes varies in relation to position on the shore. There is a gradation in maximum size of frond for plants not in pools, between the upper and lower limits of distribution. Specimens growing near high-water of neap tides have small erect fronds, usually less than 10 mm in height, while fronds from the low-water of spring tides reach a maximum size of 100 mm. Growth in the plants at the upper limit occurs only intermittently while the maximum rate at the lower level is of the order of 60 mm per annum. There is a second gradient at the edges of pools with larger plants in pools grading to smaller plants on adjacent rock, out of water. Animal grazing can modify the size of plants. This species is an effective sand-binder and under heavy grazing the axes are cut back repeatedly to the level of accumulated sand. Abundant regeneration produces dense caespitose axes.

During active growth of the axes, the tips are narrow and cylindrical; when growth slows down or stops the tips become broader and flatter. Thus, the tips are cylindrical from March/April to October/November and more or less spathulate during the remainder of the year.

The degree of branching of the axes is also determined by position on the shore. Fronds from the upper limit have axes which are simple or infrequently branched, although damage from animal grazing can produce some bizarre effects. The frequency of branching increases from the upper to the lower limits, the fronds thereby becoming increasingly complex. There is also a gradation between specimens growing in and out of pools in terms of the frequency of branching which is minimal in the latter.

The variation is the result of interaction of these three major variables. Large numbers of species and varieties have been delimited from this assemblage. The smallest specimens from the upper limits of distribution have been described under *'G. pusillum'*, whilst the most luxuriant specimens from the lowest part of the range have been referred to *'G. pulchellum'*. In general, these names have been applied to material in the summer state, with cylindrical tips to the axes. Those specimens in the winter state, with flattened spathulate tips have been assigned varietal status under such names as 'var. *claviferum*' or 'var. *claviger*'.

The life history is largely unknown. In the British Isles, gametangial plants are exceedingly rare; they are completely lacking from certain parts of the morphological range. Germlings cannot be found in the field and large-scale perennation of prostrate axes is the most significant feature of the life history.

The smaller plants of the upper shore are often mixed with or capable of being confused with *Catenella*. Axes in *Catenella* are often hollow and the single apical cell divides obliquely, whereas in *Gelidium pusillum* the axes are always solid and the apical cells divide transversely.

Gelidium pusillum has been used as a source of agar, but not now to any extent because of its small size and the difficulties of collection.

Gelidium sesquipedale (Clemente) Thuret in Bornet & Thuret (1876), p. 61.

Provisional lectotype: LD (Herb. Alg. Agardh 32975). Spain (Algeciras).

Fucus corneus var. *sesquipedalis* Clemente (1807), p. 317.

Thallus consisting of erect fronds and creeping axes; erect fronds to 0·4 m, red or blackish-red in colour, often in large tufts, much branched, usually dendroid; branching variable, but with major axes of one or more orders and ultimate axes short, forming a parallel fringe to the penultimate axes, axes parallel-sided, 1–3 mm in breadth, 100–200 μm in thickness, rigid, with broad, blunt apices; medulla made up of tightly compacted large cells with an outer cortex of small cells containing chloroplasts, internal thick-walled rhizoids occurring in a narrow band at the outer part of the medulla, immediately beneath the cortex, cortical cells small in surface view, measuring 3–5 μm in diameter.

Fertile material unknown in Britain; elsewhere, gametangial plants dioecious; spermatangia formed in superficial colourless patches of oval or irregular outline, formed behind the apices but often extending for some distance along the axes; carpogonia formed in terminal colourless areas immediately behind the apices; carposporophytes developing internally, the outer cortex lifting on both sides of the frond to form a relatively large, ovoid bilocular cystocarp, 0·9–1·2 mm in length, 0·8–1·0 m in diameter, formed terminally on the fringing ultimate axes, occasionally with a small, spike-like projection; tetrasporangia ovoid 30–40 μm in length, 25–35 μm in diameter, formed in groups in the ultimate fringing

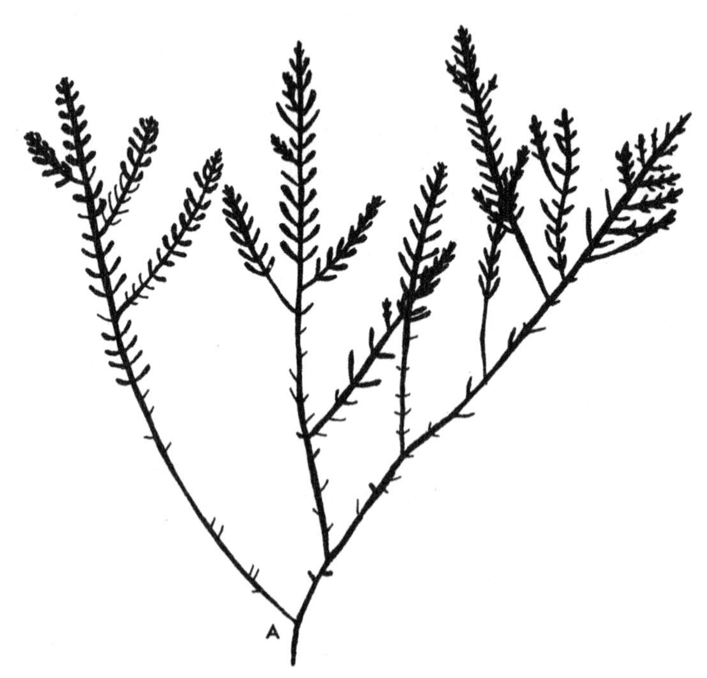

Fig. 49 *Gelidium sesquipedale*.
A. Habit ×1.

lateral axes or in clusters of short axes which develop adventitiously from these fringing axes, tetraspores arranged cruciately or irregularly.

Epilithic; in Britain, restricted to the upper sublittoral in places exposed to severe wave action. Occurs in similar situations in Finistère; further south, on the French Basque coast, occurs extensively throughout the upper sublittoral and, in pools, well into the lower littoral.

Restricted to Cornwall and Devon, in the British Isles.

French Channel coast to Mauretania; western Mediterranean.

Erect fronds survive for 2–3 years in the British Isles, with maximum growth between April and June. In Europe and North Africa, fronds persist for longer than this. Fertile material never collected in Britain; on the French Basque coast, cystocarps occur in August/September; tetrasporangia between June and September.

Data from the whole range indicates that *G. sesquipedale* is one of the least variable species of the genus, with the principal variation being only in overall size, a reflection of frond longevity. The largest specimen from the British Isles is about 200 mm while the largest observed, from the Basque coast, is about 400 mm.

Examination of records and identified material collected prior to 1950 showed that all but three specimens were misidentifications and these, from their appearance, were of drift origin (Dixon, 1958). Suggestions that the species should be removed from the British Flora were negated by the subsequent collection of material *in situ* from localities in Cornwall and Devon.

The absence of any reproductive structures in British specimens suggests that perennation on a clonal basis must occur although there is no evidence to indicate how initial recruitment takes place.

The species serves as a commercial source of agar in France, Spain and Morocco although far too rare ever to be used in the British Isles.

Fucus corneus Hudson (1762) has been the subject of much controversy. Typification (Dixon, 1967), undertaken in the hope of resolving these problems, has only added to them. It would seem best, therefore, to dismiss the epithet *corneum* as a longstanding source of confusion and error and revert to the use of *sesquipedale* for this taxon.

Excluded species

Gelidium melanoideum Bornet

This taxon was detected by Schousboe, although not formally described until later, by Bornet (1892). The species was first included in the British flora by Batters (1902) listing it from 'Sussex (Hastings) and Northumberland (Alnmouth)', with an additional locality (Dorset) added by Newton (1931). Feldmann (1939) questioned the occurrence of this taxon in the British Isles.

According to Feldmann & Hamel (1936) the distinguishing features are the blackish-violet colour, the sparse occurrence of the internal rhizoids and the V-shaped distribution of tetrasporangia. Investigation showed that the specimens on which the published records were based are referable to *G. crinale.* The specimens in question bear a slight resemblance in certain respects to *G. melanoideum,* although the two taxa are quite distinct.

See Dixon & De Valéra (1961) for full details.

Gelidium torulosum Kützing

G. torulosum was described by Kützing (1868) from material collected in Brazil. The addition of this entity to the British flora resulted from an identification by Lyle (1920) of a single specimen collected in Guernsey, on the basis only of the original description and illustration. This specimen is a narrow, somewhat aberrant example of *Gelidium latifolium*.
See Dixon & De Valéra (1961) for full details.

Gelidium versicolor (S. G. Gmelin) Lamouroux (generally referred to as *'G. cartilagineum')*.

Luxuriant specimens of *'Gelidium cartilagineum'* have been collected widely but sparsely over the past 250 years from the British Isles and other parts of western Europe (Belgium, Helgoland, Schleswig-Holstein, etc). Harvey accepted *Gelidium cartilagineum* as a British species, but with reluctance, while Batters (1902) excluded it deliberately. Examination indicates that all identifications are correct, as one might expect for such a very distinctive species, but there is no evidence to indicate that any was collected in an attached state and all specimens appear to have been obtained from the drift. There is no reason for the species to be retained in the British flora, or European flora. The Canary Islands represent the nearest location at which attached specimens are known to occur.

PTEROCLADIA J. Agardh

PTEROCLADIA J. Agardh (1851), p. XI.

Type species: *P. lucida* (Turner) J. Agardh (1852), p. 483.

Thallus differentiated into erect fronds and prostrate axes; prostrate axes cylindrical, attached to the substrate by lateral protuberances; erect frond cylindrical or compressed or markedly flattened, the lower parts usually flat, with the terminal portions either cylindrical or flattened; of uniaxial construction; medulla of large colourless cells with a cortex of smaller cells containing chloroplasts; internal thick-walled rhizoids distributed in the centre of the frond. Gametangial plants dioecious or monoecious; spermatangia formed in superficial colourless patches; carpogonia sessile; arising in groups in the ultimate divisions of the frond, carposporophytes developing internally, the outer cortex lifting on one face of the frond to form a unilocular cystocarp with one or more pores; tetrasporangia distributed in groups or singly in the ultimate divisions of the frond, tetraspores with cruciate, tetrahedral or irregular arrangement.

One species in the British Isles:

Pterocladia capillacea (S. G. Gmelin) Bornet & Thuret (1876), p. 57.

Lectotype: original illustration (Gmelin, 1768, pl. 15, fig. 1) in the absence of material. 'Mediterranean Sea'.

Fucus capillaceus S. G. Gmelin (1768), p. 146.
Fucus corneus ε *capillaceus* (S. G. Gmelin) Turner (1819), p. 146.
Gelidium corneum ε *capillaceum* (S. G. Gmelin) Greville (1830), p. 143.
Gelidium capillaceum (S. G. Gmelin) Kützing (1868), p. 18.
Fucus corneus var. δ *uniforme* Goodenough & Woodward (1797), p. 181.
Gelidium corneum var. δ *uniforme* (Goodenough & Woodward) Greville (1830), p. 143.
Gelidium corneum ρ *proliferum* Kützing (1849), p. 765.
Gelidium proliferum (Kützing) Kützing (1868), p. 19, non *G. proliferum* Harvey (1855), p. 551.

Thallus differentiated into erect fronds and prostrate axes; prostrate axes extensive, cylindrical, 1–2 mm in diameter, attached to substrate by lateral protuberances; erect frond to 300 mm, black, purple, red or pink in colour, simple in smaller specimens but much branched in larger, often dendroid with age, cylindrical or flattened, small simple fronds may be either but basal parts are always markedly flattened in larger specimens, to 4 mm in breadth, terminal portions of frond cylindrical during spring, barely 1 mm in diameter, but becoming increasingly broad and flat during the autumn; branching variable, some axes unbranched, or branched once, twice or three times, opposite or alternate, producing a frond with markedly triangular outline; medulla of large colourless cells and cortex of smaller cells containing chloroplasts, with internal thick walled rhizoids restricted to the centre of the frond.

Fertile material unknown in Britain; elsewhere, gametangial plants dioecious; spermatangia formed in superficial colourless patches of irregular outline, usually behind the apices of axes, the carposporophytes developing internally, the outer cortex lifting on one face of the frond to form a unilocular cystocarp 500–700 μm in diameter in which there are usually several pores; tetrasporangia ovoid or spherical 20–30×20–30 μm, distributed in the tips of ultimate divisions of the frond often, but not always spathulate, tetraspores cruciately, tetrahedrally or irregularly arranged.

Epilithic; in the British Isles, occurring in deep pools in the lower littoral, but never emerging above the water surface, to 2 m.

Southern and western shores, extending eastwards to Dorset, northwards to Anglesey and the Isle of Man; in Ireland from Wexford to Donegal.

Atlantic coasts of France, Spain and Portugal; western Mediterranean; West Indies; Indian and Pacific Oceans.

Erect fronds perennial, persisting for up to 7 years although the average is only 2–3 years. Growth commences in the early spring; increase in length ceases by June, but increase in breadth continues until December. Reproductive structures unknown in British material, although on the Atlantic coasts of France and Spain carpogonia and spermatangia develop in June/July with cystocarps present from July to October; tetrasporangia from June to October.

The marked seasonality of growth in length and breadth results in fronds having two distinct forms. In spring and early summer, the ultimate portions are narrow and terete; the epithet '*capillacea*' has been applied to this form. Later, the continued increase in breadth of these terete portions results in apices which are more uniformly strap-shaped; the epithet '*pinnata*' has been applied to this, albeit incorrectly.

The absence of reproductive structures throughout the British Isles raises problems with respect to life history. Perennation on a clonal basis has been shown to occur for more than 40 years in a single site although it is not clear whether recruitment results from vegetative propagation or by long-distance spore transport.

Numerous taxa described from tropical and subtropical waters differ from *P. capillacea*

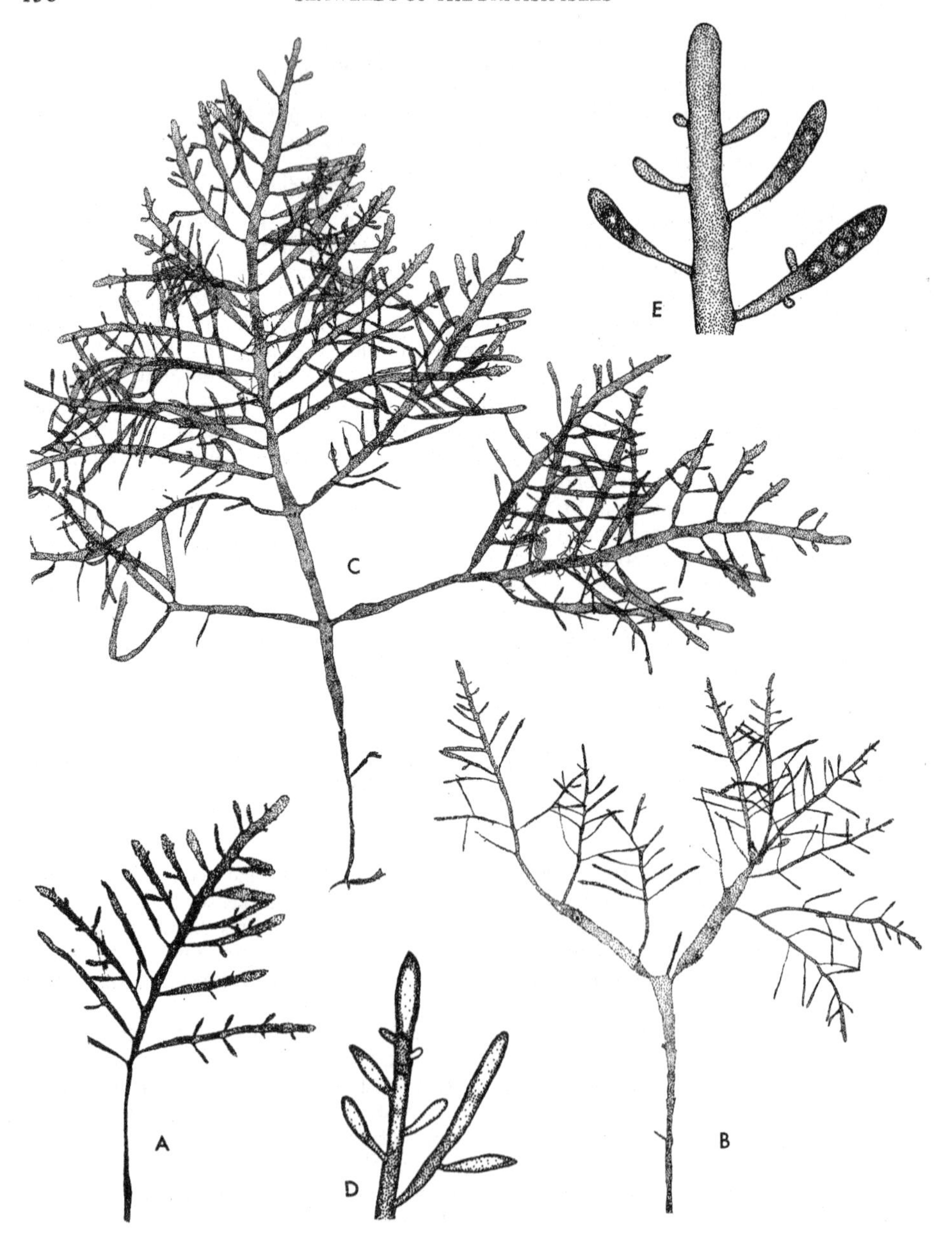

Fig. 50 *Pterocladia capillacea*
A. Two-year plant (Dec.) ×1·5; B. Three-year plant (June) ×1·5; C. Five-year plant (Dec.) ×1·5; D. Apex of spermatangial plant (non-British material) ×10; E. Apex of cystocarpic plant (non-British material) ×10.

only in minor variations of external form which could result from slight changes in the patterns of cell division and cell enlargement.

Pterocladia capillacea serves as a commercial source of agar in various parts of the world, particularly New Zealand. During the Second World War, an attempt to develop in Ireland an industry based largely on this species failed after a relatively successful harvest in the first year.

HELMINTHOCLADIACEAE J. Agardh

HELMINTHOCLADIACEAE J. Agardh (1852), p. 410 [as Helminthocladeae].

Gametangial thallus erect, terete, branched or unbranched, lubricous or calcified, multiaxial; gonimoblasts developing internally, not externally obvious, producing carposporangia or carpotetrasporangia terminally. Other phase in life history filamentous, *Audouinella*-like, sometimes forming tetrasporangia.

Three genera of the Helminthocladiaceae occur in the British Isles – *Helminthocladia, Helminthora* and *Nemalion.* Although there are difficulties with the attribution of species to these genera in certain parts of the world (Abbott, 1965; Womersley, 1965) the limits are perfectly clear for British material.

HELMINTHOCLADIA J. Agardh **nom. cons.**

HELMINTHOCLADIA J. Agardh (1852), p. 412.

Type species: *H. purpurea* (Harvey) J. Agardh (1852), p. 414 (=*H. calvadosii* (Lamouroux ex Duby) Setchell (1915), no. 2035).

Gametangial plant erect, lubricous or mucilaginous, becoming firmer with age, differentiated into basal disc and one or more erect axes; axes with central core of narrow axial filaments intricately entangled; filaments of limited growth loosely aggregated but becoming more compact with age; apical cells of filaments of limited growth enlarging once divisions have ceased; carpogonial branch 3–4 celled, lateral; carposporophyte arising from one or more segments cut off from the carpogonium at its upper end, when mature with only a few enveloping filaments; each carposporangium producing a single carpospore in the one British species.

Tetrasporangial phase filamentous or discoid in culture, not known in the field.

One species in the British Isles: *H. calvadosii* (Lamour. ex Duby) Setch. Older works also list *H. hudsonii,* but this must be rejected as based on misidentifications (see p. 139 under *H. agardhiana*).

Helminthocladia calvadosii (Lamouroux ex Duby) Setchell (1915), no. 2035.

Provisional lectotype: CN. France (Calvados).

Dumontia calvadosii Lamouroux ex Duby (1830), p. 941.
Mesogloia purpurea Harvey (1833), p. 386.
Helminthocladia purpurea (Harvey) J. Agardh (1852), p. 414.

Gametangial plant erect, to 400 mm in length, red, brown or purple in colour, lubricous or mucilaginous when young but becoming firmer with age; differentiated into basal disc and one or more erect axes; axes solid, terete, but becoming hollow and distorted with age, 2–5 mm in diameter, simple or branched; branching occasionally dichotomous but usually lateral and adventitious; central strand of axial filaments not sharply differentiated from the cortex; axial cells narrow, elongate, colourless; ultimate cells of filaments of limited growth short, 15–25 μm in length, almost quadrate, with deeply pigmented chloroplasts; following cessation of divisions, the apical cells enlarge considerably, becoming clavate, 45–60 μm in length, with conspicuous chloroplasts; monoecious; spermatangia in hemispherical clusters at the tips of the filaments of unlimited growth; carpogonium formed by the transformation of the apical cell of a 3–4 celled filament, arising laterally on

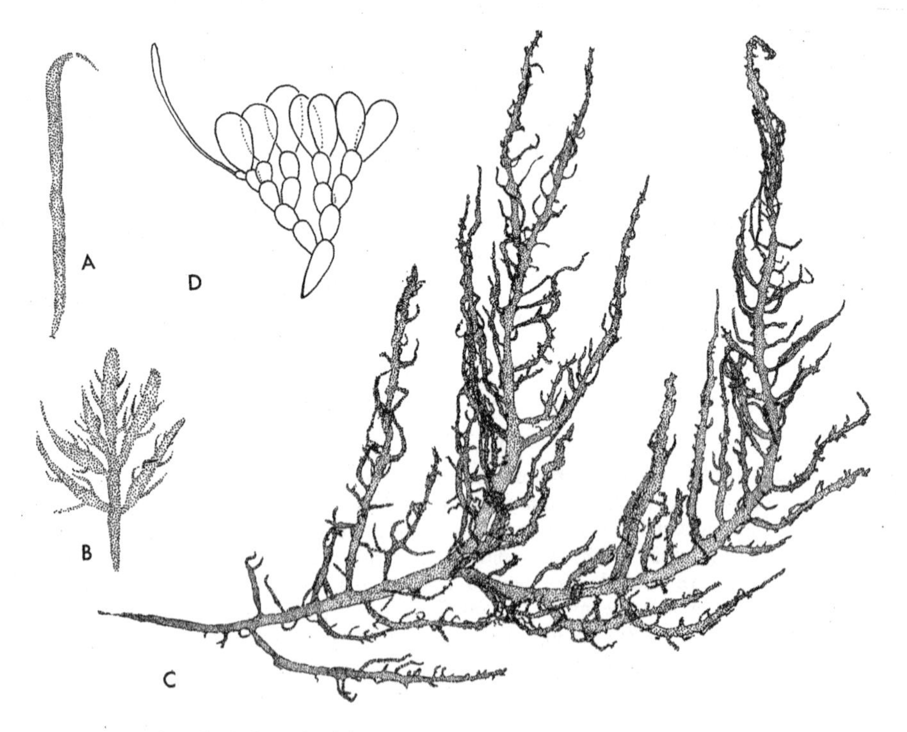

Fig. 51 *Helminthocladia calvadosii*
A – C. Habit, showing range of form ×0·5; D. Filaments of limited growth, showing inflated apical cells ×175.

the filaments of limited growth; carposporophyte arising by the oblique division of the carpogonium and subsequent elaboration, 100–270 μm in diameter when mature; gonimoblast composed of elongate cells with carposporangia formed terminally.

Tetrasporangia unknown; each carposporangium produces a single carpospore which germinates to produce an *Audouinella*-like filamentous growth known to be diploid; cells elongate, 15–20 × 7–9 μm.

Gametangial plant epilithic, in pools and channels, particularly those with a layer of sand at the bottom, in the upper sublittoral and to 5 m. The product of carpospore germination not yet recognized in the field.

Widely distributed in the British Isles, but highly sporadic; recorded from the Channel Islands, Cornwall, Devon, Sussex, Durham, Northumberland, Clare, Galway, Mayo, Dublin.

Denmark to Canary Islands.

Gametangial plants arise in May/June and persist until September/October; spermatangia and carpogonia present by early July, with cystocarps liberating carpospores by late August. Nothing is known of the presumed tetrasporangial phase in the field.

Variation of the gametangial phase not very great, being determined by the occurrence or absence of adventitious lateral branches. Such branches are present in most specimens; in their absence the plant is a simple tubular thallus of very different aspect.

Structures somewhat similar in many respects to spermatangia have been regarded as monosporangia (Rosenvinge, 1909).

Although usually quite distinct from *Nemalion helminthoides* (Vell. in With.) Batt. in gross morphology, some simple or little branched specimens may be difficult to distinguish. The terminal cells of the filaments of limited growth are inflated in *Helminthocladia calvadosii* and not in *Nemalion helminthoides;* hairs arise terminally in *N. helminthoides* whereas in *Helminthocladia calvadosii* the terminal position is obscured as a result of the small size of the hair-bearing cell and the large size of the inflated cells.

Excluded species

Helminthocladia agardhiana Dixon *(H. hudsonii pro parte).*

In the course of an attempt to locate authentic British specimens of what has been accepted in various floras as *H. hudsonii,* it was discovered that there had been a gross misapplication of the epithet *hudsoni* and that there was sufficient evidence to doubt all the previous records of the occurrence of this taxon not only in Britain but in France as well. The J. Agardh specimen from Tangier, referred by him to *H. hudsonii,* is not conspecific with the alga to which the basionym *Mesogloia hudsonii* was applied. The former has been renamed *H. agardhiana,* while the latter must be referred to *Halarachnion ligulatum* (Woodw.) Kütz. For details see Dixon (1962). As far as can be ascertained, the alga known previously as *Helminthocladia hudsonii* is restricted to a small area of Morocco and adjacent parts of North Africa.

HELMINTHORA J. Agardh **nom. cons.**

HELMINTHORA J. Agardh (1852), p. 415.

Type species: *H. divaricata* (C. Agardh) J. Agardh (1852), p. 416.

Gametangial thallus erect, lubricous or mucilaginous, differentiated into basal disc and one or more erect axes; axes terete, much branched, of multiaxial construction, with central core of enlarged rectangular cells arranged in series and compacted, not intricately entangled; filaments of limited growth very loosely aggregated in the early stages of growth and not becoming compact for some time; carpogonial branch 3–4 celled, lateral; carposporophyte arising from the upper cell formed by the transverse division of the carpogonium, when mature surrounded by a group of enveloping filaments, associated with a tuft of downgrowing rhizoidal filaments.

Tetrasporangial plants not known in the field; in one species, in culture, the carpospores germinate to produce a filamentous growth with large cells, bearing tetrasporangia terminally; tetraspores cruciately arranged.

One species in the British Isles:

Helminthora divaricata (C. Agardh) J. Agardh (1852), p. 416.

Lectotype: LD (Herb. Alg. Agardh. 31980). Spain (Cadiz).

Mesogloia divaricata C. Agardh (1824), p. 51.
Nemalion ramosissimum Zanardini (1847), p. 38.

Gametangial plant erect, to 250 mm in length, red, pinkish-red or brownish-red in colour, soft, lubricous or mucilaginous when young, but becoming firmer and more crisp with age, differentiated into basal disc and one or more erect axes; axes terete, 0·1–3 mm in diameter, usually profusely branched, of multiaxial construction, with a compact central core of axial filaments sharply differentiated from the cortex; axial cells broad, rectangular; filaments of limited growth arising at right angles, made up of rounded cells, most of which contain chloroplasts, much smaller in size than the cells of the axial core; monoecious or dioecious; spermatangia in spherical clusters at the tips of the filaments of limited growth; carpogonium with a very long trichogyne; carpogonial branch 3–4 celled, arising laterally close to the central axial core; carpogonium dividing transversely, only the upper cell giving rise to the carposporophyte; carposporophyte small, 50–175 μm in diameter when mature, surrounded by more or less compacted group of enveloping filaments and firmly attached to the gametangial thallus by downgrowing rhizoidal filaments; carposporangia formed from the terminal cells.

Tetrasporangial plant not reported in the field; in culture, the carpospores produce a discoid or filamentous growth with large cells, 50 μm in length, 15 μm in breadth, bearing tetrasporangia terminally on upright filaments, tetrasporangia 30 μm in length, 15 μm in breadth; tetraspores cruciately arranged.

The gametangial plant is epilithic, or on gravel, shells and coralline algae, very rarely epiphytic on other algae; in the sublittoral, where it can occur from deep pools to depths of 30 m. The product of germination of the carpospores has not been recognized in the field.

Generally distributed throughout the British Isles.

Denmark to French Atlantic coast; western Mediterranean.

Gametangial plants arise in May/June and persist until September/October; spermatangia are present by July, with mature cystocarps by August. Nothing is known of the presumed tetrasporangial phase in the field.

Variation in form of the gametangial phase is not very great, the major influencing factor being light intensity. Plants growing at depth or in areas where silt particles occur are attenuated but the characteristic appearance of *H. divaricata* is not lost.

Svedelius (1917) described monosporangia in *H. divaricata;* such structures have not been observed in British specimens.

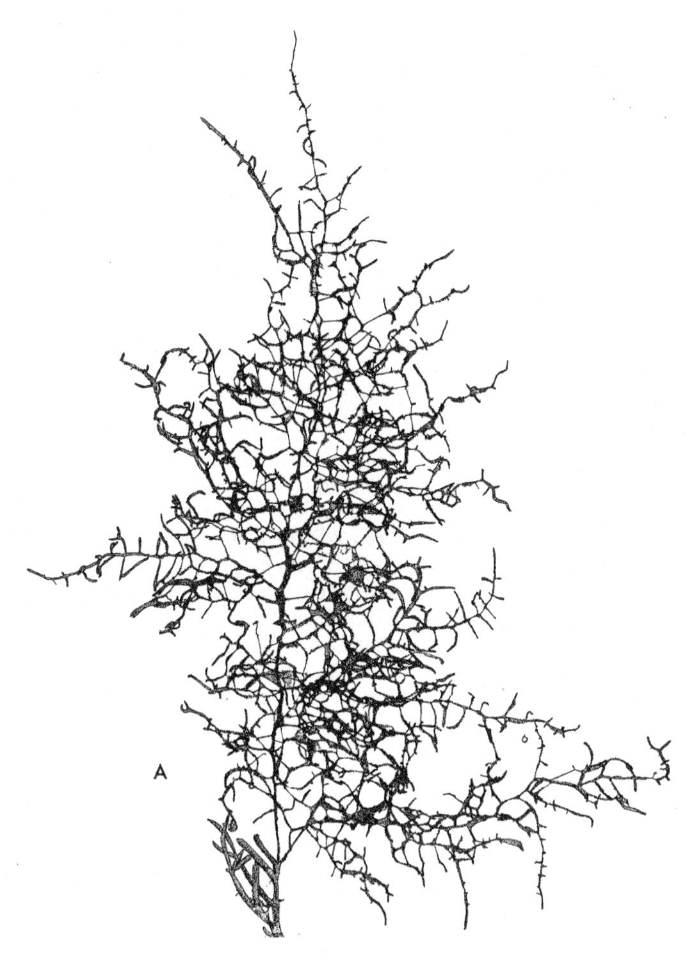

Fig. 52 *Helminthora divaricata*
A. Habit ×1.

NEMALION Duby

NEMALION Duby (1830), p. 959.

Type species: *N. lubricum* Duby (1830), p. 959, (=*N. helminthoides* (Velley in Withering) Batters (1902), p. 59).

Helminthora Fries (1825), p. 341, non *Helminthora* J. Agardh (1852), p. 415.

Gametangial plant erect, lubricous and mucilaginous, differentiated into basal disc and one or more erect axes; axes solid, terete, simple or branched, of multiaxial construction, with central core of narrow axial filaments intricately entangled; filaments of limited growth loosely aggregated throughout the whole life of the plant; carpogonial branch 4–8 celled, with little differentiation; carposporophyte arising from the upper cell formed by the transverse division of the carpogonium, when mature with only a very few enveloping filaments.

Filamentous growths resembling *Audouinella* formed in culture by the germination of carpospores, bearing tetrasporangia; gametangial plant arising in the field directly.

One species in the British Isles:

Nemalion helminthoides (Velley in Withering) Batters (1902), p. 59.

Provisional lectotype: LIV. Dorset (Portland).

Fucus elminthoides Velley in Withering (1792), p. 255.
Rivularia multifida Weber & Mohr (1804), p. 193.
Nemalion multifidum (Weber & Mohr) J. Agardh (1841), p. 453.
Nemalion lubricum Duby (1830), p. 959.

Gametangial plant erect, to 0·4 m in length, red, brown or black in colour; lubricous or mucilaginous, differentiated into basal disc and one or more erect axes; axes solid, terete, 2–20 mm in diameter, simple, branched or much-branched; central strand of axial filaments not sharply differentiated in size or form from the cortex; axial cells elongate, narrow, colourless; apical cells of filaments of limited growth short, almost quadrate, with deeply-pigmented chloroplasts; following the cessation of divisions, the apical cell not conspicuously enlarged, often giving rise to a hair; usually monoecious although occasionally dioecious; spermatangia in clusters at the tips of the filaments of limited growth; carpogonium formed by the transformation of the apical cell of a 4–8 celled filament arising sub-terminally on a filament of limited growth; carpogonium divides transversely, the carposporophyte arising from the upper cell; carposporophyte spherical, 100–300 μm in diameter, surrounded only by a few vegetative filaments; carposporangia formed terminally, rarely in an intercalary position.

In culture, carpospores germinate to produce a mass of branched filaments resembling *Audouinella* on which monosporangia and tetrasporangia have been reported, with direct development of the gametangial thallus in the field; cells elongate, 30–65 μm in length, 15–20 μm in breadth; monosporangia ovoid, 12–18 × 8–10 μm; tetrasporangia ovoid, 22–28 × 15–18 μm; tetraspores cruciately arranged.

Fig. 53 *Nemalion helminthoides*
A – E. Habit, showing range of form ×1; F. Filaments of limited growth ×175.

A
B
C
D
E
F

Gametangial phase epilithic, throughout the midlittoral, also on *Balanus, Patella, Mytilus,* etc. Growths similar to the product of germination of carpospores in culture occur on *Patella* and in *Balanus.*

Generally distributed throughout the British Isles.

Central Norway to Morocco; Mediterranean; U.S.A. (New York) to Canada (Nova Scotia).

Gametangial plants arise in March/April and persist until October in southern England. Spermatangia and carpogonia are present by early May, with carposporophytes liberating carpospores by early August. In Anglesey, development begins somewhat later, in May, with gametangia in June/July and the gametangial thalli disappear a little earlier, in mid-September. The product of germination of carpospores, or growths similar to this, has been reported as occurring during the winter months in Anglesey.

It has been traditional to regard this species as two entities, with the degree of branching as the criterion for discrimination. The completely unbranched or sparsely branched thalli were referred to *N. helminthoides,* or *N. lubricum* in the older literature, while the much-branched specimens were attributed to *N. multifidum.* Detailed investigation of this species in Europe, Australia (Womersley, 1965) and California (Hollenberg & Abbott, 1966) has finally indicated the unsatisfactory nature of this distinction. The degree of branching does not appear to correlate with any ecological factor although Hamel (1930) claimed that the unbranched thalli occurred at higher levels on the shore than the branched specimens.

The life history of *Nemalion helminthoides* is still not fully established. Carpospore germination and the early stages of development of the product have been known for many years and several investigators have shown that extensive growths of an *Audouinella*-like filamentous growth may be produced, with reproduction by monospores. Martin (1967, 1969) has demonstrated the occurrence in the field, on *Patella* and in *Balanus,* of filamentous growths of similar aspect. These growths become eroded during the late autumn so that only certain basal cells persist, which become characteristically rounded and remain in this condition through the winter. In the following spring, these rounded cells put out filamentous protuberances which in a short time produce the characteristic gametangial phase of *Nemalion.* Fries (1967), on the other hand, has reported the occurrence of tetrasporangia on material grown in culture from carpospores as well as on filaments obtained from the field and placed in culture in the laboratory.

The spelling of the specific epithet causes confusion. The form used in the original description *(elminthoides)* is often used although it is more correct to change this to *helminthoides,* as here, in accordance with nomenclatural practice.

CHAETANGIACEAE Kützing

CHAETANGIACEAE Kützing (1843), p. xxiii [as Chaetangieae].

Gametangial thallus erect, terete, hollow, dichotomously branched, soft, multiaxial; gonimoblasts developing in cortex, visible externally. Other phase in life history filamentous, *Audouinella*-like, sometimes forming cruciate tetrasporangia.

Only one genus of the Chaetangiaceae, *Scinaia,* occurs in the British Isles.

SCINAIA Bivona

SCINAIA Bivona (1822), p. 232.

Type species: *S. forcellata* Bivona (1822), p. 232, (see Dixon & Irvine, 1970).

Ginnania Montagne (1841), p. 162.
Myelomium Kützing (1843), p. 398.

Gametangial plant erect, differentiated into basal disc and erect axis; axes tubular and turgid when young but becoming squashed with age, branched dichotomously, with central core of axial filaments; some apical cells of filaments of limited growth colourless interposed with cells which retain chloroplasts and which give rise to hairs, monosporangia and spermatangia. Carpogonial branch 3-celled; carposporophyte arising from the carpogonium with numerous enveloping filaments formed from the lowermost cell of the carpogonial branch to produce a spherical or pyriform cystocarp, opening to the exterior through a pore.

Filamentous growth resembling *Audouinella* formed in culture by the germination of carpospores, bearing tetrasporangia, or giving rise directly to the gametangial phase; not known in the field.

KEY TO SPECIES

1 Gametangial plant to 100 mm in height, 1–3 mm in diameter, rarely with constrictions; multiaxial core not usually visible in herbarium specimens; thallus surface with clusters of pigmented cells; cystocarps 100–150 μm in diameter, present in large numbers . . . . *S. forcellata*

Gametangial plant to 150 mm in height, 2–4 mm in diameter, usually with constrictions; multiaxial core usually visible in herbarium specimens; thallus surface with isolated pigmented cells; cystocarps 150–200 μm in diameter, widely dispersed . . . . *S. turgida*

Scinaia forcellata Bivona (1822), p. 232.

Lectotype: original illustration (Bivona, 1822), in the absence of material. Italy.

Ulva furcellata Turner (1801), p. 301.
Scinaia furcellata (Turner) J. Agardh (1852), p. 422.
Ulva interrupta De Candolle (1807), p. 232.
Dumontia interrupta (De Candolle) Duby (1830), p. 133.

Gametangial plant erect, to 100 mm in height, dull red in colour, differentiated into basal disc and a single erect axis; axes tubular, branched dichotomously and irregularly, 1–3 mm in diameter, rarely with constrictions; multiaxial core rarely visible in herbarium specimens; surface composed of almost colourless cells among which clusters of pigmented cells are distributed; monoecious; spermatangia 2–5 μm in diameter, formed in superficial clusters of variable size; carpogonia scattered, giving rise to pyriform or spherical cystocarps, 100–150 μm in diameter, sunk in the thallus wall, pore not usually protruding, formed in large numbers; monosporangia 5–9 μm in diameter, formed singly or in small groups on thallus surface.

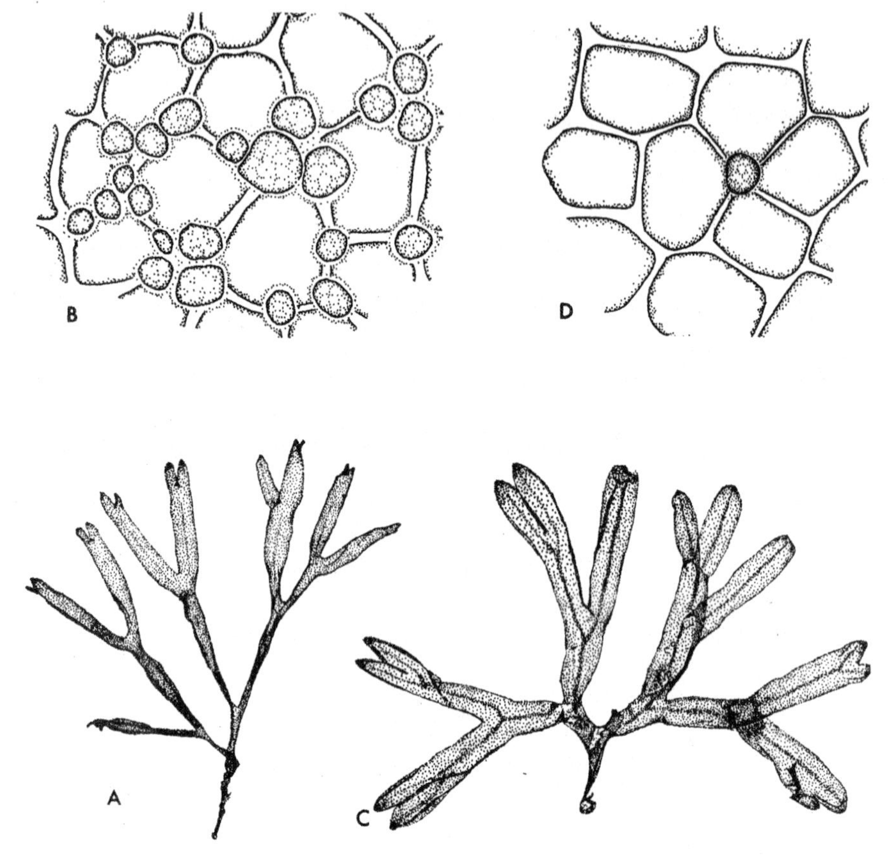

Fig. 54 *Scinaia forcellata*
A. Habit ×1; B. Surface view showing cell pattern ×625.
Scinaia turgida
C. Habit ×1; D. Surface view showing cell pattern ×625.

Tetrasporangial plant not known in the field; germination of carpospores in culture giving rise to an *Audouinella*-like filamentous growth on which tetrasporangia have been reported; cells elongate, 40–60 μm in length, 5–7 μm in diameter; tetrasporangia 10 μm in diameter, tetraspores cruciately arranged, have been described as well as reports of direct development of the gametangial thallus from these growths.

Gametangial phase epilithic, restricted to the upper sublittoral where it occurs in pools and channels (particularly those with a layer of sand at the bottom) to 8 m.

Generally distributed in the British Isles, although not reported for the east coast between Norfolk and Caithness.

Southern Norway to Morocco; Canary Islands; Mediterranean.

Gametangial plants arise in April/May and can persist until December; spermatangia and carpogonia present by May, with cystocarps in quantity by early August.

S. forcellata exhibits little morphological variation other than in overall size; it is one of the most characteristic and easily recognized red algae.

The relationship of the *Audouinella*-like growth, formed in culture by the germination of carpospores to the gametangial phase, is still not clear. Boillot (1968) described the formation of tetrasporangia and, subsequently (Boillot, 1969), the germination of the tetraspores to form the gametangial phase, although Jones & Smith (1970) have mentioned briefly the direct development of the gametangial phase from the filamentous growth.

The occurrence of monosporangia in *S. forcellata,* first reported by Svedelius (1915), is quite widespread, although the function of these structures and the product of their germination are unknown.

Scinaia turgida Chemin (1926a), p. 102.

Lectotype: LD (Herb. Alg. Agardh. 32180). Tangier.

Halymenia furcellata var. *subcostata* J. Agardh (1842), p. 98.
Scinaia furcellata var. *γ subcostata* (J. Agardh) J. Agardh (1852), p. 422.
Scinaia subcostata (J. Agardh) Chemin ex Hamel (1930), p. 85.

Gametangial plant erect, to 150 mm in height, carmine red in colour, differentiated into basal disc and erect axis; axes tubular, branched dichotomously, often regularly and in a single plane, 2–4 mm in diameter, constrictions usually present at dichotomies; surface layer made up of almost colourless cells among which isolated pigmented cells are distributed; multiaxial core usually visible in herbarium specimens; monoecious; spermatangia 2–3 μm in diameter, formed in superficial clusters; carpogonia scattered, giving rise to globular or pyriform cystocarps, 150–200 μm in diameter, sunk in thallus wall, pore protruding and raised above thallus surface; cystocarps relatively few in number; monosporangia not recorded.

Germination of carpospores in culture gives rise to *Audouinella*-like filamentous growths, on which tetrasporangia have been reported; tetraspores cruciately arranged.

Gametangial phase occurs predominantly in the sublittoral, to 30 m; nothing known of the tetrasporangial phase in the field.

Restricted to southern and western shores of England; (north Devon to Dorset) and Ireland (Wexford to Donegal).

Atlantic coast of France and northern Spain; western Mediterranean.

Fully grown gametangial plants with mature cystocarps known from July and August, but with no knowledge of when development and reproduction commence.

Little morphological variation other than in overall size.

NACCARIACEAE Kylin

NACCARIACEAE Kylin (1928), p. 11.

Gametangial thallus erect, terete, profusely and irregularly branched, soft, uniaxial; gonimoblasts developing internally, producing swollen subterminal cystocarps, visible externally. Other phase in the life history prostrate, compactly filamentous, sometimes forming tetrasporangia.

Two genera occur in the British Isles, *Naccaria* and *Atractophora*. Although confused by the earlier authors, the points of distinction were clearly outlined by the brothers Crouan (1848) and the characters discussed by them are still perfectly acceptable.

ATRACTOPHORA Crouan frat.

ATRACTOPHORA Crouan frat. (1848), p. 371.

Type species: *A. hypnoides* Crouan frat. (1848), p. 372.

Gametangial thallus erect, lubricous and mucilaginous, differentiated into basal disc and erect frond; axes much branched, with the axial cells of considerable diameter, branching somewhat irregular; uniaxial, each axial cell giving rise to four primary pericentral cells; short filaments of limited growth radiate from and invest the axial filament, apical cells of the filaments of limited growth frequently converted into hairs; monoecious; spermatangia formed in superficial clusters; carpogonial branch 3-celled; carposporophyte arising from the carpogonium after fusion with the hypogynous cell, gonimoblast envelops a length of thallus and forms a complex cystocarp.

Carpospores on germination form a disc, from which direct development of the gametangial phase has been reported; tetrasporangia unknown.

One species in the British Isles:

Atractophora hypnoides Crouan frat. (1848), p. 372.

Provisional lectotype: CO. France (Brest).

Naccaria hypnoides (Crouan frat.) J. Agardh (1863), p. 712.

Gametangial thallus erect, gelatinous, red in colour, to 100 mm in height, attached by a small basal disc; axes uniaxial, to 1 mm in diameter but usually much less (100 μm), much branched, downgrowing filaments enclosing the axial filament in a compact cortex, spreading filaments incompletely compacted to give the thallus a loose, diffuse, 'hairy' aspect; monoecious; spermatangia in superficial clusters of varying size; carposporophytes spreading over a length of thallus, cystocarps markedly swollen, to 200 μm in length, to 150 μm in diameter, with an elongate tip; carposporangia externally visible.

Carpospores on germination form a compact disc with direct development of the gametangial phase; tetrasporangia unknown.

Epiphytic on coralline algae in the sublittoral to 20 m.

Cornwall, Devon, Dorset, Channel Islands, Isle of Man. The species is known only from a few isolated collections.

Atlantic coast of France.

Fully-grown plants of the gametangial phase known from July–September.

Data on form variation inadequate for comment.

Germination of the carpospores produces a discoid germling (Chemin, 1927) from which the direct development of the gametangial phase has been reported (Boillot, 1967). The occurrence of tetrasporangia on this discoid germling has not been noted.

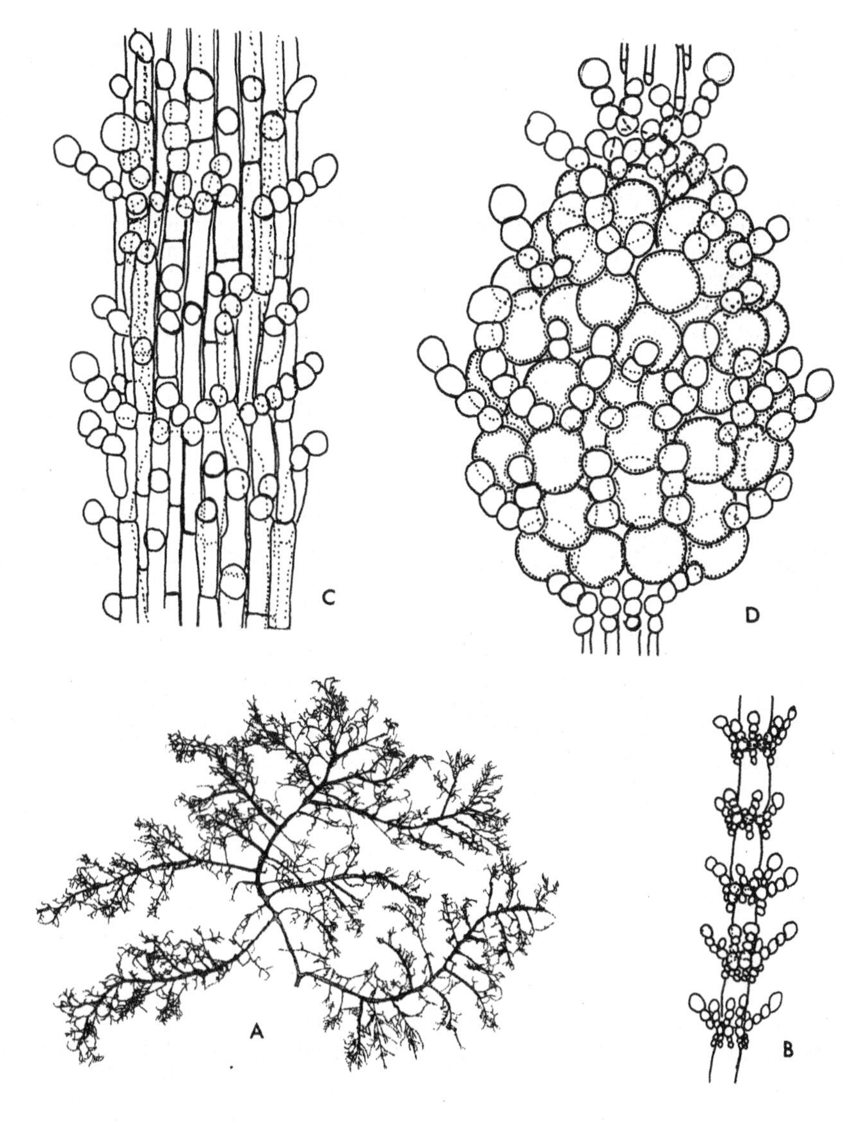

Fig. 55 *Atractophora hypnoides*
A. Habit ×1·25; B. Young axis ×200; C. Mature axis ×200; D. Cystocarp ×200.

NACCARIA Endlicher

NACCARIA Endlicher (1836), p. 6.

Type species: *N. wiggii* (Turner) Endlicher (1836), p. 6.

Chaetospora C. Agardh (1824), p. 146, non *Chaetospora* R. Brown (1810), p. 99.

Gametangial thallus erect, lubricous and mucilaginous, differentiated into basal disc and erect frond; axes much branched, axial cells somewhat narrow and enveloped by the cortication; branching somewhat irregular; uniaxial, each axial cell giving rise to two primary pericentral cells arranged on opposite sides, although when mature the pericentral cells also lie at opposite ends; filaments of limited growth closely invest the axial filament; dioecious; spermatangia formed in superficial clusters with a spiral arrangement around the axis; carpogonial branch 2–3 celled; çarposporophyte arising from segment cut off from the carpogonium after fusion with hypogynous cell, gonimoblast envelops a length of thallus and forms a complex cystocarp.

Carpospores germinate to form a filamentous growth on which tetrasporangia have been reported and from which direct development of the gametangial phase has also been noted.

One species in the British Isles:

Naccaria wiggii (Turner) Endlicher (1836), p. 6.

Lectotype: BM-K. Norfolk (Yarmouth).

Fucus wiggii Turner (1802), p. 135.
Naccaria vidovichii Meneghini (1844), p. 298.
Naccaria gelatinosa J. Agardh (1863), p. 713.

Gametangial thallus erect, lubricous, red or pinkish-red in colour, to 250 mm in height, attached by a small basal disc; axes much branched, branches arranged in a spiral, compact, cylindrical, to 800 μm in diameter, clothed in hairs; dioecious; spermatangia grouped in superficial clusters, often spirally arranged; carposporophytes spreading over a length of thallus, cystocarps slightly swollen, to 1 mm in diameter, with a small sterile tip; carposporangia enveloped by cystocarp wall and barely visible.

Carpospores in culture germinate to produce an *Audouinella*-like filamentous growth, not recognised in the field; cells 15–20 μm in length, 4–6 μm in diameter; direct development of the gametangial phase reported as well as tetrasporangia; tetrasporangia 20–25 × 16–20 μm, tetraspores cruciately arranged.

Gametangial phase epiphytic, restricted to the sublittoral, to 20 m.

Widely distributed but highly sporadic in occurrence on southern and western shores, extending northwards to Norfolk, Isle of Man, Clare.

Atlantic coasts of France and Spain; Mediterranean.

Fully grown gametangial plants collected between June and October.

There appears to be little variation in form in the British Isles. Material from the Mediterranean is often more elongate and filiform, with the filaments of limited growth less compacted than in British specimens. These attenuated specimens have been regarded as distinct species *(N. vidovichii, N. gelatinosa)* although there is no justification for this separation.

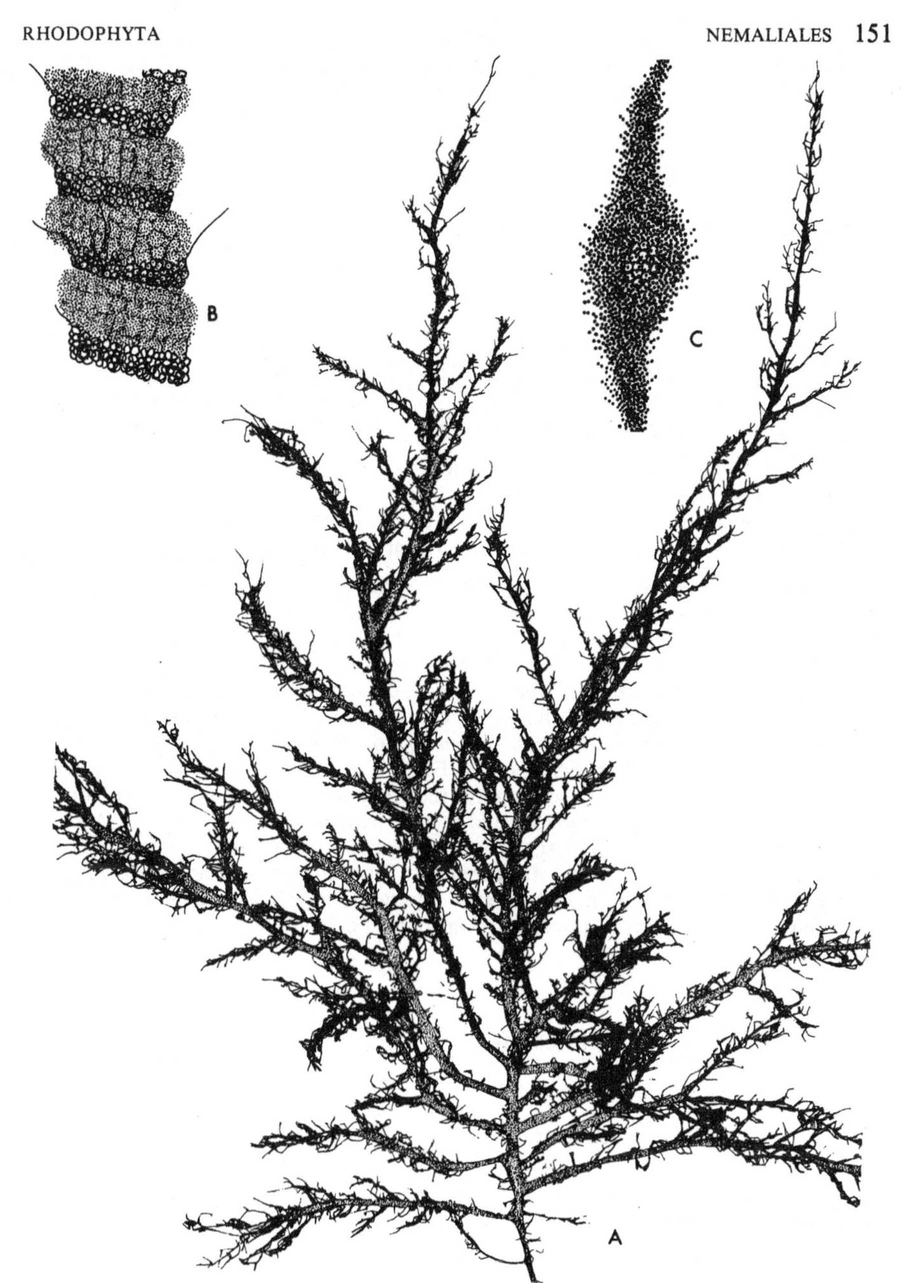

Fig. 56 *Naccaria wiggii*
A. Habit ×1; B. Spermatangial axis ×40; C. Cystocarp ×10.

Germination of the carpospores to produce a filamentous *Audouinella*-like growth was reported by Chemin (1927), while Boillot (1967) reported the development from this of the erect gametangial phase. Jones & Smith (1970) have described the occurrence of structures regarded as tetrasporangia on this product of carpospore germination.

The gametangial phase of *Naccaria wiggii* sometimes resembles *Gloiosiphonia capillaris* (Huds.) Carm. ex Berk. Mature cystocarps are usually present in specimens of both; the cystocarps of *Naccaria* completely envelop the axis, occurring singly in a subterminal position while those of *Gloiosiphonia* occur to one side of the axis over its entire length.

BONNEMAISONIACEAE Schmitz

BONNEMAISONIACEAE Schmitz in Engler (1892), p. 20.

Gametangial thallus erect, terete or flattened, profusely and regularly branched, soft, uniaxial; gonimoblasts developing terminally in short lateral branches, producing swollen terminal flask-shaped cystocarps. Other phase in the life history prostrate or filamentous, of various morphologies, sometimes forming cruciate or tetrahedral tetrasporangia.

Two genera, *Bonnemaisonia* and *Asparagopsis,* occur in the British Isles. Although the species are often distinctive, there is little agreement as to criteria by which these genera should be distinguished. Svedelius (1933) based his distinction on the origin of the carposporophyte, attributing to *Bonnemaisonia* those species in which this arose from the carpogonium and to *Asparagopsis* those in which the hypogynous cell served as the point of origin. Okamura (1921) and Kylin (1956) characterized the two genera on the arrangement of branches, referring distichous species to *Bonnemaisonia* and those with a spiral arrangement to *Asparagopsis.* Feldmann & Feldmann (1943) used the development and extent of the filaments of limited growth, the secretory cells, the arrangement and number of carposporangia within the pericarp and the life history. Hudson & Wynne (1969) indicated that the position of the axial cell from which the procarp develops can also be used to distinguish *Bonnemaisonia* and *Asparagopsis.* There is little correlation between these criteria and it might be questioned whether separation is justified. At present, it would seem best to retain the two genera as defined by Feldmann & Feldmann (1943) and Hudson & Wynne (1969).

ASPARAGOPSIS Montagne

ASPARAGOPSIS Montagne (1841), p. XIV.

Type species: *A. delilei* Montagne (1841), p. XIV (=*A. taxiformis* (Delile) Trevisan (1845), p. 45).

Lictoria J. Agardh (1841), p. 22.
Falkenbergia Schmitz (1897), p. 479.

Gametangial phase (=*Asparagopsis*) erect, much branched; branches opposite, with unequal development of the two components of each pair, one lateral branch being short and simple, the opposite branch long, with one or more further orders of branching, the two kinds of branch each following a spiral arrangement; axes uniaxial, cylindrical or slightly compressed, with three pericentral cells associated with each axial cell and, in most specimens, additional cortication formed from filaments of limited growth; secretory cells numerous; dioecious; spermatangia formed in dense clusters, ovate or clavate, several clusters to each long branch, carpogonia formed in short and simple fertile branches which represent a replacement of the whole long branch of a pair, usually only two such branches are converted at the lower part of a fertile system; cystocarps spherical, pedicellate, filled with an irregularly arranged mass of spores when mature.

Tetrasporangial phase (=*Falkenbergia*) polysiphonous, each segment of the uniaxial thallus with three pericentral cells, sparsely branched, branching irregular; secretory cells present; tetrasporangia without a pedicel, formed by the transformation of one of the three pericentral cells of a segment, 40–70×25–40 μm, tetraspores cruciately or irregularly arranged.

One species in the British Isles:

Asparagopsis armata Harvey (1855), p. 544.

Lectotype: TCD. Australia (Garden Island, Western Australia).

Polysiphonia rufolanosa Harvey (1855), p. 540.
Falkenbergia rufolanosa (Harvey) Schmitz (1897), p. 479.
Polysiphonia hillebrandii Bornet in Ardissone (1883), p. 376.
Falkenbergia hillebrandii (Bornet) Falkenberg (1901), p. 689.
Polysiphonia vagabunda Harvey (1857), p. 300.
Falkenbergia vagabunda (Harvey) Falkenberg (1901), p. 690.
Falkenbergia doubletii Sauvageau (1925), p. 18.

Gametangial thallus (=*Asparagopsis*) erect, to 300 mm, of a rosy-pink, yellowish-pink, or whitish-pink colour; basal attachment unknown, at least in British material; thallus with prominent major axes, naked in the lower parts, but densely tufted in the upper; apical tufts pyramidal in outline, composed of spirally-arranged branches arising in pairs, falling away with age to give the naked basal axes; certain axes converting into short spiny axes, usually in pairs, to 100 mm in length, 2–4 mm broad, provided with short reflexed laterals; ultimate lateral branches polysiphonous, with three spirally-arranged pericentral cells for each axial cell; further division gives rise to additional cortication; secretory cells prominent in most peripheral cells, particularly in the major axes. Dioecious; spermatangia in clavate swollen lateral axes, to 2 mm in length; cystocarps flask-shaped, to 2 mm in length, spores releasing through a terminal pore. Tetrasporangial thallus (=*Falkenbergia*) rose-pink in colour, formed of irregularly-branched polysiphonous axes, to 40–50 μm in diameter when fully mature, usually aggregated into spherical or hemispherical masses, 10–30 mm in diameter, although also occurring as isolated axes; axes attached to substrate by multicellular haptera; each axial cell associated with three pericentral cells orientated spirally and closely appressed; pericentral cells with prominent secretory cells; tetrasporangia ovoid 70×30 μm, formed by the transformation of a pericentral cell; tetraspores cruciately or irregularly arranged.

Gametangial phase sublittoral or occasionally in deep pools in the littoral, attached to other algae by the barbed axes. Tetrasporangial phase usually sublittoral, epiphytic, rarely epilithic, often free floating.

Southern and western shores of the British Isles. The gametangial phase extends in

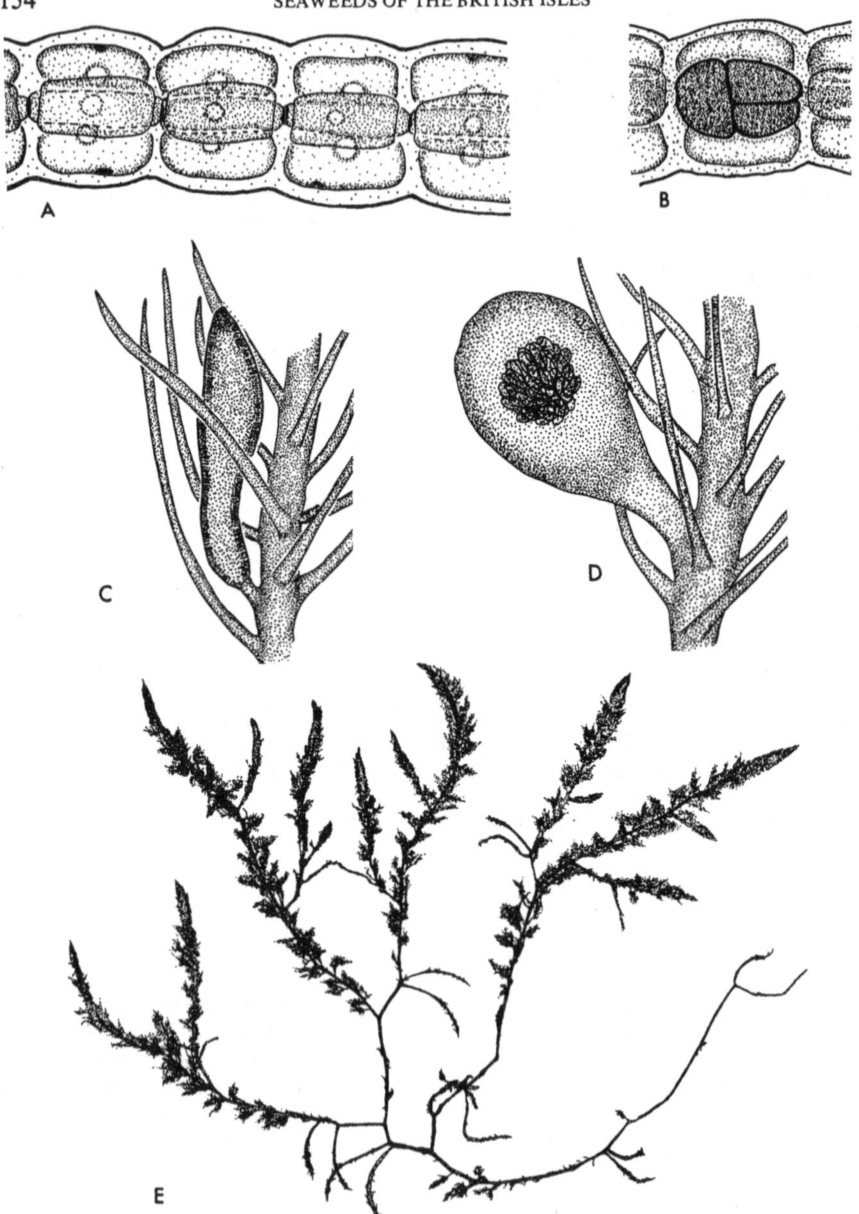

Fig. 57 *Asparagopsis armata*
A. *Falkenbergia* phase ×525; B. Tetrasporangium in *Falkenbergia* phase ×525; C. Spermatangial axis ×20; D. Cystocarp ×20; E. Habit ×0·5.

Ireland from Mayo to Cork and in England from North Devon to Dorset. The tetrasporangial phase is more widely distributed although absent from eastern coasts of Ireland, England and Scotland.

Atlantic coast of France, Spain, Portugal, Mediterranean; Australia, Pacific and Indian Oceans.

Fragments of both gametangial and tetrasporangial phases occur at all times of the year. Regeneration of whole plants from such fragments occurs most rapidly between July and October. De Valéra & Folan (1964) have suggested that the size of plants of the gametangial phase has diminished since the initial introduction into the British Isles.

Tetrasporangia form during the winter months on the Atlantic coasts of France, Spain and Portugal, but so far there has been only one report of their occurrence in the British Isles (McLachlan, 1967). Most gametangial plants are sterile in the British Isles, although fertile material has been collected during summer and early autumn. De Valéra & Folan (1964) suggest that there has been a decrease in fertility since the first introduction.

A. armata is of very constant appearance both in the tetrasporangial and gametangial phases.

First detected in the British Isles in November 1939 at Galway, the species has spread extensively and is widely distributed on southern and western shores.

The life history is unlikely to be a sequence of gametangial, carposporangial and tetrasporangial phases. Tetrasporangia have been detected once while carpospore production appears to be diminishing and there is good evidence that spread and survival are determined by vegetative propagation.

The relationship between *A. armata* and *A. taxiformis* is in need of clarification. There are indications in the British Isles of plants with characteristics of *A. taxiformis* while the *Falkenbergia* phases of the two entities are indistinguishable.

Falkenbergia rufolanosa is often infected with the fungus *Olpidiopsis feldmannii* Aleem (Lagenidiales), particularly in the early autumn.

BONNEMAISONIA C. Agardh

BONNEMAISONIA C. Agardh (1822), p. 196.

Type species: *B. asparagoides* (Woodward) C. Agardh (1822), p. 197.

Trailliella Batters (1896), p. 10.

For comments on *Trailliella* Batters and *Hymenoclonium* Crouan frat. as generic synonyms of *Bonnemaisonia*, see p. 156.

Gametangial thallus *(=Bonnemaisonia)* erect, much branched; branches opposite, distichous or spirally-arranged, with unequal development of the two components of each pair, one lateral branch being short and simple, the opposite branch long, the two kinds of branch either alternating in successive pairs or arranged spirally; axes uniaxial, cylindrical or slightly compressed, with three pericentral cells associated with each axial cell and additional cortication formed from the filaments of limited growth; secretory cells numerous. Monoecious or dioecious; spermatangia formed in dense clusters, ovoid, clavate or elongate, formed by the transformation of the short branch of a pair; carpogonia also formed in the short branch of a pair; cystocarps flask-shaped, pedicellate, containing relatively few sporangia, radially arranged from the base.

Tetrasporangial thalli of various forms, either filamentous *(=Trailliella)* or a single-layered crust *(=Hymenoclonium)*. *Trailliella* phases filamentous, uniseriate, attached to substrate by unicellular rhizoids or multicellular haptera at intervals along the thallus; thallus branched, branching irregular and highly variable, even in different parts of the same thallus; numerous secretory cells present, located laterally between adjacent cells of the thallus; tetrasporangia sunk into the cell of the filament from which the primordium was formed; tetraspores tetrahedrally arranged. *Hymenoclonium* phases prostrate, made up of uniseriate fiaments appressed to the substrate, densely branched; branching frequently anastomosing to give a flat sheet of cells; numerous secretory cells present; tetrasporangia borne on a short pedicel; tetraspores cruciately or irregularly arranged.

KEY TO SPECIES

A. Gametangial phases

1 Branches spirally arranged; crozier-shaped hooks present . . . *B. hamifera*
Branches distichous; crozier-shaped hooks absent 2

2 Plant monoecious; spermatangial clusters ovoid, less than 100 μm diameter; cystocarp sub-spherical . . . . *B. asparagoides*
Plant dioecious; spermatangial clusters elongate, visible to the naked eye, 1 mm long × 300 μm diameter; cystocarp ovoid . . . . *B. clavata*

B. Tetrasporangial phases

1 Filamentous, sparsely branched, with secretory cells prominent, lying between adjacent cells . . . . *Trailliella* phase
Prostrate, much branched, forming a dense cell mass; secretory cells obscure, each lying on the upper side of a single cell. . *Hymenoclonium* phase

The tetrasporangial phase of *Bonnemaisonia hamifera* was recognized in Britain as *Trailliella intricata* Batt., prior to the association of the two entities. The tetrasporangial phase of the Pacific *Bonnemaisonia nootkana* (Esper) Silva is also recognized as *Trailliella intricata* and is indistinguishable from the British material. Plants identified as *Trailliella* have not been associated with any genus other than *Bonnemaisonia*. For this reason, *Trailliella* may be cited as a generic synonym of *Bonnemaisonia*. It would be incorrect to cite *Trailliella intricata* in the synonymy of *Bonnemaisonia hamifera*, in view of its association with a second species of this same genus. The position with respect to *Hymenoclonium serpens* (Crouan frat.) Batt. is even more complex. It has been customary to cite this as a synonym of *Bonnemaisonia asparagoides* with *Hymenoclonium* cited as a generic synonym of *Bonnemaisonia*. The *Hymenoclonium* phase of *Bonnemaisonia clavata* is indistinguishable from that of *B. asparagoides* and there are now *Hymenoclonium* phases known for several genera both of the Cryptonemiales and Nemaliales. These are morphologically indistinguishable. It would not be appropriate to assign *Hymenoclonium* as a generic synonym of *Bonnemaisonia*, or *H. serpens* as a synonym of any of the species for which the life histories contain such phases.

Bonnemaisonia asparagoides (Woodward) C. Agardh (1822), p. 197.

Lectotype: original illustration (Woodward, 1794, pl. 6), in the absence of material. Norfolk (North Yarmouth).

Fucus asparagoides Woodward (1794), p. 29.
Bonnemaisonia adriatica Zanardini (1847), p. 20.

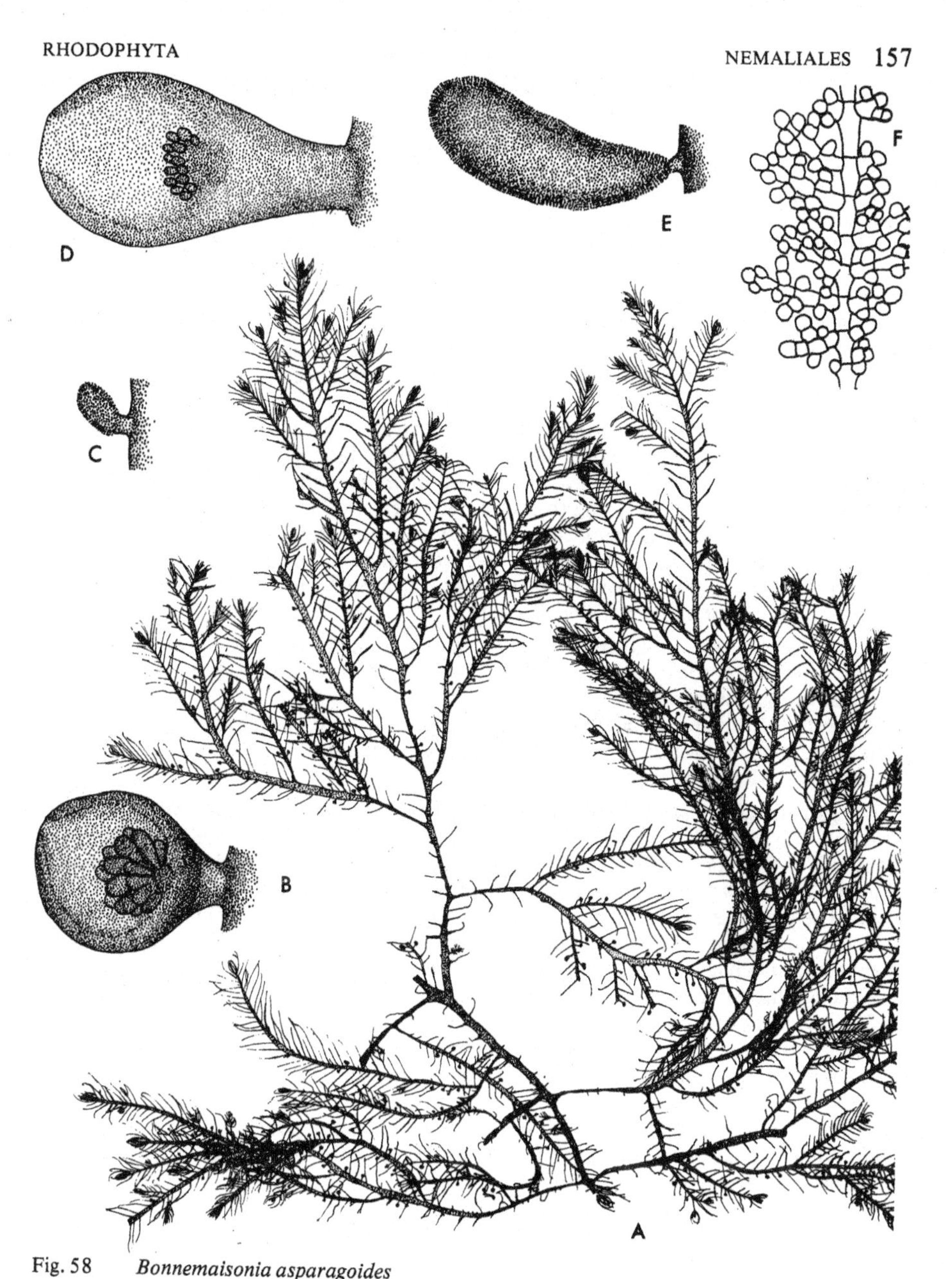

Fig. 58 *Bonnemaisonia asparagoides*
A. Habit ×1; B. Cystocarp ×100; C. Spermatangial axis ×100.
Bonnemaisonia clavata
D. Cystocarp ×100; E. Spermatangial axis ×100; F. *Hymenoclonium* phase ×200.

Gametangial phase *(=Bonnemaisonia asparagoides)* erect, bright red in colour, to 400 mm, main axis to 800 μm in diameter, much branched, attached by a small basal disc; branching distichous, opposite, with unequal and alternate development of the two components of each pair; the longer branch of a pair to 3 mm in length, 200 μm in diameter when fully developed, incurved at the apex, but straightening with age, orientated to the axis at an angle of 45–60°; the shorter branch of a pair minute, usually converted into male or female reproductive structures, or into indeterminate axes; secretory cells present, usually in abundance, formed by the transformation of a cortical cell, colourless; monoecious; spermatangia formed in spherical or ovoid clusters, 100 μm in diameter, replacing the shorter lateral branch of a pair; carpogonial branches formed in the same position; cystocarps stipitate, spherical to subspherical, 300–400 μm in diameter when mature, with a terminal pore; carposporangia arranged radially at the base of the cystocarp, ovoid, 100×40–50 μm.

Tetrasporangial *(Hymenoclonium)* phase a prostrate crust made up of uniseriate filaments appressed to the substrate, densely branched; branching usually opposite, but occasionally alternate or irregular, with second order branches reduced to a single cell, branches frequently anastomosing to give a flat sheet of cells; cells of principal filaments 20–30 μm long, 10–15 μm in diameter; secretory cells numerous, arising on the upper face of each cell of the principal filaments and first order laterals, colourless, refringent, 4–6 μm in diameter; tetrasporangia ovoid, 50×40 μm, tetraspores tetrahedrally or irregularly arranged, not present in the British collections.

Gametangial phase sublittoral, to 15 m, or occasionally in deep pools in the lower littoral, growing on rocks, shells, coralline algae, *Zostera,* etc.; tetrasporangial phase, to 30 m, on various substrates.

Generally distributed throughout the British Isles, although somewhat sporadic in the Irish Sea and on the eastern coasts of England and Scotland.

Norway to Morocco; Mediterranean.

The gametangial phase is found from February to September, with growth most active in April and May. Fragments and denuded thalli sometimes occur through the winter months. The first stages of cystocarp formation are visible in April and May, with mature examples in June and continuing until October. Nothing is known of growth and development, or seasonal aspects of reproduction of the *Hymenoclonium* phase in the field.

Under low light intensities, thalli of the gametangial phase become markedly attenuated; such plants correspond to *B. asparagoides β teres* Harv.

Although the gametangial phase is widespread and easily recognizable, plants identified as *Hymenoclonium serpens,* which may or may not represent the *Hymenoclonium* phase of *Bonnemaisonia asparagoides,* have been collected only very rarely in the British Isles.

The relation between the gametangial phase and the supposed *Hymenoclonium* phase is controversial. The latter may produce tetrasporangia or give rise directly to the gametangial phase, the first suggesting that it should be regarded as a tetrasporangial phase while the second suggests that it is merely a protonemal base. It is possible to reconcile these two views (Dixon, 1970).

Bonnemaisonia clavata Hamel (1930), p. 104.

Lectotype: PC (see Dixon, 1962). France (Marseille).

Gametangial thallus erect, reddish-pink in colour, to 150 mm, main axis to 500 μm in diameter, much branched, attached by a small basal disc; branching distichous and opposite, with unequal and alternate development of the two components of each pair, although occasionally somewhat irregular in which case the difference between the opposite branches of a pair is less marked; the longer branch of a pair to 15 mm in length, 100 μm in diameter when fully developed, slightly incurved at the apex, but straightening with age and often becoming reflexed, orientated to the axis at an angle of 80–100°; the shorter branch of a pair minute, usually converted into male or female reproductive structures, or into indeterminate axes; secretory cells present, although sometimes difficult to find, formed by the transformation of a cortical cell, colourless. Dioecious; spermatangia formed in elongate, clavate clusters, 1 mm long and 300 μm in diameter, replacing the shorter lateral branch of a pair; carpogonia formed in the same position; cystocarps stipitate, ovoid, 700–800 μm long and 500–600 μm in diameter when mature, with a terminal pore; carposporangia arranged radially at the base of the cystocarp, ovoid, 50–60×20–25 μm.

Tetrasporangial phase unknown. Germination of carpospores in culture gives rise to a prostrate crust resembling a *Hymenoclonium* phase, but nothing is known of its occurrence in the field or of the reproductive structures which it forms.

Gametangial phase restricted to the sublittoral, to 20 m, on rock or coralline algae.
Cornwall.
Atlantic coast of France; western Mediterranean.

Fertile plants of the gametangial phase occur between May and June.
Data on form variation insufficient for comment.

B. clavata is indistinguishable from *B. asparagoides* when sterile, and reproductive structures are essential for recognition. *B. clavata* is dioecious; *B. asparagoides* is monoecious. Both spermatangial clusters and cystocarps of *B. clavata* are more massive than those of *B. asparagoides* and clearly visible to the naked eye. Germination of the carpospores of *B. clavata* produces a plant which is indistinguishable from the *Hymenoclonium* phase of *B. asparagoides*. Knowledge of the life history is far from complete but it is likely to be similar to that of the latter.

The occurrence of this species in the British Isles rests on two specimens collected at Plymouth Sound and Padstow in the last century. These specimens are referred to by Hamel (1930) in the original treatment of the taxon and are now preserved in the Thuret-Bornet herbarium (PC).

Bonnemaisonia hamifera Hariot (1891), p. 223.

Lectotype: PC. Japan (Yokosuka [as Yokoska]).

Asparagopsis hamifera (Hariot) Okamura (1921), pls. 183, 4.

For comments on the status of *Trailliella intricata* Batt. as a synonym of this taxon, see p. 156.

Gametangial thallus erect, deep to blackish red in colour, to 200 mm, main axis to 1 mm in diameter, much branched, attachment structure not known, at least in British material, branching opposite and spirally arranged, with unequal development of the two components of a pair; the longer branch of a pair to 2 mm in length, 200 μm in diameter when fully developed, incurved at the apex, but straightening with age, orientated to the main axis at an angle of 60°; the shorter branch of a pair a small protuberance, 100 μm in

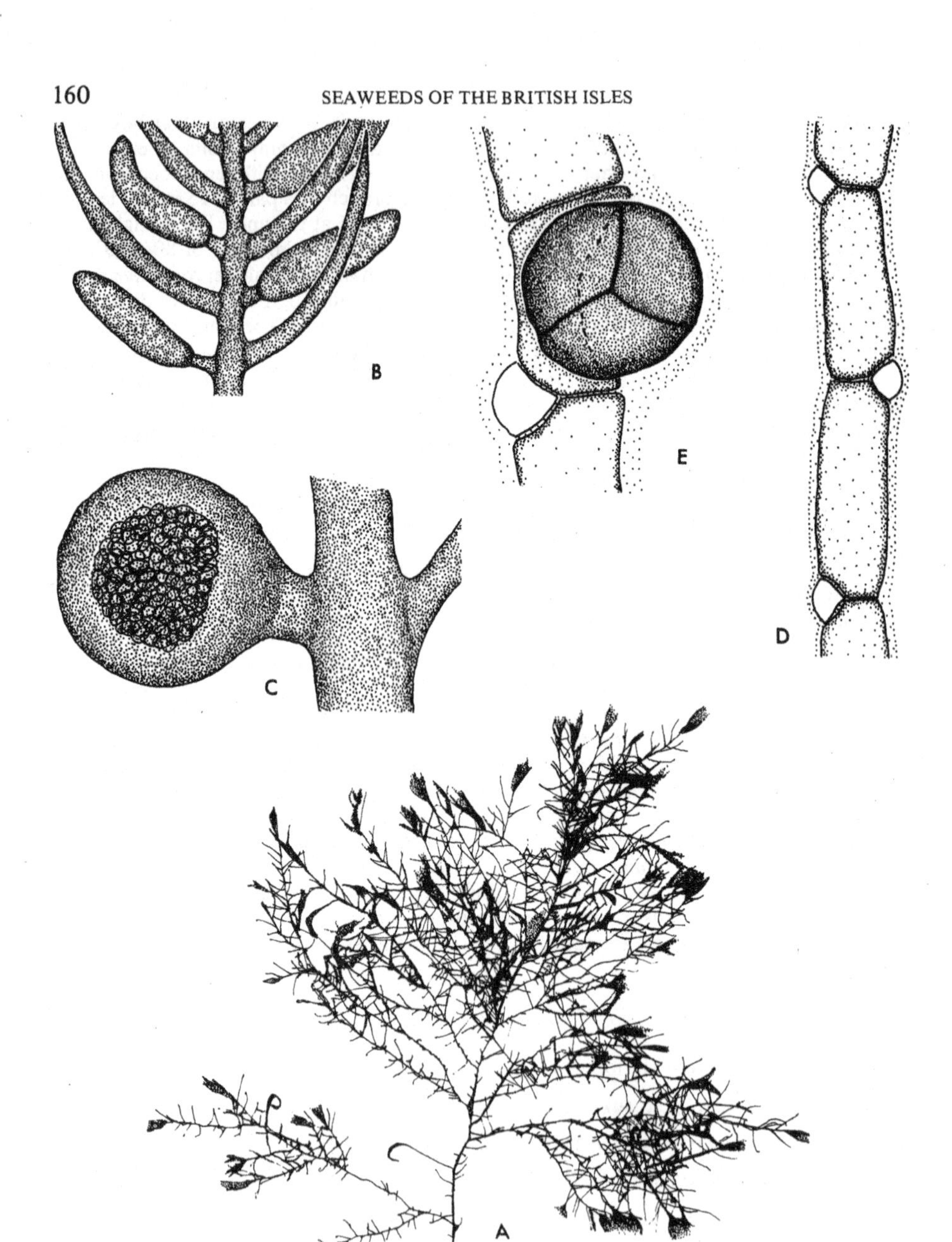

Fig. 59 *Bonnemaisonia hamifera*
A. Habit ×1; B. Spermatangial axes ×100; C. Cystocarp ×100; D. *Trailliella* phase, vegetative structure ×525; E. *Trailliella* phase, tetrasporangium ×525.

diameter, only some smaller branches converted into reproductive structures or indeterminate axes, a few modified to form the prominent reflexed hooks from which the species derives its name; dioecious; spermatangia formed in elongate, clavate clusters to 1·5 mm in length, 400–600 μm in diameter, replacing the shorter lateral branch of a pair; carpogonial branches formed in the same position; cystocarps stipitate, spherical to subspherical, 2 mm in diameter, with a terminal pore but no evidence of carposporangium formation in British material.

Tetrasporangial phase a uniseriate filament *(=Trailliella)* attached to the substrate by unicellular rhizoids or multicellular haptera formed at intervals; thallus branched, branching irregular and highly variable; cells slightly inflated, or barrel shaped, 40–90 μm in length, 20–30 μm in diameter; secretory cells numerous, 10–15 μm in diameter, formed at the upper end of each cell of the thallus, colourless, highly refractive; tetrasporangia ovoid, 50–70×30–40 μm, each sunk partially or completely into the cell from which it has been formed; tetraspores cruciately arranged.

Gametangial phase sublittoral or occasionally in deep pools in the lower littoral, attached to other algae by the hooked axes. Tetrasporangial phase usually epiphytic, although occasionally epilithic in the lower littoral and sublittoral, to 8 m.

Southern and western shores of the British Isles, with the tetrasporangial phase recorded further to the north (Shetland) than the gametangial (Argyll), both extending eastwards to Sussex.

Gametangial phase from Norway to Atlantic France; Japan; tetrasporangial phase from Denmark to the Canary Islands.

The gametangial phase is found throughout the year, with growth most active in spring and early summer. The tetrasporangial phase occurs as dense, bright red clumps in summer and as isolated filaments throughout the remainder of the year.

Spermatangial plants reported only infrequently in the British Isles and it would appear that viable carpospores are not formed. Tetrasporangia are rare, the few records suggesting their formation in the autumn.

There are no indications that thallus form undergoes much modification in relation to seasonal or environmental factors.

Bonnemaisonia hamifera is a species of Japanese origin, which became established in the North Atlantic Ocean towards the end of the last century. The date of the first collection in the British Isles is not certain because of confusion between the tetrasporangial phase and a species of *Spermothamnion.* The first positive collection was made in the Isle of Wight in September 1890, by E. M. Holmes (Batters, 1896), while the gametangial phase was first detected by T. H. Buffham at Falmouth (Cornwall) in August 1893 (Buffham, 1896). The species was reported subsequently from various localities in Europe and eventually from the eastern coast of North America in 1927 (Lewis & Taylor, 1928).

The life history of *B. hamifera* in the North Atlantic shows various peculiarities and it is highly improbable that the sequence of gametangial, carposporangial and tetrasporangial phases found in Japan occurs in this area. The initial reports of the gametangial phase in the British Isles and other parts of the North Atlantic made no reference to male plants and the cystocarps were said not to contain carposporangia. Furthermore, tetrasporangia appear to be of infrequent occurrence. This evidence suggests that independent vegetative propagation of both gametangial and tetrasporangial phases (Chemin, 1929) has probably provided the only means for the survival and spread of *B. hamifera* in the North Atlantic. However, male plants have now been reported from France, Helgoland and Nova Scotia and, at the same places, some cystocarps containing carposporangia. Male plants and carposporangia are still extremely scarce so that although there may now be some reproduction by spores occurring, vegetative propagation is still the major means of reproduction. Life history studies in culture are still incomplete, with the curious observations that germination of tetraspores in German material produces only female plants, Canadian only male. The *Trailliella* phase has always been interpreted as being tetrasporangial, but direct development from it of the gametangial phase has been reported.

The records of *B. hamifera* from the Pacific coast of North America are referable to *B. nootkana* (Esper) Silva, previously known as *B. californica*.

Most gametangial thalli of *B. hamifera* contain the endophyte, *Audouinella asparagopsis* Chemin (1926). Specimens on the eastern coast of North America are said to contain *Acrochaetium americanum* Jao which is virtually indistinguishable.

The fungus *Olpidiopsis feldmannii* Aleem (Lagenidiales) has been reported in the tetrasporangial phase of *B. hamifera*.

REFERENCES FOR NEMALIALES

ABBOTT, I. A. (1965). *Helminthora* and *Helminthocladia* from California. *Hydrobiologia* **25:** 88–98.

AGARDH, C. A. (1822). *Species algarum* **1**(2). Lundae.

—— (1824). *Systema algarum.* Lundae.

AGARDH, J. G. (1841). In historiam algarum symbolae. *Linnaea* **15:** 1–50; 443–457.

—— (1842). *Algae Maris Mediterranei et Adriatici.* Parisiis.

—— (1851). *Species Genera et Ordines Algarum.* **2**(1). Lundae.

—— (1852). *Species Genera et Ordines Algarum.* **2**(2). Lundae.

—— (1863). *Species Genera et Ordines Algarum.* **2**(3). Lundae.

ARDISSONE, F. (1883). *Phycologia Mediterranea.* Varese.

—— & STRAFFORELLO, J. (1877). *Enumerazione delle Alghe di Liguria.* Milano.

BATTERS, E. A. L. (1896). Some new British marine algae. *J. Bot., Lond.* **34:** 6–11.

—— (1896a). New or critical British marine algae. *J. Bot., Lond.* **34:** 384–390.

—— (1897). New or critical British marine algae. *J. Bot., Lond.* **35:** 433–440.

—— (1902). A catalogue of the British marine algae. *J. Bot., Lond.* **40** (Suppl.): 1–107.

BIVONA-BERNARDI, A. (1822). *Scinaia,* algarum marinarum novum genus. *Iride* **1:** 232–234.

BOILLOT, A. (1967). Sur le développement des carpospores de *Naccaria wiggii* (Turner) Endlicher et *d'Atractophora hypnoides* Crouan (Naccariacées, Bonnemaisoniales). *C. r. hebd. Séanc. Acad. Sci., Paris,* sér. D **264:** 257–260.

—— (1968). Sur l'existence d'un tétrasporophyte dans le cycle de *Scinaia furcellata* (Turner) Bivona, Nemalionales. *C. r. hebd. Séanc. Acad. Sci., Paris,* sér. D **266:** 1831–1832.

—— (1969). Sur le développement des tétraspores et l'édification du gamétophyte chez *Scinaia furcellata* (Turner) Bivona, Rhodophycées (Nemalionales). *C. r. hebd. Séanc. Acad. Sci., Paris,* sér. D **268:** 273–275.

—— & MAGNE, F. (1973). Le cycle biologique de *Kylinia rosulata* Rosenvinge (Rhodophycées, Acrochaetiales). *Bull. Soc. phycol. Fr.* **18:** 47–53.

BORNET, E. (1892). Les algues de P.-K.-A. Schousboe. *Mém. Soc. natn. Sci. nat. math. Cherbourg* **28:** 165–376.

—— (1904). Deux *Chantransia corymbifera* Thuret. *Acrochaetium* et *Chantransia.* *Bull. Soc. bot. Fr.* **51** (Suppl.): xiv–xxiii.

—— & THURET, G. (1876). *Notes algologiques.* **1.** Paris.

BORSJE, W. J. (1973). The life history of *Acrochaetium virgatulum* (Harv.) J. Ag. in culture. *Br. phycol. J.* **8:** 204–205.

BORY DE ST. VINCENT, J. B. M. (1823). Ceramiaires. *In: Dictionnaire classique d'Histoire Naturelle.* 3. Paris.

BROWN, R. (1810). *Prodromus Florae Novae Hollandiae et Insulae Van-Diemen.* London.

BUFFHAM, T. H. (1892). *Chantransia trifila:* a new marine alga. *J. Quekett microsc. Club,* ser. 2 **5:** 24–26.

—— (1896). On *Bonnemaisonia hamifera,* Hariot, in Cornwall. *J. Quekett microsc. Club,* ser. 2 **6:** 177–182.

CHEMIN, E. (1926). Une nouvelle espèce de *Colaconema* sur *Asparagopsis hamifera* Okam. *C. r. hebd. Séanc. Acad. Sci., Paris* **183:** 900–902.

—— (1926a). Sur le développement des spores dans le genre *Scinaia* et sur la nécessité d'une espèce nouvelle: *Scinaia turgida. Bull. Soc. bot. Fr.* **73:** 92–102.

—— (1927). Sur le développement des spores de *Naccaria wiggii* Endl. et *Atractophora hypnoides* Crouan. *Bull. Soc. bot. Fr.* **74:** 272–277.

—— (1929). *L'Asparagopsis hamifera* (Hariot) Okamura et son mode de multiplication. *Revue algol.* **4:** 29–42.

CHRISTENSEN, T. (1967). Two new families and some new names and combinations in the algae. *Blumea* **15:** 91–94.

CLEMENTE Y RUBIO, S. de R. (1807). *Ensayo sobre las Variedades de la Vid Comun que vegetan en Andalucia, con un Indice Etimológico y Tres Listas de Plantas en que se Caracterizan Varias Especias Nuevas.* Madrid.

COLLINS, F. S. (1906). *Acrochaetium* and *Chantransia* in North America. *Rhodora* **8:** 189–196.

—— & HERVEY, A. B. (1917). The algae of Bermuda. *Proc. Am. Acad. Arts Sci.* **53:** 1–195.

CROUAN, P. L. & H. M. (1848). Sur l'organisation, la fructification et la classification du *Fucus wigghii* de Turner et de Smith, et de *l'Atractophora hypnoides. Annls Sci. nat.,* sér. 3, Bot. **10:** 361–376.

DARBISHIRE, O. V. (1899). *Chantransia endozoica* Darbish., eine neue Florideen-Art. *Ber. dt. bot. Ges.* **17:** 13–17.

—— (1910). *Chantransia sanctae-mariae.* A new British species. *Rep. scient. Invest. Northumb. Sea Fish. Comm.* **1901/1910:** 40–41.

DE CANDOLLE, A. P. (1807). Rapport sur une voyage botanique et agronomique dans les Départements de l'Ouest. *Mém. Agric.* **10:** 228–292.

DE VALÉRA, M. & FOLAN, A. (1964). Germination *in situ* of carpospores in Irish material of *Asparagopsis armata* Harv. and *Bonnemaisonia asparagoides* (Woodw.) Ag. *Br. phycol. Bull.* **2:** 332–338.

DILLWYN, L. W. (1802–1809). *British Confervae.* London.

DIXON, P. S. (1958). The occurrence of *Gelidium sesquipedale* (Clem.) Thur. in the British Isles. *Phycol. Bull.* **1**(6): 47–48.

—— (1962). Taxonomic and nomenclatural notes on the Florideae, III. *Bot. Notiser* **115:** 245–260.

—— (1967). The typification of *Fucus cartilagineus* L. and *F. corneus* Huds. *Blumea* **15:** 55–62.

—— (1970). The Rhodophyta: some aspects of their biology. II. *Oceanogr. mar. Biol. ann. Rev.* **8:** 307–352.

—— (1976). Appendix. *In:* Parke, M. & Dixon, P.S. Check-list of British marine algae – third revision. *J. mar. biol. Ass. U.K.* **56:** 527–594.

—— (1977). The *Acrochaetium/Rhodochorton* complex in the British Isles. (in preparation).

—— & DE VALÉRA, M. (1961). A critical survey of the evidence for the occurrence of *Gelidium torulosum* Kütz. and *G. melanoideum* Schousb. ex Born. in Britain and Ireland. *Br. phycol. Bull.* **2:** 67–71.

—— & IRVINE, L. M. (1970). Miscellaneous notes on algal taxonomy and nomenclature, III. *Bot. Notiser* **123:** 474–487.

—— & —— (1977). Miscellaneous notes on algal taxonomy and nomenclature, IV. (in preparation).

DREW, K. M. (1928). A revision of the genera *Chantransia, Rhodochorton,* and *Acrochaetium. Univ. Calif. Publs Bot.* **14:** 139–224.

DUBY, J. E. (1830). *Botanicon Gallicum*, ed. 2. **2.** Paris.

DUFOUR, L. (1863). *Erbario Crittogamico Italiano.* **20.** Genova.

ENDLICHER, S. L. (1836–1840). *Genera plantarum.* Vindobonae.

ENGLER, A. (1892). *Syllabus der Vorlesungen über specialle und medicinisch-pharmaceutische Botanik.* Berlin.

FALKENBERG, P. (1901). Die Rhodomelaceen des Golfes von Neapel und der angrenzenden Meeresabschnitte. *Fauna Flora Golf. Neapel* **26:** 1–754.

FELDMANN, J. (1939). Les algues marines de la Côte des Albères. IV – Rhodophycées. *Revue algol.* **11:** 247–330.

—— (1958). Le genre *Kylinia* Rosenvinge (Acrochaetiales) et sa réproduction. *Bull. Soc. bot. Fr.* **105:** 493–500.

—— (1962). The Rhodophyta order Acrochaetiales and its classification. *Proc. 9th Pacific Sci. Congr.* **4:** 219–221.

FELDMANN, J. & G. (1939). Additions à la flore des algues marines de l'Algérie. Fascicle 2. *Bull. Soc. Hist. nat. Afr. N.* **30:** 453–464.

—— (1943). Recherches sur les Bonnemaisoniacées et leur alternance de générations. *Annls Sci. nat.*, sér. 11, Bot. **3:** 75–175.

FELDMANN, J. & HAMEL, G. (1936). Floridées de France VII Gélidiales. *Revue algol.* **9:** 85–140.

FRIES, E. M. (1825). *Systema orbis vegetabilis, I Plantae homonemeae.* Lund.

FRIES, L. (1967). The sporophyte of *Nemalion multifidum* (Weber et Mohr) J. Ag. *Svensk bot. Tidskr.* **61:** 457–462.

FRITSCH, F. E. (1944). Present-day classification of algae. *Bot. Rev.* **10:** 233–277.

GIBSON, R. J. H. (1891). On the development of the sporangia in *Rhodochorton Rothii*, Näg., and *R. floridulum*, Näg.; and on a new species of that genus. *J. Linn. Soc., Bot.* **28:** 201–205.

GOODENOUGH, S. & WOODWARD, T. J. (1797). Observations on the British Fuci, with particular descriptions of each species. *Trans. Linn. Soc.* **3:** 84–235.

GMELIN, S. G. (1768). *Historia Fucorum.* Petropoli.

GREVILLE, R. K. (1830). *Algae Britannicae.* Edinburgh.

HAMEL, G. (1927). *Recherches sur les genres* Acrochaetium *Naeg. et* Rhodochorton *Naeg.* St. Lo.

—— (1930). Floridées de France VI. *Revue algol.* **5:** 61–109.

HARIOT, P. (1891). Liste des algues marines rapportées de Yokoska (Japon) par M. le Dr Savatier. *Mém. Soc. natn. Sci. nat. math. Cherbourg* **27:** 211–230.

HARVEY, W. H. (1833). Div. II Confervoideae. Div. III Gloiocladeae. *In:* Hooker, W. H., *The English Flora of Sir James Edward Smith.* **5.** London.

—— (1836). Algae. *In:* Mackay, J. T., *Flora Hibernica.* **2.** Dublin.

—— (1855). Some account of the marine botany of the colony of western Australia. *Trans. R. Ir. Acad.* **22:** 525–566.

—— (1857). Nat. Ord. VIII Algae. *In:* Hooker, J. D., *The Botany of the Antarctic Voyage.* III *Flora Tasmaniae.* II *Monocotyledons & Acotyledons.* London.

HAUCK, F. (1885). *Die Meeresalgen Deutschlands und Oesterreichs. In:* Rabenhorst, L. *Kryptogamen-Flora von Deutschland, Oesterreich und der Schweiz.* ed. 2. **2.** Leipzig.

HEYDRICH, F. (1892). Beiträge zur Kenntniss der Algenflora von Kaiser-Wilhelms-Land (Deutsch-Neu-Guinea). *Ber. dt. bot. Ges.* **10:** 458–485.

HOLLENBERG, G. J. & ABBOTT, I. A. (1966). *Supplement to Smith's Marine Algae of the Monterey Peninsula.* Stanford.

HOWE, M. A. & HOYT, W. D. (1916). Notes on some marine algae from the vicinity of Beaufort, North Carolina. *Mem. N. Y. bot. Gdn* **6:** 105–123.

HOYT, W. D. (1920). Marine algae of Beaufort, N. C. and adjacent regions. *Bull. Bur. Fish., Wash.* **36:** 367–556.

HUDSON, P. R. & WYNNE, M. J. (1969). Sexual plants of *Bonnemaisonia geniculata* (Nemaliales). *Phycologia* **8:** 207–213.

HUDSON, W. (1762). *Flora anglica.* London.

—— (1778). *Flora Anglica.* ed. 2. London.

JAO, C. C. (1936). New Rhodophyceae from Woods Hole. *Bull. Torrey bot. Club* **63**: 237–257.

JONES, W. E. & SMITH, R. M. (1970). The occurrence of tetraspores in the life history of *Naccaria wiggii* (Turn.) Endl. *Br. phycol. J.* **5**: 91–95.

JÓNSSON, H. (1901). The marine algae of Iceland. (1. Rhodophyceae). *Bot. Tidsskr.* **24**: 127–155.

KJELLMAN, F. R. (1875). Om spetsbergens marina Klorofyllförende Thallophyter 1. *Bih. K. svenska VetenskAkad. Handl.* **3** (7): 1–34.

—— (1883). The algae of the Arctic Sea. *K. svenska VetenskAkad. Handl.* **20** (5): 1–350.

—— (1906). Zur Kenntnis der marinen Algenflora von Jan Mayen. *Ark. Bot.* **5** (14): 1–30.

KÜTZING, F. T. (1843). *Phycologia generalis.* Leipzig.

—— (1849). *Species Algarum.* Lipsiae.

—— (1868). *Tabulae Phycologicae.* **18.** Nordhausen.

KYLIN, H. (1906). Zur Kenntnis einiger schwedischen *Chantransia*-Arten. *In: Botaniska Studier til F. R. Kjellman.* pp. 113–126. Upsala.

—— (1907). *Studien über die Algenflora der schwedischen Westküste.* Upsala.

—— (1928). Entwicklungsgeschichtliche Florideenstudien. *Acta Univ. lund.*, Ny Följd, Avd. 2 **24** (4): 1–127.

—— (1944). Die Rhodophyceen der schwedischen Westküste. *Acta Univ. lund.*, Ny Följd, Avd. 2 **40** (2): 1–104.

—— (1956). *Die Gattungen der Rhodophyceen.* Lund.

LAMOUROUX, J. V. F. (1813). Essai sur les genres de la famille des thalassiophytes non articulées. *Annls Mus. Hist. nat. Paris.* **20**: 21–47; 115–139; 267–293. [reprint pp. 84].

LE JOLIS, A. (1863). Liste des algues marines de Cherbourg. *Mém. Soc. natn. Sci. nat. math. Cherbourg* **10**: 6–168.

LEVRING, T. (1937). Zur Kenntnis der Algenflora der norwegischen Westküste. *Acta Univ. lund.*, Ny Följd, Avd. 2 **33** (8): 1–147.

LEWIS, I. F. & TAYLOR, W. R. (1928). Notes from the Woods Hole Laboratory – 1928. *Rhodora* **30**: 193–198.

LIGHTFOOT, J. (1777). *Flora Scotica.* **2.** London.

LUND, S. (1959). The marine algae of East Greenland. I. Taxonomical part. *Meddr Grønland* **156** (1): 1–247.

LYLE, L. (1920). The marine algae of Guernsey. *J. Bot., Lond.* **58** (Suppl. 2): 1–53.

—— (1929). Marine algae of some German warships in Scapa Flow and of the neighbouring shores. *J. Linn. Soc., Bot.* **48**: 231–257.

LYNGBYE, H. C. (1819). *Tentamen Hydrophytologiae Danicae.* Hafniae.

MAGNUS, P. (1875). Die botanischen Ergebnisse der Nordseefahrt 1872. *Jhber. Comm. wiss. Unters. Meeres* **2**: 61–75.

MARTIN, M. T. (1967). Observations on the life-history of *Nemalion helminthoides* (Vell. in With.) Batt. *Br. phycol. Bull.* **3**: 408.

—— (1969). A review of life-histories in the Nemalionales and some allied genera. *Br. phycol. J.* **4**: 145–158.

MARTIUS, K. (1817). *Flora Cryptogamia Erlangensis.* Norimbergae.

MCLACHLAN, J. (1967). Tetrasporangia in *Asparagopsis armata. Br. phycol. Bull.* **3**: 251–252.

MENEGHINI, G. (1844). Algarum species novae vel minus notae. *G. bot. ital.* **1**: 296–306.

MONTAGNE, C. (1841). Plantes cellulaires. *In:* Webb, P. & Berthelot, S., *Histoire naturelle des îles Canaries* III *Botanique* (2) *Phytographia Canariensis.* Paris.

NAGELI, C. (1861 [1862]). Beiträge zur Morphologie und Systematik der Ceramiaceen. *Sber. bayer. Akad. Wiss.* **1861** (2): 297–415.

NEWTON, L. (1931). *A Handbook of the British Seaweeds.* London.

NEWTON, L. M. (1953). Marine Algae. *Scient. Rep. John Murray Exped.* **9**: 395–420.

OKAMURA, K. (1916–1923). *Icones of Japanese Algae.* **4.** Tokyo.

—— (1934). On *Gelidium* and *Pterocladia* of Japan. *J. imp. Fish. Inst., Tokyo* **29**: 47–67.

PAPENFUSS, G. F. (1945). Review of the *Acrochaetium-Rhodochorton* complex of the red algae. *Univ. Calif. Publs Bot.* **18**: 299–334.

—— (1947). Further contributions toward an understanding of the *Acrochaetium-Rhodochorton* complex of the red algae. *Univ. Calif. Publs Bot.* **18:** 433–447.
PRINGSHEIM, N. (1862 [1863]). Beiträge zur Morphologie der Meeres-Algen. *Abh. preuss. Akad. Wiss.* **1862:** 1–37.
ROSENVINGE, L. K. (1893). Grønlands havalger. *Meddr Grønland* **3:** 765–981.
—— (1894). Les algues marines du Groenland. *Annls Sci. nat.*, sér. 7, Bot. **19:** 53–164.
—— (1900). Note sur une Floridée aérienne (*Rhodochorton islandicum* nov. sp.). *Bot. Tidsskr.* **23:** 61–81.
—— (1909). The marine algae of Denmark. Contributions to their natural history. Part I Introduction. Rhodophyceae I. (Bangiales and Nemaliales). *K. danske Vidensk. Selsk. Skr.*, 7 Raekke, Nat. Math. Afd. **7:** 1–151.
—— (1924). The marine algae of Denmark. Contributions to their natural history. Part III Rhodophyceae III. (Ceramiales). *K. danske Vidensk. Selsk. Skr.*, 7 Raekke, Nat. Math. Afd. **7:** 285–488.
ROTH, A. W. (1797). *Catalecta Botanica.* 1. Lipsiae.
SAUVAGEAU, C. (1925). Sur une Floridée (*Polysiphonia doubletii* mscr.) renfermant de l'iode à l'état libre. *C. r. hebd. Séanc. Acad. Sci., Paris* **181:** 293–295.
SCHIFFNER, V. (1931). Neue und bemerkenswerte Meeresalgen. *Hedwigia* **71:** 139–205.
SCHMITZ, F. (1897). *In:* Engler, A. & Prantl, K., *Die natürlichen Pflanzenfamilien.* **1** (2). Leipzig.
SETCHELL, W. A. (1915). *In:* Collins, F. S., Holden, I. & Setchell, W. A., *Phycotheca Boreali-Americana.* 41. Malden.
SIRODOT, S. (1876). Le *Balbiania investiens* – étude organogénique et physiologique. *Annls Sci. nat.*, sér. 6, Bot. **3:** 146–174.
STACKHOUSE, J. (1795). *Nereis Britannica.* Fasc. 1. Bathoniae & Londini.
—— (1809). Tentamen marino-cryptogamicum. *Mém. Soc. Nat. Moscou* **2:** 50–97.
SUHR, J. N. (1839). Beiträge zur Algenkunde. *Flora, Regensburg* **22:** 65–75.
SVEDELIUS, N. (1915). Zytologisch-entwicklungsgeschichtliche Studien über *Scinaia furcellata. Nova Acta R. Soc. Scient. upsal.*, ser. 4 **4**(4): 1–55.
—— (1917). Die Monosporen bei *Helminthora divaricata* nebst Notiz über die Zweikernigkeit ihres Karpogons. *Ber. dt. bot. Ges.* **35:** 212–224.
—— (1933). On the development of *Asparagopsis armata* Harv. and *Bonnemaisonia asparagoides* (Woodw.) Ag. *Nova Acta R. Soc. Scient. upsal.*, ser. 4 **9** (1): 1–61.
TREVISAN, V. B. A. (1845). *Nomenclator Algarum.* Padova.
TURNER, D. (1801). *Ulva furcellata* et *multifida. J. Bot., Göttingen* **1800** (1): 300–302.
—— (1802). Descriptions of four new species of *Fucus. Trans. Linn. Soc.* **6:** 125–136.
—— (1819). *Fuci.* 4. London.
TURTON, W. (1806). *A General System of Nature.* **3.** London.
WEBER, F. & MOHR, D. M. H. (1804). *Naturhistorische Reise durch einen Theil Schwedens.* Göttingen.
WEBER VAN BOSSE, A. (1921). Liste des algues du Siboga. II Rhodophyceae. Part 1. Protoflorideae, Nemalionales, Cryptonemiales. *Siboga-Exped.* **59B:** 187–310.
WEST, J. A. (1969). The life history of *Rhodochorton purpureum* and *R. tenue* in culture. *J. Phycol.* **5:** 12–21.
WITHERING, W. (1792). *An Arrangement of British Plants.* ed. 2. **3.** Birmingham & London.
WOELKERLING, W. J. (1971). Morphology and taxonomy of the *Audouinella* complex (Rhodophyta) in Southern Australia. *Aust. J. Bot. Suppl. Ser.* **1:** 1–91.
—— (1973). The morphology and systematics of the *Audouinella* complex (Acrochaetiaceae, Rhodophyta) in northeastern United States. *Rhodora* **75:** 529–621.
WOMERSLEY, H. B. S. (1965). The Helminthocladiaceae (Rhodophyta) of Southern Australia. *Aust. J. Bot.* **13:** 451–487.
WOODWARD, T. J. (1794). Descriptions of two new British Fuci. *Trans. Linn. Soc.* **2:** 29–31.
ZANARDINI, G. (1847). Notizie intorno alle cellulari marine delle lagune e de'litorali di Venezia. *Atti Ist. veneto Sci.* **6:** 185–262.

Gigartinales

GIGARTINALES Schmitz

GIGARTINALES Schmitz in Engler (1892), p. 18.
Nemastomales Kylin (1925), p. 39.
Sphaerococcales Sjöstedt (1926), p. 75.

Thalli crustose and discoid and/or erect and frondose, pseudoparenchymatous, the degree of aggregation of the constituent filaments ranging from loose to compact; of uniaxial or multiaxial construction.

Carpogonium arising from the apical cell of an undifferentiated filament of the thallus; carposporophyte development following transfer of the zygote nucleus to an auxiliary cell which is always an unspecialized vegetative cell of the thallus; it may be the supporting cell (or its daughter cell) of the branch bearing the carpogonium or a cell at a distance from this; carposporangia liberating one carpospore or four carpotetraspores; gametangial plant and tetrasporangial plant of similar or totally dissimilar organization.

An order not sharply separated from the Cryptonemiales, with representatives of the following families occurring in marine situations in the British Isles:

Calosiphoniaceae
Gymnophlaeaceae
Polyideaceae
Furcellariaceae
Cruoriaceae
Solieriaceae
Rhabdoniaceae
Rhodophyllidaceae
Plocamiaceae
Sphaerococcaceae
Gracilariaceae
Phyllophoraceae
Gigartinaceae

CALOSIPHONIACEAE Kylin

CALOSIPHONIACEAE Kylin (1932), p. 5.

Thallus irregularly laterally branched, soft and lubricous; uniaxial with a transversely dividing apical cell, cortex composed of repeatedly branched fascicles arranged in whorls around axial filament, loosely arranged inwards, more compact outwards; auxiliary cells remote from carpogonial branches, indistinguishable before fertilization, gonimoblasts developing outwards, embedded in cortex without enveloping filaments, all cells becoming carposporangia arranged in groups, tetrasporangia unknown except in culture.

Genera of the Gymnophlaeaceae in which the structure is uniaxial were removed to the Calosiphoniaceae by Kylin (1932). One such genus, *Calosiphonia,* has been recorded for the British Isles.

CALOSIPHONIA Crouan frat.

CALOSIPHONIA Crouan frat. (1852), no. 181.

Type species: *C. finisterrae* Crouan frat. (1852), no. 181 (=*C. vermicularis* (J. Agardh) Schmitz (1889), p. 453 (reprint p. 19)).

Lygistes J. Agardh (1876), p. 118.

Thallus consisting of small crust giving rise to an erect frond; fronds profusely laterally branched, slightly compressed, soft, lubricous; structure uniaxial, each axial cell with four branches of limited growth disposed in a cross and forming whorls, these branches repeatedly branched and composed of cells diminishing in size towards periphery where they are very small, ellipsoid, moniliform and embedded in mucilage to form a continuous layer, internal rhizoids arising from the innermost cells of these branches, clothing axial cells with long simple downgrowing filaments.

Gametangial plants monoecious; spermatangia at ends of whorled branches; carpogonial branches 3-celled, curved, auxiliary cells numerous in cortex, carpogonium fusing with supporting cell before producing connecting filament, gonimoblasts lying in inner cortex, developing outwards and consisting of several successively formed gonimolobes, cystocarps with a pore; tetrasporangia unknown except in cultures of Syrian material.

One species in the British Isles:

Calosiphonia vermicularis (J. Agardh) Schmitz (1889), p. 453 (reprint p. 19).

Lectotype: LD (Herb. Alg. Agardh. 34787). Spain (Cadiz).

Halymenia floresia var. *angusta* C. Agardh (1822), p. 209.
Nemastoma vermicularis J. Agardh (1851), p. 163.
Calosiphonia finisterrae Crouan frat. (1852), no. 181.

Thallus initially a small crust consisting of a basal layer of radially elongated cells 4–8×2–3 μm, each with one or two erect filaments of cells 5–6×4–5 μm; later with a frond produced by the continued growth of an erect filament; frond much-branched, compressed, rose, to 120 mm in length but often much smaller, main axes up to 2 mm broad, soft and lubricous, branches alternate, tapering at apices, appearing somewhat banded throughout because of whorled arrangements of branches of limited growth.

Structure uniaxial, apical cells dividing transversely to form an axial filament whose cells elongate rapidly to about 200–250×20–30 μm 10 mm behind the apex, and produce a whorl of four filaments of limited growth, themselves repeatedly branched and composed of cells diminishing in size towards the periphery where they are embedded in mucilage to form a loose cortex of moniliform cells, the outermost cells being 5–7 μm in surface view; axial filament becoming surrounded by a compact layer of unbranched rhizoidal filaments produced by the lowermost cells of the whorled branches.

Gametangial plants monoecious; spermatangia grouped on terminal and subterminal cells of whorled branches, 2 μm in diameter; carpogonium with a long, spirally twisted trichogyne, gonimoblast about 85 μm in diameter when mature, consisting of successively produced gonimolobes, all the cells of which form carposporangia 15–25 μm in diameter, cystocarps surrounded by an envelope of stratified mucilage and pushing out the cortex to form swellings visible externally; cortex eventually separating to form a pore; tetrasporangial plants unknown except in cultures of Syrian material.

Known in Britain from a very few specimens which were probably collected unattached. Recorded from the French Channel coast at depths of about 15 m, occurring in sheltered areas on small stones and shells associated with *Dudresnaya verticillata* (With.) Le Jol. and *Scinaia* spp., and from the Mediterranean at 20 m where it is occasionally epiphytic on *Cystoseira* spp. etc.

Dorset and Channel Isles.

British Isles to Tangier; Mediterranean eastwards to Syria.

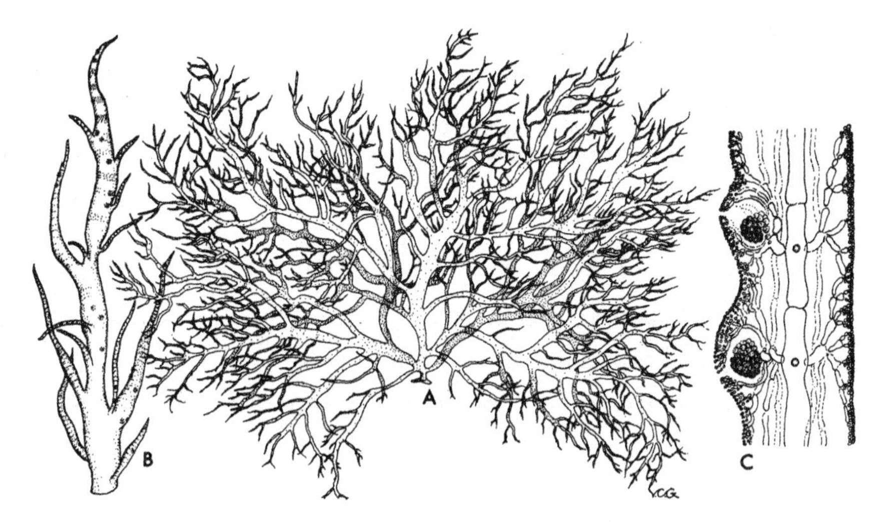

Fig. 60 *Calosiphonia vermicularis*
A. Habit (Aug.) $\times\frac{4}{3}$; B. Branch with cystocarps (Aug.) ×8; C. L.S. same (July) ×80 (after Bornet & Thuret).

Recorded for the months of July and August; gametangial plants only. Mayhoub (1974) reported that in Syrian material in culture erect gametangial fronds developed directly from the crust produced by the carpospore without the intervention of a tetrasporangial phase. He postulated the occurrence of meiosis during the initiation of the erect frond. More recently, (1975) he has described the production of tetrasporangia on carposporelings subjected to 8 hours or less of daily illumination. The tetrasporangia are $45 \times 20\ \mu m$ and occur terminally either on the erect filaments or on the prostrate filaments at the margins.

Information on form variation scanty; type specimen 120 mm long, British specimens to 80 mm, Mediterranean specimens usually only 10 mm (Feldmann, 1954).

The occurrence of this species in the British Isles is accepted on the basis of collections made by Batters at Weymouth in 1883 and by Batters and Dyke-Poore at Jersey in 1883 and 1885. It has not been reported subsequently and newly collected material has not been available for study.

According to Feldmann (1954), some of the reports of *Calosiphonia vermicularis* from northern France are based on misidentifications of *Schmitzia neapolitana* (Berth.) Lagerh. ex Silva (=*Bertholdia neapolitana* (Berth.) Schmitz), a closely similar species; the British specimens have not been re-examined by a specialist.

The gametangial phase of *C. vermicularis* bears some resemblance to *Crouania attenuata* (C. Ag.) J. Ag. and *Dudresnaya verticillata* (With.) Le Jol. In *C. attenuata* the filaments of limited growth are strongly curved and forwardly directed. Those of *C. vermicularis* are patent, relatively short and much branched, a branch usually arising from each cell. Those of *D. verticillata* are longer with branches occurring at intervals of 3–7 cells. *Gloiosiphonia capillaris* (Huds.) Carm. ex Berk. is closely similar in anatomical construction but the cortical cells are more firmly united so that the thallus is much less soft, becoming hollow in older plants. The terminal cells of the branches of limited growth are 8–13 μm in diameter (5–7 μm in *C. vermicularis*).

In *Naccaria wiggii* (Turn.) Endl. and *Atractophora hypnoides* Crouan frat., species with a similar habit, each cystocarp develops all round an axis and does not protrude only on one side. The terminal cells of the branches of limited growth of *N. wiggii* are cylindrical and about 10 μm in diameter whilst those of *A. hypnoides* are spherical and up to 15 μm in diameter.

GYMNOPHLAEACEAE Kützing

GYMNOPHLAEACEAE Kützing (1843), p. 389.
Nemastomataceae (J. Agardh) Engler (1892), p. 22 [as Nemastomaceae].
Nemastomeae J. Agardh (1842), p. 89.

Thallus terete or flattened, either foliose and undivided, irregularly lobed, or irregularly dichotomously divided; mucilaginous, fleshy, occasionally calcified; multiaxial, structure distinctly filamentous; carpogonial branches 3–5 celled, borne laterally on inner cortical cells, auxiliary cells remote from carpogonial branches, more or less recognisable before fertilisation, enveloping filaments absent, gonimoblasts developing outwards but remaining largely embedded in thallus, most cells becoming carposporangia; tetrasporangia scattered in cortex, cruciate.

This family is represented in Britain by the genera *Schizymenia* and *Platoma*. These genera are obviously closely related and their limits are at present rather uncertain. Kylin (1956) distinguished *Schizymenia* by the presence of secretory cells in the cortex (though these are frequently invisible in British material) and by the cystocarps which he considered protrude above the thallus more than in *Platoma*. *Nemastoma dichotomum* J. Agardh, formerly included in treatments of the marine algae of the British Isles (Parke 1953, Parke & Dixon 1964, 1968), has been excluded (see p.178).

PLATOMA Schmitz

PLATOMA Schmitz (1894), p. 627.

Type species: *P. cyclocolpa* (Montagne) Schmitz (1894), p. 627.

Thallus with a discoid holdfast and erect fronds, soft and lubricous, terete, compressed or flattened, dichotomously or irregularly branched, sometimes proliferous from margins; multiaxial with a distinctly filamentous structure, medulla thick and fairly compact, medullary filaments narrow, accompanied by internal rhizoids, inner cortex rather loose, interspersed with internal rhizoids, outer cortex very compact, composed of small cells in distinct radial fascicles, embedded in mucilage, secretory cells sometimes reported but not known in British species.

Spermatangia where known superficial, scattered; 3–5 celled carpogonial branches and auxiliary cells scattered in inner cortex, gonimoblast developing outwards, cystocarps with a pore; tetrasporangia scattered in cortex, cruciate.

Plants attributed to *P. bairdii* from the western Atlantic, including the type, are often considerably larger than any found in European waters, up to 140 mm long and 20 mm broad. The type specimen, illustrated by Taylor (1957), bears a strong resemblance to some plants of *P. marginifera*, e.g. isotype material illustrated by the Crouans (1867). The characters used in the key and description are based on European specimens only.

Key to Species

1 Frond terete or slightly compressed, maximum breadth not more than 5 mm; carpogonial branches 3 celled, cystocarps 65–75 μm in diameter. . *P. bairdii*
Frond compressed or (usually) flattened, maximum breadth at least 15 mm; carpogonial branches 4–5 celled, cystocarps *c.* 160 μm in diameter *P. marginifera*

Platoma bairdii (Farlow) Kuckuck (1912), p. 190.

Holotype: FH (see Kuckuck, 1912; Taylor, 1957). U.S.A. (Gay Head, Martha's Vineyard, Mass.)

Nemastoma ? bairdii Farlow (1875), p. 372.

Thallus consisting of a compact basal disc giving rise to a group of erect fronds which are terete or slightly compressed, lubricous, to 50 mm in length and up to 2 mm broad, rose to carmine; up to three times dichotomously divided, tapering above.

Structure multiaxial, medulla composed of comparatively few thick-walled filaments embedded in mucilage, cylindrical, *c.* 8 μm in diameter, cortical filaments arranged in fascicles, also embedded in mucilage, about 10 cells deep, the inner cells producing downgrowing internal rhizoids, the outermost layers consisting of small moniliform cells, the superficial cells *c.* 4·5 μm in diameter and widely separated in surface view.

Spermatangia unknown; carpogonial branches 3-celled, arising from inner cortical cells, auxiliary cells similarly placed; several connecting filaments arising from each carpogonium, each fusing with several auxiliary cells, most gonimoblast cells forming carposporangia, cystocarps very small, lying amongst the cortical filaments, 65–75 μm in diameter, with a few enveloping filaments and pore, carpospores 7–9 μm in diameter; tetrasporangia occurring both in the erect fronds and in the encrusting base, borne laterally near the base of cortical or crust filaments, 17–20×11–18 μm, with cruciately arranged spores.

Nothing is known of its habitat in Britain. In Helgoland (Kuckuck, 1912), Denmark (Rosenvinge, 1917), and eastern U.S.A. (Taylor, 1957) it has been reported growing on stones in the lower littoral and sublittoral to 20 m.

Northumberland.

Helgoland and Denmark; Canada (Nova Scotia) to U.S.A. (Massachusetts).

Data on seasonal and form variation in the British Isles too inadequate for comment; the only specimens were found in July 1853, and bore tetrasporangia. In Helgoland the earliest erect fronds arise in May, with tetrasporangia occurring in June and ripe cystocarps in July; littoral plants are largest in August and September (Kuckuck, 1912).

This species was recorded for the British Isles by Batters (1900) as *Helminthocladia hudsonii* J. Ag. (see Dixon, 1964); newly collected material has not been available for study.

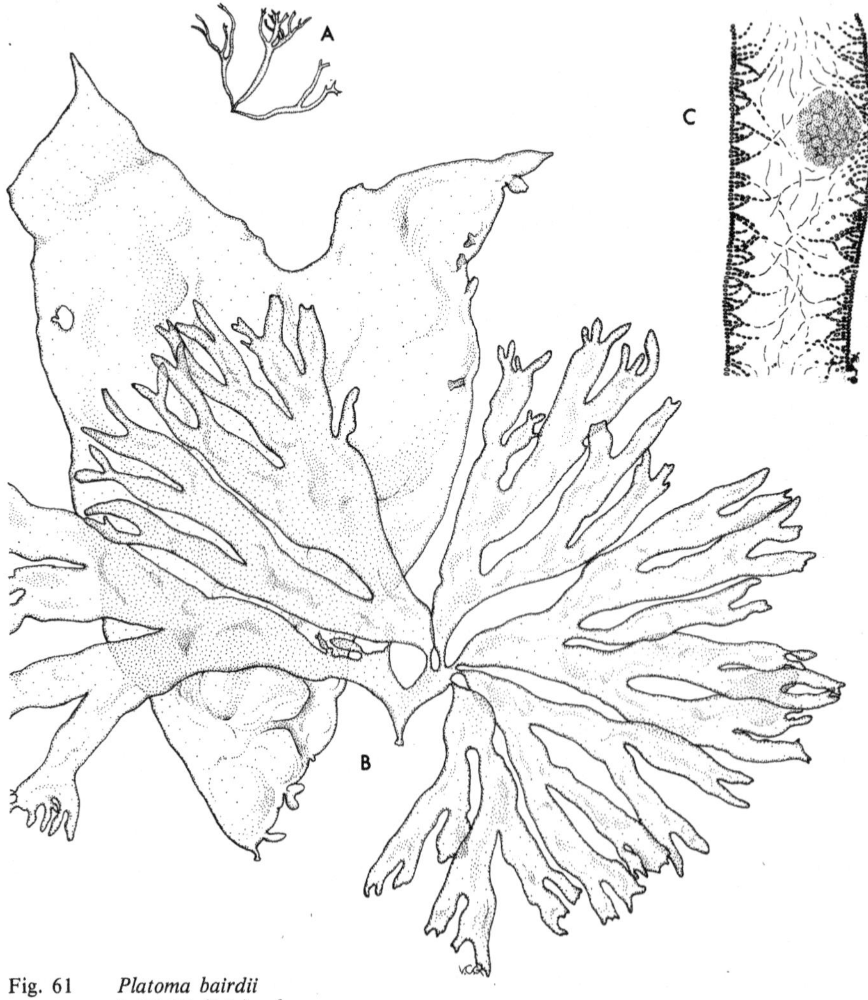

Fig. 61 *Platoma bairdii*
A. Habit (July) ×$\frac{2}{3}$.
Platoma marginifera
B. Habit of two plants (Sep.) ×$\frac{2}{3}$; C. T.S. blade with cystocarp (undated) ×80.

Platoma marginifera (J. Agardh) Batters (1902), p. 94.

Holotype: LD (Herb. Alg. Agardh. 22125). France (Brest).

Nemastoma marginifera J. Agardh (1851), p. 165.

Thallus consisting of erect fronds arising in groups from a small discoid base, stipe short, expanding gradually into a compressed or flattened blade, lubricous, fleshy and elastic, to 240 mm in length, and up to 100 mm broad, and 300–500 μm thick, rose to carmine, repeatedly dichotomously or proliferously branched sometimes with angle keyhole-shaped, frequently narrowed above a division.

Structure multiaxial, medulla distinctly filamentous, composed of comparatively few cylindrical filaments embedded in mucilage, 4–8 μm in diameter, with a conspicuous sheath; cortical filaments in fascicles up to 10 cells deep, embedded in mucilage, the inner cells producing down-growing internal rhizoids, the outermost layers consisting of small moniliform cells, 3–4·5 μm in diameter in surface view and widely separated.

Gametangial plants monoecious; spermatangia produced from cells of cortical fascicles, small, not united into large groups; carpogonial branches 4–5 celled, borne at the base of the cortical fascicles, auxiliary cells similarly placed, cystocarps very small, *c.* 160 μm, gonimoblast lobed, surrounded by a delicate transparent envelope, immersed in unmodified cortex, enveloping filaments absent, all gonimoblast cells transformed into carposporangia, spores remarkably small, about 5 μm in diameter, escaping through a narrow pore; tetrasporangia unknown.

In the British Isles found only in deep rock pools in the upper sublittoral; in France recorded by Bornet & Thuret (1876) from midlittoral to sublittoral in rock crevices and on stones.

Cornwall, Kerry, Clare (drift) and the Channel Isles.

British Isles to the Basque coast of France; Tangier.

Except for the first record of the species for Britain (Whitsand Bay, Nov. 1848), specimens have only been collected in August and September. The Crouans (1867) reported the species for the autumn in Brittany but it apparently occurs earlier (June and July) in southwestern France (Bornet & Thuret 1876).

This species is very imperfectly known in the British Isles; the few specimens available for examination have varied widely in size and extent of subdivision. For a discussion of form variation in plants in the vicinity of Biarritz, where the species is fairly common, see Bornet & Thuret (1876).

Bornet & Thuret comment on the similarity between this species and *Halarachnion ligulatum* (Woodw.) Kütz., q.v. It is also similar to *Schizymenia dubyi* (Chauv. ex Duby) J. Ag., q.v.

SCHIZYMENIA J. Agardh

SCHIZYMENIA J. Agardh (1851), p. 169.

Type species: *S. dubyi* (Chauvin ex Duby) J. Agardh (1851), p. 171.

Thallus with a small discoid holdfast and erect fronds, fronds soft and fleshy, flattened, irregularly lobed or split; structure multiaxial, obviously filamentous, medulla fairly thick with loosely arranged, narrow, branched filaments, interspersed with internal rhizoids, inner cortex somewhat looser, also with rhizoids, outer cortex compact, composed of small cells in radial fascicles, with scattered large secretory cells.

Gametangial plants monoecious; spermatangia aggregated in superficial sori; 3 celled carpogonial branches and auxiliary cells scattered in inner cortex, gonimoblasts developing outwards, small, elevating cortex slightly, enveloping filaments absent, most cells becoming carposporangia, spores shed through a comparatively large, well-defined pore; tetrasporangia unknown in the British Isles.

In some Pacific species tetrasporangia occur in foliose fronds similar to those of the gametophyte and secretory cells are more regularly seen. In the European species secretory cells are too infrequently seen for conclusions regarding identity to be drawn from their absence.

One species in the British Isles:

Schizymenia dubyi (Chauvin ex Duby) J. Agardh (1851), p. 171.

Lectotype: STR. France (Cherbourg).

Halymenia dubyi Chauvin ex Duby (1830), p. 944.
Turnerella atlantica Kylin (1930), p. 40.

Erect frond arising from a small attachment disc; stipe very short, 2–3 mm long, and comparatively stout and fleshy, expanding gradually into a foliose blade which is entire or variously split but not normally proliferous from the margin, usually flat but occasionally somewhat cucullate or with an undulate margin, ovate or obovate, firmly mucilaginous, translucent, brownish-red, to 500 mm long, 250 mm broad and 600 μm thick.

Structure multiaxial, medulla composed of a thick layer of compact filaments interwoven with rhizoids, filaments thick-walled, to 15 μm in diameter, further enlarged in the region of a cross-wall (i.e. 'bone-shaped'); inner cortex of large, rounded cells arranged more loosely, merging into an outer cortex of radial fascicles of smaller cells about 4–5 μm in diameter in surface view, secretory cells with highly refractive contents sometimes visible amongst the cortical cells, 25–45 × 8–12 μm.

Gametangial plants monoecious; spermatangia in large superficial sori, radially elongate, spermatia *c.* 3 μm in diameter; auxiliary cells and 3 celled carpogonial branches in the inner cortex, cystocarps embedded in the medulla, elevating the cortex slightly to give the frond the texture of sandpaper, scattered, very small, *c.* 150 μm, containing comparatively few (10–30) carposporangia 15–35 μm in diameter; tetrasporangia unknown in the British Isles.

Epilithic, in pools exposed to some wave action, midlittoral to shallow sublittoral, tolerant of some sand cover.

St Kilda, Ireland, Wales and the western coasts of England, extending eastwards along the Channel to Dorset.

Iceland; British Isles to Morocco; Mediterranean.

Little is known of the seasonal behaviour of this species; fronds probably annual, recorded between February and October with a peak in May; spermatangia occur in March and ripe cystocarps from May onwards.

There is comparatively little variation in external appearance apart from the degree of subdivision of the blades; the lobes are always elongated and the angle between them is very narrow so that the lobes may even overlap.

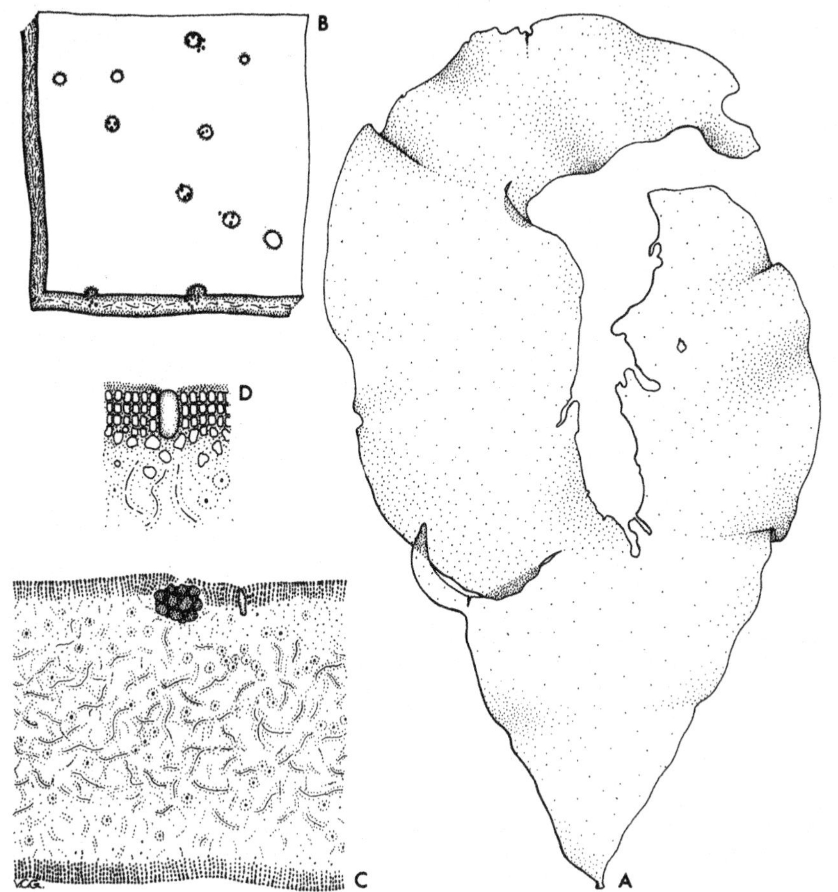

Fig. 62 *Schizymenia dubyi*
A. Habit (Aug.) $\times\frac{2}{3}$; B. Part of blade with carpospores emerging from cystocarps (Sep.) ×25; C. T.S. same ×80; D. T.S. blade showing secretory cell ×360.

Tetrasporangia were illustrated by the Crouans (1867) and Newton (1931) but are otherwise unknown. It is possible that the fronds arise from a persistent encrusting base, and that the tetrasporangia occur in a similar crust. This type of life history is already known to occur in *Halarachnion ligulatum* (Woodw.) Kütz., for example. Difficulties are frequently encountered in identifying specimens of foliose red algae, especially when sterile. In *Schizymenia* the presence of secretory cells is diagnostic although in the British Isles these are visible in only a minority of specimens. A squash preparation can be made after brief treatment with a dilute acid such as 10 per cent HCl. This will show fascicles of rounded cortical cells, occasional secretory cells and medullary filaments with 'bone-shaped' cells. In similar preparations of foliose members of the Grateloupiaceae and

Kallymeniaceae the cortical cells do not hang together in fascicles and 'stellate' cells of various kinds would be seen. The internal structure of *Dilsea carnosa* (Schmidel) O. Kuntze is similar to that of *Schizymenia dubyi* but more compact and the blade is at least 1 mm in thickness. *Platoma marginifera* (J. Ag.) Batt. is much softer, the cortical cells being more loosely embedded in mucilage, and the inner as well as the outer cells of the cortical fascicles are very small.

Excluded species

Nemastoma dichotomum J. Agardh.

The record of this species for the British Isles is based on a single specimen collected from a rock pool in Guernsey (Lyle, 1920). The identification is doubtfully correct as this species is not otherwise known outside the Mediterranean. Lyle's specimen has not been available for study but the illustration she gives of the thallus in transverse section is closely similar to that of *Platoma marginifera* J. Ag., q.v.

POLYIDEACEAE Kylin

POLYIDEACEAE Kylin (1956), p. 166.

Thallus with erect fronds, either terete and repeatedly dichotomously branched *(Polyides),* or compressed, segmented and proliferously branched; multiaxial, medulla composed of longitudinal filaments accompanied by numerous internal rhizoids, cortex compact, inner large-celled, interspersed with rhizoids, outer small-celled; carpogonial branches and intercalary auxiliary cells developing in special outgrowths from thallus surface, gonimoblasts developing from connecting filaments, terminal cells becoming carposporangia; tetrasporangia scattered in outer cortex, cruciate.

This small family, created by Kylin in 1956, was removed from the Cryptonemiales by Papenfuss (1966) and placed in the Gigartinales because the auxiliary cell is 'an undifferentiated intercalary cell of a nemathecial (vegetative) filament'. The family has two unusual features, namely the localization of the female organs in special outgrowths and the development of the gonimoblast from the connecting filament after the latter has fused with the auxiliary cell, rather than from the auxiliary cell itself. One genus, *Polyides,* occurs in Britain.

POLYIDES C. Agardh

POLYIDES C. Agardh (1822), p. 390.

Type species: *P. lumbricalis* C. Agardh (1822), p. 392 (=*P. rotundus* (Hudson) Greville (1830), p. 70).

Spongiocarpus Greville (1824), p. 286.

Thallus with a discoid holdfast and erect fronds, fronds terete, repeatedly dichotomously branched, cartilaginous; structure multiaxial, medulla fairly thick, composed of longitudinal filaments interspersed with numerous internal rhizoids, inner cortex with large cells and rhizoids, outer cortex of small cells in distinct radial rows.

Gametangial plants dioecious, spermatangia superficial on upper dichotomies; carpogonial branches 5–7 celled, auxiliary branches longer; tetrasporangia scattered in outer cortex of upper branches, cruciate.

One species in the British Isles:

Polyides rotundus (Hudson) Greville (1830), p. 70.

Lectotype: Hudson's description, based on OXF (Herb. Sherard 1906) (see Drew, 1958, pl. 61). Cornwall (St Ives).

Fucus rotundus Hudson (1762), p. 471.
Fucus caprinus Gunnerus (1766), p. 96.
Polyides caprinus (Gunnerus) Papenfuss (1950), p. 194.
Polyides lumbricalis C. Agardh (1822), p. 392, nom. superfl.

Thallus with erect fronds arising from a perennial disc of variable size, often about 20 mm in diameter, terete (rarely slightly compressed at the nodes), bushy, cartilaginous, translucent, up to 200 mm long and about 2 mm in diameter; dark purplish red to black (red in transmitted light), repeatedly dichotomously branched in more or less the same plane, axils rounded, branches not patent, often abruptly tapering, apices obtuse or acute.

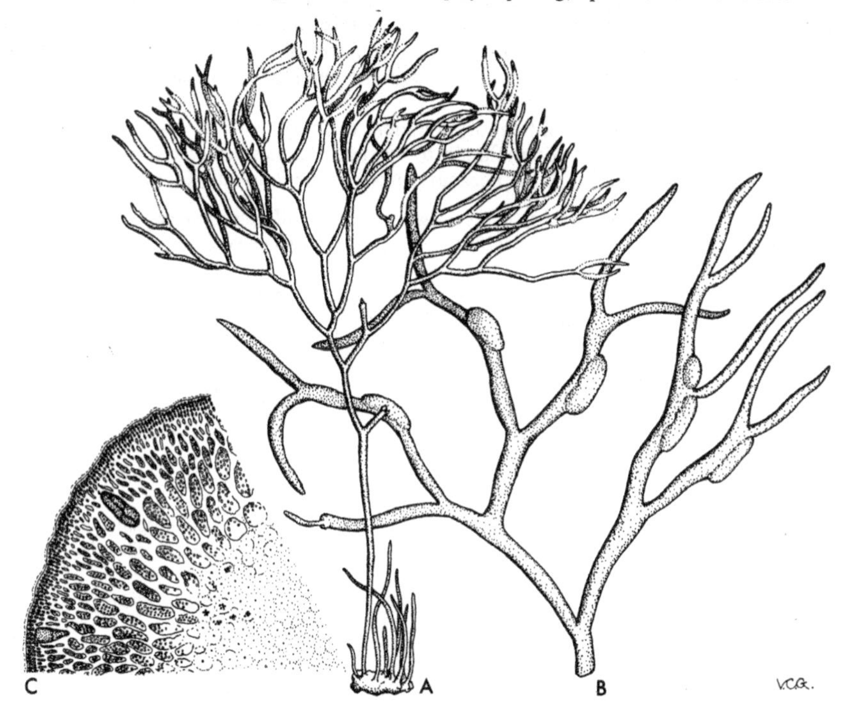

Fig. 63 *Polyides rotundus*
A. Habit of plant with spermatangia (Sep.) ×1; B. Part of plant with carposporangial outgrowths (Sep.) ×4; C. T.S. branch with tetrasporangia (Oct.) ×80.

Structure multiaxial, medulla of longitudinal filaments 25–33 μm in transverse diameter, interspersed with rhizoids, inner cortex composed of a zone of large radially elongated cells about 60–80×30–40 μm, outer cortex of much smaller cells in radial rows, about 5–7 μm in diameter in surface view.

Gametangial plants dioecious; spermatangia developing superficially over the entire surface of the upper dichotomies, spermatangial mother cells lateral on branched filaments, producing 1–4 spermatangia each liberating a single spermatium about 6 μm in diameter; female outgrowths spreading centrifugally, sometimes fusing and extending completely around thallus, cystocarps up to 300 μm in diameter, enveloping filaments absent, central tissue remaining small-celled and sterile, carposporangia formed terminally on the gonimoblast filaments, obconical, 45–55 μm long, 20–35 μm in diameter at the base, pore absent; tetrasporangia in immersed sori in the younger parts, thickening the branch somewhat, 70–100×35–55 μm, tetraspores cruciately arranged.

Epilithic, littoral in pools and sublittoral to at least 12 m, tolerating sand cover.

Generally distributed throughout the British Isles.

Arctic Ocean (Kara Sea) to northern Spain; Baltic; Canada (Hudson Straits) to U.S.A. (New York).

Fronds perennial, rapid growth occurring in February and March (Austin 1960a). Spermatangia produced in August and September; female outgrowths begin to develop in August as small pale oval patches in the younger parts of the erect frond; carpospores released in December and January after which the outgrowths deeay. Tetrasporangia recorded between October and March.

A species with comparatively little morphological variation.

This species bears a close resemblance to *Furcellaria lumbricalis* (Huds.) Lamour. Morphological and anatomical differences are shown in the following table:

	Furcellaria lumbricalis	*Polyides rotundus*
Attachment	branching holdfast	expanded disc
Frond	occasionally compressed at dichotomies, otherwise terete and regular	more completely terete but often showing slight irregularities in growth
Apices	gradually tapering, not paler	abruptly tapering, paler
Colour in incandescent transmitted light	brown	red
L.S. cortex – outer	1–3 layers of radially cylindrical cells	2–5 layers of vertically subcylindrical cells
L.S. cortex – inner	cells irregularly arranged at 90° to axis	cells regularly arranged at 45° to axis
L.S. medulla	cells 15–20×longer than diameter	cells 8–15×longer than diameter

P. rotundus is said to support fewer epiphytes than *F. lumbricalis*, q.v.
Gigartina pistillata (S. G. Gmel.) Stack. q.v. is also rather similar in habit.

FURCELLARIACEAE Greville

FURCELLARIACEAE Greville (1830), p. 66 [as Furcellarieae].

Thallus with either a discoid or branched holdfast, with or without erect fronds, erect fronds terete, compressed or flattened, dichotomous or irregularly branched; multiaxial, medulla distinctly filamentous, cortex compact, inner large-celled, outer small-celled; auxiliary cells remote from carpogonia, easily distinguished before fertilization, gonimoblast developing inwards, embedded in thallus, surrounded only by displaced cortical filaments, pore present or absent, most cells becoming carposporangia, tetrasporangia embedded in cortex of erect fronds or among erect filaments of an encrusting plant, zonate.

This is a small family in which several carpogonial branches are borne on each supporting cell. Two genera, *Furcellaria* and *Halarachnion*, occur in Britain; they differ in morphology and life history but correspond closely in the pattern of carposporophyte development.

FURCELLARIA Lamouroux **nom. cons.**

FURCELLARIA Lamouroux (1813), p. 45 (reprint p. 25).

Type species: *F. lumbricalis* (Hudson) Lamouroux (1813), p. 46 (reprint p. 26).

Fastigiaria Stackhouse (1809), p. 59, 90.

Thallus with a branched holdfast and erect fronds, terete, repeatedly dichotomously branched, cartilaginous; multiaxial, medulla fairly compact, composed of longitudinal filaments interwoven with numerous internal rhizoids, cortex very compact with larger cells inwards and smaller cells in distinct radial rows outwards.

Gametangial plants dioecious; spermatangia in superficial sori restricted to apical regions; carpogonial branches 3–5 celled, arising from inner cortical cells, usually several attached to each supporting cell, auxiliary cells in large numbers in inner cortex, gonimoblasts developing inwards in inner cortex and often closely appressed, fairly small, each with an obscure pore, developing almost completely into carposporangia; tetrasporangia scattered in outer cortex, zonate.

One species in the British Isles:

Furcellaria lumbricalis (Hudson) Lamouroux (1813), p. 46 (reprint p. 26).

Lectotype: BM. An undated unlocalized Hudson specimen considered provisionally to be of this status. (See Dixon & Irvine, 1977). England.

Fucus lumbricalis Hudson (1762), p. 471.
Fucus fastigiatus Linnaeus (1753), p. 1162.
Fucus lumbricalis var. *fastigiatus* Turner (1808), p. 11.
Furcellaria fastigiata (Turner) Lamouroux (1813), p. 46 (reprint p. 26).

Thallus consisting of a branched entangled prostrate holdfast and erect fronds up to 300 mm or more in length; brownish-black (brown in transmitted light), cartilaginous, repeatedly dichotomously branched, fastigiate, branches terete except in region of dichotomies, about 1–2 mm in diameter, gradually tapering, apices acute, axils acute.

Structure multiaxial; consisting of a medulla of filaments 11–15 μm in transverse diameter, interwoven with narrower rhizoids, and a cortex of rather irregularly arranged filaments composed internally of ellipsoidal cells about 70 μm in radial length and externally of one or two rows of radially elongated cells about 25 × 8 μm (4–7 μm in surface view).

Gametangial plants dioecious; spermatangial mother cells developing superficially in much swollen apical regions, the whole surface becoming covered with yellowish cells slightly elongated radially and cutting off spermatangia containing spermatia about 9·5 × 3·5 μm which escape through the well-marked cuticle by a definite pore; cystocarps developing internally in the apical regions, gonimoblast producing a few sterile cells and gonimolobes up to 800 μm in diameter, with displaced cortical filaments appearing as enveloping filaments, almost all cells becoming carposporangia (about 200), carpospores 35–50 μm when shed, a tract of cells disintegrating to form an ill-defined pore to the exterior; tetrasporangia in markedly thickened apical regions, each arising from the distal end of an inner cortical cell, increasing to a maximum length of about 110 μm, tetraspores zonately arranged, slightly smaller than carpospores, liberated by disintegration of the thallus surface.

Epilithic in the lower littoral and sublittoral to at least 12 m, tolerating sand cover.

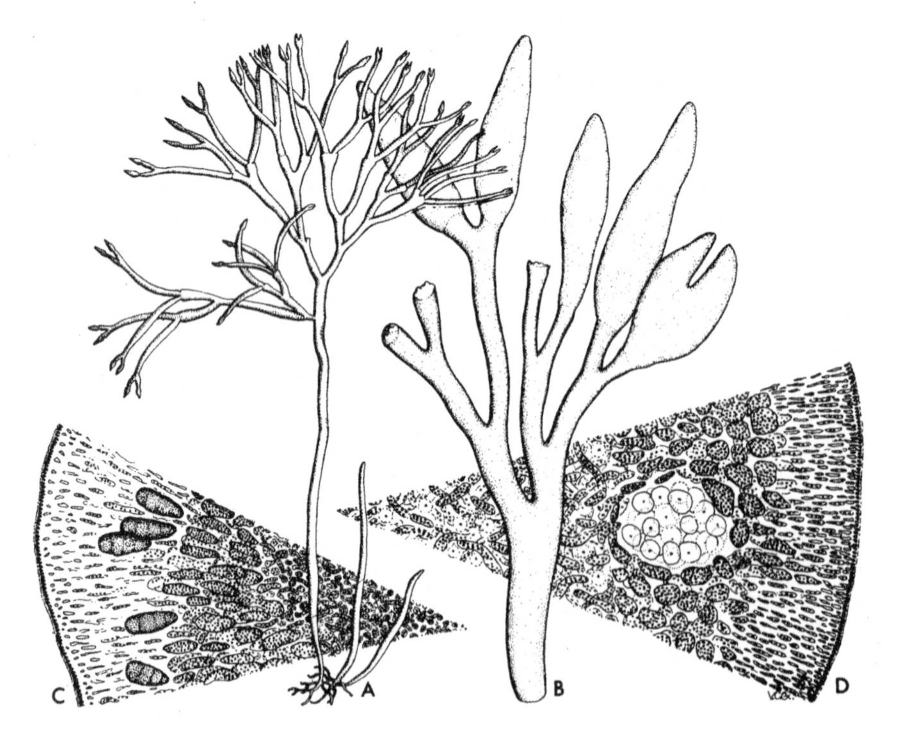

Fig. 64 *Furcellaria lumbricalis*
A. Habit of plant with spermatangia (Feb.) ×1; B. Same ×3; C. T.S. branch apex with tetrasporangia (Dec.) ×80; D. T.S. branch apex with cystocarp (Dec.) ×80.

Generally distributed throughout the British Isles.
Northern Russia, Iceland, Faeroes, Norway to France (Basque coast); Baltic; Greenland; Canada (Arctic to Nova Scotia).

Fronds perennial, growth occurring throughout the year with a maximum in March/April; becoming fertile when full-sized (this varies between 90 mm and 300 mm according to habitat) during the fourth to seventh year (Austin 1960, 1960a). Spermatangia produced continuously from December–April, female sexual organs developing in December/January, the gonimoblast gradually maturing from January until the following December; in plants from northerly and very exposed places development may be somewhat slower; carpospores shed in late December and January; tetrasporangia initiated in April with meiosis occurring in November and spore liberation during late December and early January. Tetrasporangial plants slightly more robust and plentiful than gametangial ones.

Plants show comparatively little variation from the dichotomous pattern since, when damage occurs, growth is resumed from the surface of the wounded apex rather than by the production of lateral branches (Austin 1960a).

The species penetrates into parts of the Baltic where the salinity is about 5 parts per thousand.

Polyides rotundus (Huds.) Grev., q.v. and *Gigartina pistillata* (S. G. Gmel.) Stackh., q.v. bear a close resemblance to this species.

Older plants are usually covered with epiphytic Bryozoa and encrusting corallines; galls produced by nematode worms are also known (Barton, 1901).

Floating populations several metres in thickness occur in Denmark and are the basis of the Danish phycocolloid industry (Levring *et al.* 1969). Small populations of similar unattached plants might occur in Scottish and Irish lochs.

HALARACHNION Kützing

HALARACHNION Kützing (1843), p. 394.

Type species: *H. ligulatum* (Woodward) Kützing (1843), p. 394.

Thallus of gametangial phase composed of erect fronds, compressed or flattened, undivided or dichotomously to irregularly branched, frequently proliferous from margins or frond surface; structure multiaxial, medullary filaments usually few, very loose, interspersed with narrow rhizoids, cortex thin, more compact, composed of 2–3 layers with large cells inwards and somewhat smaller cells outwards.

Monoecious; spermatangia in small superficial sori; carpogonial branches on inner side of cortex, curved, usually 3 celled, auxiliary cells very numerous among inner cortical cells; gonimoblast rounded or lobed, developing inwards, lying in medulla, without enveloping filaments, carpospores discharging through cortex which is pierced by a pore.

Thallus of tetrasporangial phase crustose and closely similar to plants described as *Cruoria rosea* (Crouan frat.) Crouan frat., tetrasporangia scattered among erect filaments, zonate.

In the European species of *Halarachnion* the gametangial phase produces a delicate but often quite large proliferous frond in summer. The tetrasporangial phase, however, consists exclusively of a persistent crustose plant so far known to occur only sublittorally and fertile in winter. The gametangial plants are probably crustose at first and produce the fertile erect fronds when conditions are suitable. The Japanese species *H. latissimum* Okamura appears to have tetrasporangial fronds similar to those bearing gametangia.

One species in the British Isles:

Halarachnion ligulatum (Woodward) Kützing (1843), p. 394.

Lectotype: BM-K. An undated, unlocalized specimen labelled 'from Mr Woodward' accepted provisionally as of lectotype status. Norfolk (Yarmouth (drift) and Cromer).

Ulva ligulata Woodward (1797) p. 54.

Gametangial plants consisting of a small attachment disc about 1 mm in diameter from which arise one to several fronds which are compressed to flattened (sub-cylindrical when narrow) and extremely variable in dimensions; thin plants soft and diaphanous, thicker plants sub-cartilaginous; to 500 mm long and 2–200 mm broad, translucent, deep brownish red, paler in thin specimens, becoming pinkish in fresh water or when drifting, stipe short and narrow, expanding gradually into irregularly dichotomously divided blades which are frequently furnished with ligulate proliferations from the margins and blade surfaces, lobes tapering above.

Structure multiaxial; medulla ranging from rather compact to very loose and lacunose, composed of comparatively few filaments embedded in mucilage, filaments cylindrical, about 8 μm in diameter, cortex 2–3 cells thick, inner cells large and rounded, outer cells smaller, 6–11 μm in diameter in surface view, and forming a compact layer.

Monoecious; spermatangia in superficial sori, 1–4 mother cells produced by each cortical cell, spermatia *c.* 2 μm; gonimoblasts developing inwards, 100–260 μm, most cells becoming carposporangia 15–20 μm in diameter, enveloping filaments absent, pore reputedly present, obscure.

Tetrasporangial plants crustose consisting of a single basal layer of radially elongated cells *c.* 15×5 μm from which arise erect simple or branched filaments composed of 6–8 cells, 8–10 μm in diameter, filaments sometimes replaced by secretory cells *c.* 30 μm long×12 μm broad, with highly refractive contents; tetrasporangia borne laterally at the base of the erect filaments, 22–25×8 μm, with zonately arranged spores.

Erect fronds attached to stones, shells and maerl mainly in the lower sublittoral, to 17 m, occasionally extending into the upper sublittoral. In view of the problems in identifying the crustose phases, precise comments on the habitats of these are not possible.

Plants with erect fronds generally distributed throughout the British Isles.

Plants with erect fronds recorded from Norway (More) to Morocco; Helgoland, Denmark; Mediterranean. Doubtfully recorded for Florida, Barbados, Brazil; records probably based on misidentifications of *Halymenia* spp.

For comparisons and distribution of crustose phases see *Cruoria rosea* (Crouan frat.) Crouan frat. and South *et. al.* (1972).

Foliose phase short-lived, recorded only between June and September; spermatangia recorded for June–July and cystocarps for June–September; encrusting phases probably perennial, tetrasporangial plants indistinguishable from *Cruoria rosea* (Crouan frat.) Crouan frat., q.v.

The form of the foliose frond varies widely, the differences in appearance being due largely to the range in length and breadth. Those which are 1–5 mm broad have been

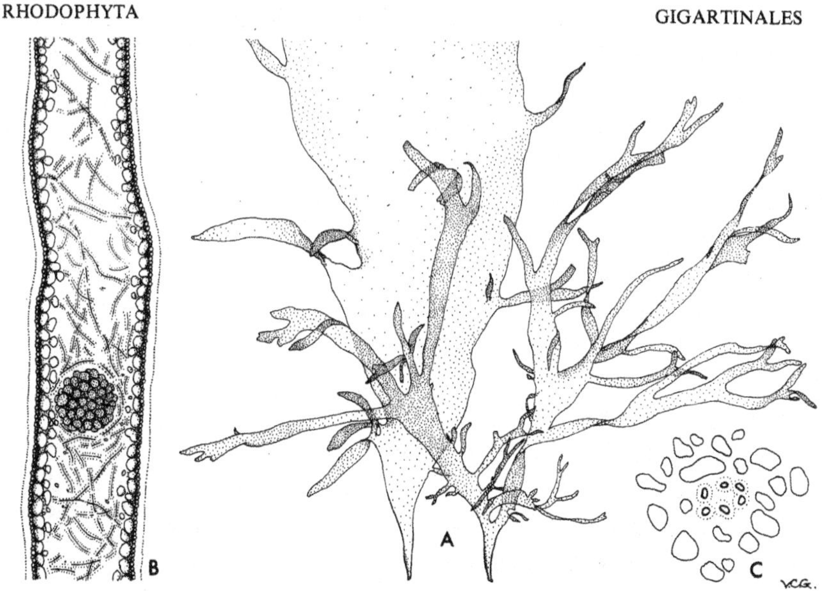

Fig. 65 *Halarachnion ligulatum*
A. Habit of two plants (Aug.) ×1; B. T.S. blade with cystocarp (Aug.) ×80; C. Surface view of blade with spermatangia (June) ×300.

called var. *aciculare* Hauck and var. *dichotomum* Harvey. The broader forms are more obviously flattened; those plants which are 15–20 mm broad have been called var. *ramentaceum* Harvey whilst var. *latifolium* Harvey has been retained for the largest plants of all, reaching 500 mm in length and 200 mm in breadth. These varieties are of doubtful taxonomic validity but no further information is available at present.

Bornet & Thuret (1876) comment on the similarity between some forms of *H. ligulatum* and *Platoma marginifera* (J. Ag.) Batt., adding that they can be distinguished by the fact that in the former all the carpospores mature and discharge together, leaving behind a group of sterile cells, whereas in the latter the carpospores mature in batches and no sterile cells are present. There are also vegetative differences; for example the cortex of *Platoma* spp. is thicker and consists of fascicles of very small moniliform cells 4–5 μm in diameter. In *H. ligulatum* the cortex consists of an outer layer of cells 6–11 μm in diameter and an inner layer of larger rounded cells.

CRUORIACEAE (J. Agardh) Kylin

CRUORIACEAE (J. Agardh) Kylin (1928), p. 29.
Cruorieae J. Agardh (1852), p. 487.

Thallus encrusting, closely adherent to substrate, composed of prostrate radiating filaments and parallel erect filaments embedded in mucilage; carpogonial branches 2–4 celled, auxiliary cells remote from carpogonia, not visible before fertilization, gonimoblasts developing among erect filaments, all cells becoming carposporangia; tetrasporangia among erect filaments, zonate.

This family is provisionally placed in the Gigartinales; carposporophytes have been described only for *Cruoria pellita* (Rosenvinge, 1917) and the evidence upon which Rosenvinge's interpretations were based requires re-examination. Parke & Dixon (1968) placed three genera, *Cruoria, Cruoriopsis* and *Petrocelis,* in this family. It now appears that *Cruoriopsis*-like plants are involved in the life-history of *Gloiosiphonia* and so the former genus has been placed provisionally in the Gloiosiphoniaceae (Cryptonemiales). For similar reasons the genus *Petrocelis* has been removed to the Gigartinaceae.

CRUORIA Fries

CRUORIA Fries (1836), p. 316.

Type species: *C. pellita* (Lyngbye) Fries (1836), p. 316.

Chaetoderma Kützing (1843), p. 326.

Thallus encrusting, ranging from very thin to comparatively thick, closely attached to substrate by undersurface, rhizoids normally absent; consisting of a monostromatic basal layer the cells of which give rise to curved or straight erect filaments which are simple or sparingly branched, embedded in mucilage and very loosely bound together except at apices.

Gametangial plants where known monoecious; spermatangia forming small lateral tufts at upper end of erect filaments; carpogonial branches 2–4 celled, gonimoblasts developing among erect filaments as spindle-shaped groups of cells each of which becomes a large carposporangium, enveloping filaments absent; tetrasporangia lateral on erect filaments, zonate.

Plants apparently indistinguishable from *C. rosea* have been found to represent the tetrasporangial phase of both *Halarachnion ligulatum* (Woodw.) Kütz. (Boillot, 1965) and *Turnerella pennyi* (Harv.) Schmitz (South *et al.,* 1972). Although not so far recorded for the British Isles, it is possible that *T. pennyi* occurs sublittorally in northern waters. Gametangial plants produce erect foliose fronds which could be confused with *Kallymenia reniformis* (Turn.) J. Ag. or *Schizymenia dubyi* (Chauv. ex Duby) J. Ag. The presence of secretory cells in the cortex distinguishes it from *K. reniformis,* but such cells are sometimes seen in *S. dubyi.* In *T. pennyi,* however, the cystocarps are larger than those of *S. dubyi,* develop towards the interior and are furnished with a large fusion cell and a much larger number of smaller carposporangia.

Key to Species

1	Secretory cells present, erect filaments usually 2–6 cells long . .	*C. rosea*
	Secretory cells absent, erect filaments 10 or more cells long . .	*C. pellita*

Cruoria pellita (Lyngbye) Fries (1836), p. 316.

Lectotype: C (see Rosenvinge, 1917). Isotype: PC (see Bornet, 1892). Faeroes (Quivig).

Chaetophora pellita Lyngbye (1819), p. 193, excl. spec. ab Hindsholm.
Cruoria adhaerens Crouan frat. ex J. Agardh (1852), p. 491.

Crust dark red, mucilaginous, up to 100 mm or more in diameter and more than 500 μm thick, rhizoids rare, apparently occurring only on uneven substrates, prostrate filaments of fairly regular elongated cells, giving rise to simple or branched erect filaments loosely united by mucilage and easily separated by gentle pressure, usually about 20 cells long, cells barrel-shaped below, 12–14 μm in diameter tapering to 6–11 μm above and then more parallel-sided, cell fusions common, sometimes in zones.

Gametangial plants monoecious; spermatangia forming small lateral tufts at the upper ends of the erect filaments; cystocarps consisting of spindle-shaped groups of carposporangia 50–70 μm in diameter; tetrasporangia lateral near the base of the erect filaments, 240–285 × 45–65 μm, tetraspores zonately arranged.

On rock, shells, *Laminaria* holdfasts and, apparently preferentially, on encrusting coralline algae from the upper sublittoral to a depth of at least 15 m.

Recorded for scattered localities, probably generally distributed throughout the British Isles.

Norway (Nordland) to Tangier; Baltic.

Little is known about the seasonal behaviour of this species; cystocarps have been recorded for December–January and tetrasporangia from October–February; spermatangia recorded for Denmark in July (Rosenvinge, 1917).

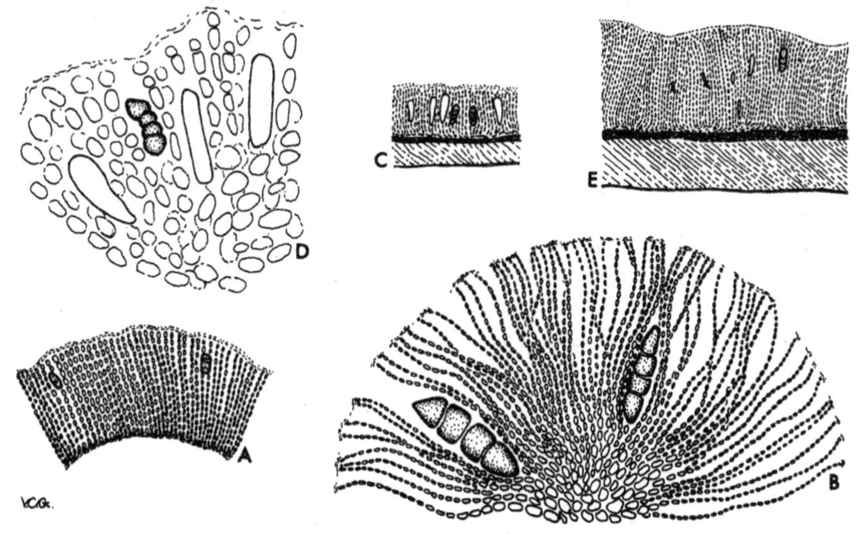

Fig. 66 *Cruoria cruoriaeformis* (Type)
A. V.S. crust with tetrasporangia (Nov.) ×80.
Cruoria pellita
B. V.S. crust with tetrasporangia (Jan.) ×80.
Cruoria rosea var. *rosea*
C. V.S. crust with tetrasporangia and secretory cells (Nov.) ×80; D. Same (Feb.) ×300.
Cruoria rosea var. *purpurea* (Type)
E. V.S. crust with tetrasporangia and secretory cells (Feb.) ×80.

The appearance of plants is affected by the frequent attacks of grazing animals; filaments are apparently also shed at other times, especially from above a zone of cell fusions.

Reports of gametangia (Rosenvinge, 1917) need re-examining in view of the peculiarities in thallus structure which arise from damage by grazing animals and the variety of adventitious inclusions possible in a thallus of this loose consistency. This species is sometimes confused with *Petrocelis* spp. In *C. pellita* the erect filaments taper markedly from base to apex whilst in *Petrocelis* the filaments are scarcely tapering and narrow (4–6 μm) throughout. The type specimen of *Cruoria adhaerens* Crouan frat. ex J. Ag. is typical *C. pellita*. The binomial has sometimes been misapplied, however: both Holmes (in herb.) and Cotton (1912) misidentified specimens of *C. rosea* (Crouan frat.) Crouan frat., whilst Arnott and colleagues applied it to sterile specimens of *Petrocelis* from the Clyde area (Johnstone & Croall 1859; Gatty 1872). Specimens sometimes contain the endophytic green algae *Codiolum petrocelidis* Kuckuck and *Chlorochytrium* sp. (see Parke & Dixon, 1976).

Cruoria rosea (Crouan frat.) Crouan frat. (1867), p. 147.

Holotype: CO. France (Brest).

Contarinea rosea Crouan frat. (1858), p. 72.

Crusts pale rose, 10 mm or more in extent, very thin, about 100 μm thick, mucilaginous, rhizoids absent, prostrate filaments with cells 1–5 times longer than broad, giving rise to short erect filaments slightly larger at the base than at the apex, 5–10 μm in diameter, simple or branched, composed of 3–4 (8) cells of which the lowest and uppermost are as long as broad and those in the centre twice as long, apices obtuse, cell fusions absent, cylindrical or clavate secretory cells interspersed among the erect filaments, about 9 (7–13) μm in diameter and up to at least 45 μm long, with highly refractive contents.

Gametangia unknown; tetrasporangia elliptical, lateral on the lowermost cell of the erect filaments, 22–44 × 9–15·5 μm, tetraspores zonately arranged.

On pottery, glass and *Solen* shells, in the sublittoral to a depth of 30 m.

Devon, Mayo, Bute and Shetland.

British Isles and northern France; Canada (Labrador and Newfoundland).

Vegetative crusts probably perennial, tetrasporangia recorded for November–April.

Data on form variation insufficient for comment.

The secretory cells which distinguish this species from other crusts such as *Rhododiscus pulcherrimus* Crouan frat., *Cruoriopsis* spp. etc., remained unrecognized until comparatively recently, probably being mistaken for undivided sporangia. The Crouans' original description refers to both elliptical and claviform sporangia: the latter adjective is more likely to apply to the secretory cells which are in fact visible in the type material.

Denizot in Boillot (1965) drew attention to the similarity of the tetrasporophyte of *Halarachnion ligulatum* (Woodw.) Kütz. to *C. rosea*. Later, South *et al.* (1972) made a similar comparison in the case of *Turnerella pennyi* (Harv.) Schmitz.

A specimen dredged from the Yealm estuary, Devon, in January 1896 by Brebner was described by Batters (1896) as *Cruoria rosea* var. *purpurea*. Batters considered it differed by having larger cells in the basal layer, a much thicker crust (200–300 μm) and tetrasporangia borne towards the apices of the erect filaments. Brebner (1896) thought that his specimen appeared to provide characters intermediate between *C. rosea* and *C. purpurea*

Crouan frat. (=*C. cruoriaeformis* (Crouan frat.) Denizot), whilst Batters thought *C. cruoriaeformis* might be identical to his *C. rosea* var. *purpurea.* A re-examination of Batters's type material (BM Slide 7958) has revealed the presence of secretory cells whilst the erect filament width and the tetrasporangium size also indicate its affinities with *C. rosea.* The status of the taxon remains uncertain, however, especially in view of the apparent involvement of these tetrasporangia-bearing crusts in the life-histories of other species.

Cruoria cruoriaeformis differs in lacking secretory cells and in having larger tetrasporangia (30–80×15–30 μm). It has been reported by Feldmann (Denizot, 1968, Cabioch, 1969) on the maerl (*Lithothamnium corallioides* Crouan frat. etc.) off the northern coast of France to a depth of 15 m. It has not so far been recorded for the British Isles but probably occurs on the maerl, particularly on southern and western coasts.

SOLIERIACEAE (Harvey) Hauck

SOLIERIACEAE (Harvey) Hauck (1885), p. 17.
Solieria J. Agardh (1842), p. 156.
Solieria chordalis (C. Agardh) J. Agardh (1842), p. 157.

This species has recently been found in the sublittoral in Cornwall (Falmouth Harbour) by Farnham & Jephson (see Parke & Dixon, 1976). A full description and illustration will be given in Part 2 of the present volume. In habit the species resembles *Gracilaria verrucosa* (Huds.) Papenf. and *Cystoclonium purpureum* (Huds.) Batt. From the former it differs by having obvious filaments in the centre of the medulla and zonately divided tetrasporangia. *C. purpureum* is similar to *S. chordalis* in these respects but is uniaxial with obvious obliquely dividing apical cells.

RHABDONIACEAE Kylin

RHABDONIACEAE Kylin (1925), p. 38.

Thallus erect or prostrate, terete or compressed, much branched; uniaxial, axial filament distinct, medulla usually obviously filamentous, cortex compact, inner large-celled, outer small-celled; auxiliary cells usually remote from carpogonia, not visible before fertilization, gonimoblast initially developing inwards or laterally, with a large fusion cell whose processes produce carposporangium-bearing filaments, enveloping filaments absent, carposporangia in rows or terminal, cystocarps immersed, protruding on one or both sides, usually with a pore; tetrasporangia scattered in outer cortex or in sori, zonate.

This family includes many genera but is represented in the British Isles by only one, *Catenella,* which is cosmopolitan. This genus is one of the few which is restricted to the upper littoral, occupying this niche in salt-marshes, mangrove swamps and shady crevices of the upper shore the world over. It is occasionally extensive enough to be used for food, e.g. in Burma (Børgesen, 1938).

CATENELLA Greville **nom. cons.**

CATENELLA Greville (1830), p. lxiii, 166.

Type species: *C. opuntia* (Goodenough & Woodward) Greville (1830), p. lxiii, 166 nom. illeg. (=*C. repens* (Lightfoot) Batters (1902), p. 69=*C. caespitosa* (Withering) L. Irvine).

Clavatula Stackhouse (1809), p. 95.

Thallus prostrate to erect, catenate, with both terete and compressed portions, irregularly branched, becoming hollow; structure uniaxial, with a 2 sided apical cell, medulla at first a group of interlacing filaments becoming lacunose later, cortex compact, inner large-celled, outer small-celled, in distinct radial rows.

Gametangial plants monoecious; spermatangia immersed in cortex of swollen segments; carpogonial branches 2–3 (–5) celled, intercalary cortical cells functioning as auxiliary cells, gonimoblasts in local thickenings of medulla, with an irregular fusion cell which produces carposporangium-bearing branch fascicles, carposporangia arranged in rows, enveloping filaments absent, cystocarps protruding, usually single in short lateral or terminal segments, with a distinct pore; tetrasporangia scattered in cortex of terminal segments, zonate.

Kylin (1928) commented on the similarity of the apical development and anatomy of the cystocarps in the genus *Cystoclonium*. The latter genus differs, however, in having a daughter cell of the supporting cell functioning as the auxiliary cell, and is therefore placed in the Rhodophyllidaceae.

One species in the British Isles:

Catenella caespitosa (Withering) L. Irvine in Parke & Dixon (1976), p. 590.

Holotype: OXF. Anglesey.

Ulva caespitosa Withering (1776), p. 735.
Fucus repens Lightfoot (1777), p. 961.
Catenella repens (Lightfoot) Batters (1902), p. 69.
Fucus opuntia Goodenough & Woodward (1797), p. 219, nom. illeg.
Catenella opuntia (Goodenough & Woodward) Greville (1830), p. lxiii, 166, nom. illeg.

Thallus composed of a discoid holdfast and a prostrate stolon-like portion which gives rise to more or less erect portions up to 20 mm in height, the whole much and irregularly branched and forming extensive mats with secondary attaching haptera; branches terete to compressed, constricted into irregular segments of variable diameter up to 2 mm, cartilaginous, almost opaque, dark brownish purple.

Structure uniaxial with a 2 sided apical cell; medulla a loose network of thick-walled filaments 11–16 μm in diameter; cortex compact, composed of 1–2 rows of tangentially elongated cells about 17×11 μm inwards and 2–3 rows of radially elongated cells outwards, 6–11 μm in diameter in surface view.

Gametangial plants monoecious; spermatangia about 6×3 μm intermixed with carposporophytes of all ages; grouped in pale sori reputedly with a withered and wrinkled appearance and a faint appearance of numerous lobes; gonimoblast developing inwards, with a large fusion cell supporting a mass of much-branched filaments whose cells become carposporangia; the whole forming a single cystocarp 200–360 μm in diameter within a

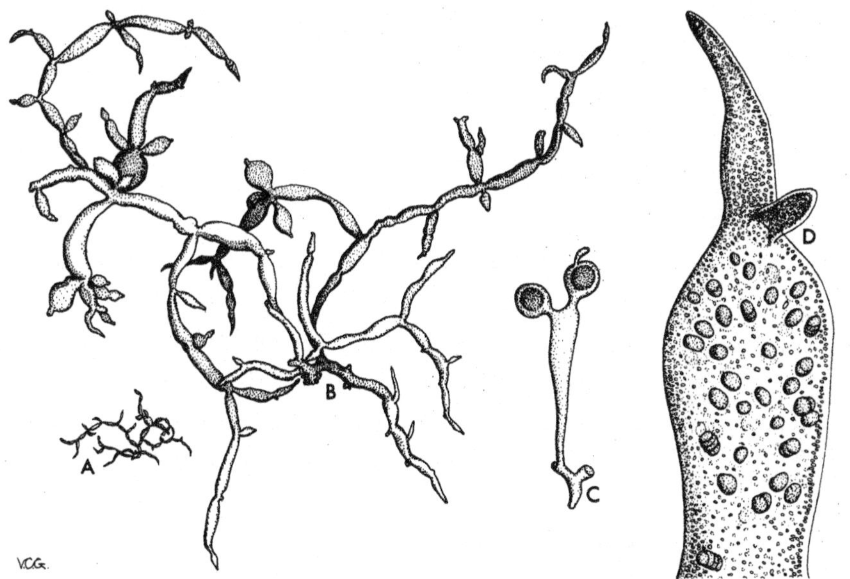

Fig. 67 *Catenella caespitosa*
A. Habit (Aug.) ×1; B. Same ×6; C. Branch with cystocarps (Aug.) ×8; D. Surface view of branch with tetrasporangia (Aug.) ×80.

swollen lateral branch; enveloping filaments absent, pore reputedly present or absent; carpospores 26–33×17·5–22 μm; tetrasporangia occurring in young segments, formed terminally on the cortical filaments amongst which they are scattered, 55–65×33–43 μm, tetraspores zonately arranged.

On rock and wood in the upper littoral in sheltered areas; tolerant of somewhat reduced salinity.

Generally distributed throughout the British Isles.

Norway (More) to Morocco; Mediterranean. Widely distributed in warmer parts of the Atlantic, Indian and Pacific Oceans.

Plants perennial, probably propagating vegetatively by re-attachment of prostrate fronds; spermatangia recorded by Buffham (1888) and Gibson (1892) for August and November, cystocarps extremely rare, recorded for June, August and November, tetrasporangia recorded for July–August.

Although the individual segments vary greatly in size and shape, the overall aspect of this moss-like creeping species remains fairly constant.

C. caespitosa is sometimes confused with *Gelidium pusillum* (Stackh.) Le Jol. In *G. pusillum* the apex is uniaxial but segmented plants are always completely flattened and never hollow. *Lomentaria articulata* (Huds.) Lyngb. is also similar but is usually reddish in colour, and more regular in the size and shape of its segments and branching pattern; it is frequently fertile, bearing cruciately arranged tetraspores in depressions in the cortex, or external cystocarps with a large prominent pore; its structure is multiaxial with the axial filaments clearly visible at the apex and forming a network lining the internal cavity.

RHODOPHYLLIDACEAE Engler

RHODOPHYLLIDACEAE Engler (1892), p. 19.

Thallus with a branched *(Calliblepharis, Cystoclonium)* or discoid *(Rhodophyllis)* holdfast and erect fronds, fronds terete, compressed or flattened, dichotomously or laterally branched, irregularly lobed; uniaxial but not showing an axial filament, medulla composed of narrow elongate filaments sometimes surrounded by a pseudoparenchymatous region, cortex compact, with large cells inwards and small cells outwards, occasionally of only one or two cell-layers; a daughter cell of supporting cell functioning as auxiliary cell, visible or not visible before fertilization, gonimoblast usually initially developing inwards, either with a large fusion cell on which carposporangium-bearing filaments are borne, or gonimoblast developing secondarily from base of cystocarp with or without sterile cells between gonimoblast and cystocarp wall, enveloping filaments absent, carposporangia in rows, cystocarps immersed in thallus, protruding on one or both sides, pore usually absent, tetrasporangia scattered in outer cortex, zonate.

This family is represented in Britain by three genera, *Calliblepharis, Cystoclonium* and *Rhodophyllis*. *Calliblepharis* and *Cystoclonium* have an extensive 2-layered medulla but in *Rhodophyllis* it consists only of a few filaments. *Calliblepharis* differs from the other two genera in lacking a fusion cell in the cystocarp.

CALLIBLEPHARIS Kützing **nom. cons.**

CALLIBLEPHARIS Kützing (1843), p. 403.

Type species: *C. ciliata* (Hudson) Kützing (1843), p. 404.

Ciliaria Stackhouse (1809), p. 54, 70.

Thallus with a branched holdfast and erect fronds, fronds flattened, dichotomous or pinnately branched, beset with simple or branched narrow proliferations on margins and also sometimes on blade surface; structure uniaxial, with a 2 or 3 sided apical cell, axial filament not visible, medulla with a few filaments, surrounded by a pseudoparenchymatous region consisting of several layers of large cells, cortex compact, small-celled, with inner cells larger than outer.

Gametangial plants dioecious; spermatangia superficial; carpogonial branches in cortex, 3 celled, a daughter cell of supporting cell functioning as auxiliary cell; gonimoblast developing inwards from a small-celled tissue, carposporangia in rows, enveloping filaments absent; cystocarps protruding strongly, pericarp formed by cortical thickening, reputedly without a pore; tetrasporangia either scattered over frond or confined to proliferations, zonate.

The two British species of *Calliblepharis* are sometimes difficult to distinguish. Young plants of *C. ciliata* are found in spring and they mature the following winter whereas young plants of *C. jubata* are found in autumn and they mature the following summer, with old plants persisting in the drift in both cases. Plants of *C. ciliata* can be distinguished from young plants of *C. jubata* in which the proliferations are still short since the latter plants have narrow lanceolate blades. It has been suggested (Newton, 1931) that the tetrasporangia are restricted to the proliferations in *C. jubata*. Tetrasporangia are rare in the British Isles but in the few cases examined they usually occurred on the blade also and a similar situation has been observed by Ardré (1969) in Portugal.

Key to Species

1 Proliferations rarely more than 5 mm long, frond normally more than 10 mm broad; mature in winter, commonly fertile; plants mainly sublittoral . *C. ciliata*
Proliferations up to 30 mm long, frond normally less than 7 mm broad; mature in summer, rarely fertile; plants mainly in lower littoral *C. jubata*

Calliblepharis ciliata (Hudson) Kützing (1843), p. 404.

Lectotype: BM. An unlocalized, undated Hudson specimen accepted provisionally as of lectotype status. England.

Fucus ciliatus Hudson (1762), p. 472.

Thallus consisting of a branched holdfast giving rise to an erect frond with a terete stipe about 0·5–1 mm in diameter and 5–10 mm long which gradually expands into a simple, dichotomous or irregularly divided blade up to 70 mm broad 300 mm long and 350–650 μm thick, with cuneate or ovate lobes, apices acuminate, margins plane, fringed

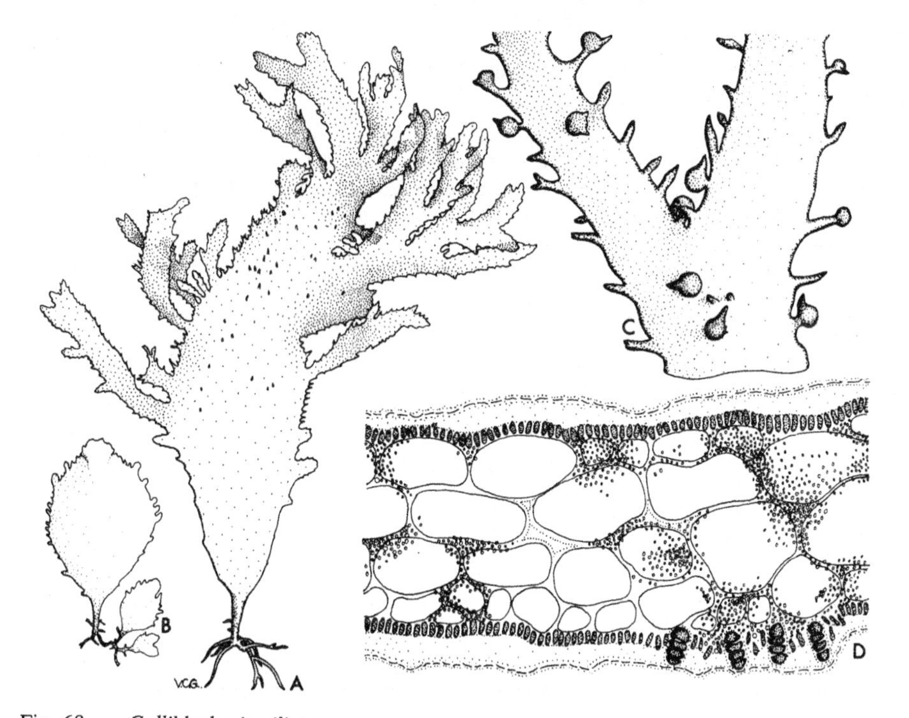

Fig. 68 *Calliblepharis ciliata*
A. Habit of mature plant (Aug.) ×⅔; B. Habit of young plant (Apr.) ×⅔; C. Part of blade with cystocarps (Oct.) ×4; D. T.S. blade with tetrasporangia (Jan.) ×80.

with short, simple or occasionally branched proliferations up to 5 mm long, proliferations also arising from blade surfaces at least in older plants; cartilaginous to coriaceous, dark red, drying blacker.

Structure uniaxial but axial filament not visible, inner medulla filamentous, outer medulla pseudoparenchymatous, cells up to 200 μm in transverse diameter and about twice as long, cortex compact, of 2 or 3 layers of cells about 7–15 μm in diameter in surface view.

Gametangial plants dioecious; spermatangia in large superficial sori on the younger parts of the blade; carpogonial branches developing in inner cortex along the margins of the proliferations, auxiliary cells producing a group of small cells from which the gonimoblast develops inwards, producing rows of carposporangia 25–35 μm in diameter; the carposporangial mass covered by a thick cortical pericarp and protruding as a lateral cystocarp 1–2 mm in diameter with the apex of the original proliferation appearing as a beak; although reputedly absent, an obscure pore can sometimes be seen, formed by disintegration of the pericarp cells; tetrasporangia grouped in large sori on the blade surface and margins, developing in the cortex, increasing in size up to 65–75×40–45 μm and raising the surface of the thallus in this region, with zonately arranged tetraspores.

Epilithic, occasionally epiphytic, known to a depth of 21 m, sometimes extending upwards into pools in the upper sublittoral.

Northwards to Argyll (Mull) and eastwards to Kent; other records from the east coasts of England and Scotland probably based on drift specimens. Reported occurrence in Orkney not confirmed. Generally distributed in Ireland.

British Isles to Mauretania; Mediterranean.

Plants annual; young broadly ovate specimens are seen in April, with growth occurring throughout the summer and autumn; old plants become detached and persist in the drift until the following summer; spermatangia recorded for September; cystocarps become visible about September and mature over the winter period; tetrasporangia recorded for October–April.

Plants vary considerably in blade width and degree of subdivision; in southeastern England plants with blades often only about 10–15 mm broad, but otherwise resembling this species, have been found. In old plants, especially in the drift, the proliferations in the apical regions may elongate and divide further.

Calliblepharis jubata (Goodenough & Woodward) Kützing (1843), p. 404.

Lectotype: BM-K. An undated specimen from Ilfracombe accepted provisionally as of lectotype status. Devonshire (Ilfracombe).

Fucus jubatus Goodenough & Woodward (1797), p. 162.
Fucus lanceolatus Withering (1976), p. 104, nom. superfl., pro parte, quoad descr., non *F. ligulatus* S. G. Gmelin.
Calliblepharis lanceolata Batters (1902), p. 71 pro parte, non *Fucus lanceolatus* Withering quoad typus; Newton (1931), p. 433.

Thallus consisting of a branched holdfast giving rise to an erect frond with a terete stipe 0·5–1 mm in diameter and 5–50 mm long, gradually expanding into a simple, dichotomous or irregularly divided blade up to 6 (15) mm broad and 300 mm long, with lanceolate lobes, apices acute; margins plane, furnished with long often branched proliferations up to 30 mm, which also arise from the blade surfaces in older luxuriant plants, sometimes hooked, especially those arising from the apex; cartilaginous, somewhat flaccid, brownish red, drying darker.

Structure uniaxial but axial filament not visible, medulla pseudoparenchymatous, cells not uniform in size, 50 μm to over 100 μm in transverse diameter, with a few cells of smaller diameter in the centre, all longitudinally elongated up to five times their diameter; cortex compact, of 2 or 3 layers of cells about 7–15 μm in surface view.

Gametangial plants probably dioecious; spermatangia recorded for the British Isles but not described; carpogonial branches arising near the apices of the proliferations, auxiliary cells producing a group of small cells from which the gonimoblast develops inwards, producing rows of carposporangia 30–50 μm in diameter; the carposporangial mass covered by a thick cortical pericarp and protruding as a lateral cystocarp 1–2 mm in diameter; although reputedly absent, an obscure pore formed by disintegration of the pericarp cells can sometimes be seen; tetrasporangia developing in the cortex, 50–55×35–45 μm with zonately arranged tetraspores.

Epilithic and epiphytic in open pools in the lower littoral in both sheltered and

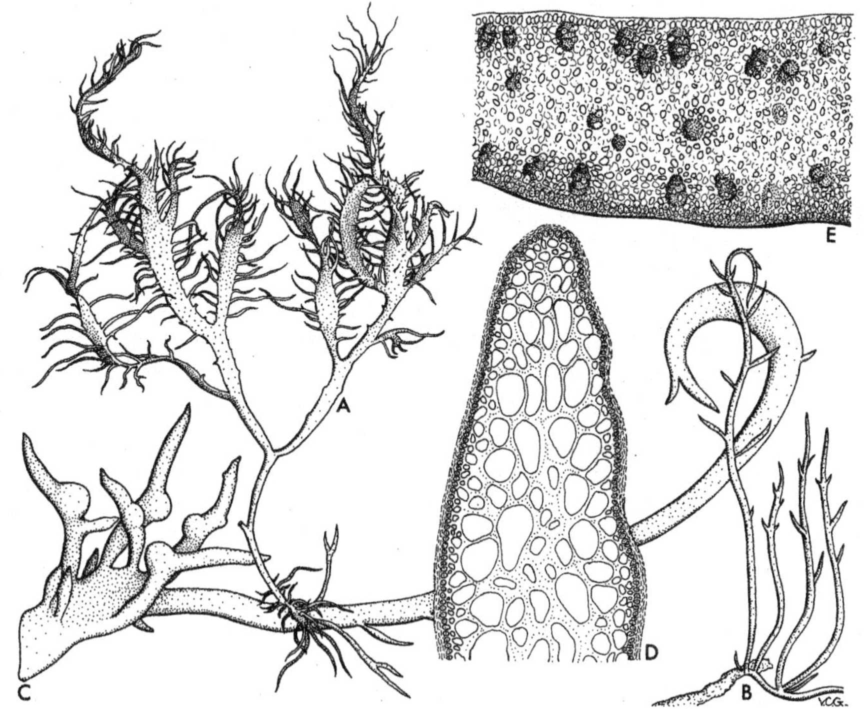

Fig. 69 *Calliblepharis jubata*
A. Habit of mature plant (June) ×1; B. Habit of young plant (Sep.) ×1; C. Branch with cystocarps and hooked apex (Aug.) ×8; D. T.S. old frond (Sep.) ×80; E. Surface view of proliferation with tetrasporangia (Aug.) ×80.

moderately exposed localities, probably not extending far into the sublittoral; tolerant of some sand cover; older plants often entangled with fronds of the same or other species.

Generally distributed on southern and western shores, northwards to Argyll (Mull) and eastwards to Isle of Wight. The report for Orkney (Traill, 1890) is based on a specimen which appears to be *Cystoclonium purpureum* (Huds.) Batt. Generally distributed in Ireland.

British Isles to Mauretania; Mediterranean.

Plants annual, young narrowly lanceolate specimens with short proliferations appear in August and September; as the season advances the blades become broader and the proliferations lengthen; old plants become detached and persist in the drift until the following winter; spermatangia recorded for June–July in the Isle of Man, possibly also occurring earlier, cystocarps becoming visible about April and maturing over the summer period; tetrasporangia rarely recorded, known for April, June and July, but no further information is available.

As in *C. ciliata,* plants vary considerably in blade width and degree of subdivision; individuals in which the blade remains narrow and quite unlike the normally accepted appearance for the genus frequently occur. In Cornwall and the Scilly Isles specimens are encountered having particularly broad fronds (up to 15 mm).

Specimens with very narrow blades can be confused with other species such as *Gigartina acicularis* (Roth) Lamour., whilst very young plants may resemble *Gelidium* spp. The cortical cells of *C. jubata* are much larger, however, (7–15 μm).

CYSTOCLONIUM Kützing

CYSTOCLONIUM Kützing (1843), p. 404.

Type species: *C. purpurascens* (Hudson) Kützing (1843), p. 404 (=*C. purpureum* (Hudson) Batters (1902), p. 68).

Thallus with a branched holdfast giving rise to erect fronds, terete, richly laterally branched; structure uniaxial with a 2 sided, obliquely dividing apical cell, axial filament visible only at apex, medulla pseudoparenchymatous and filamentous, loosely interspersed with narrow internal rhizoids, cortex compact, with large cells inwards, small cells outwards.

Gametangial plants dioecious; spermatangia in superficial sori; carpogonial branches 3 celled, auxiliary cells visible before fertilization, gonimoblast with a large irregular fusion cell, developing inwards, and lying in medulla, most cells becoming carposporangia, enveloping filaments absent, cystocarps developing in younger branches, overlying cortex becoming a thick pericarp without a pore; tetrasporangia scattered, developing in cortex of younger branches, zonate.

One species in the British Isles:

Cystoclonium purpureum (Hudson) Batters (1902), p. 68.

Lectotype: BM. An unlocalized, undated Hudson specimen accepted provisionally as of lectotype status. England.

Fucus purpureus Hudson (1762), p. 471.
Fucus purpurascens Hudson (1778), p. 589, nom. superfl.

Thallus with a branched holdfast giving rise to an erect frond, terete throughout, up to 2 mm in diameter, soft but not mucilaginous or lubricous; up to 600 mm in length, dull purplish or brownish red, repeatedly alternately branched, often becoming very bushy, branches attenuate at apices, the youngest ones quite slender (about 100–200 μm in diameter).

Structure uniaxial with an obliquely dividing apical cell, axial filament visible only in the youngest parts; inner medulla obviously filamentous, the filaments 15–25 μm in diameter, thick-walled (to 7 μm), accompanied by internal rhizoids, outer medulla pseudoparenchymatous; inner cortical cells radially elongated, 44–60×22–33 μm, outer cortex of two or three layers of cells about 7–15 μm in surface view.

Gametangial plants dioecious; spermatangia occurring in large coalescent sori on the surface of the younger parts, 4–6 mother cells developing superficially from each cortical cell each giving rise to 2 or 3 spermatangia, spermatia about 6×3 μm; gonimoblast developing inwards, producing a large irregular fusion cell from which arise fascicles of filaments the cells of which all produce carposporangia about 55–65 μm in diameter, enveloping filaments absent, cortex thickening to form a pericarp without a pore, cystocarps 500 μm in diameter, often occurring as a series of swellings along a branch; tetrasporangia scattered over the younger parts of the frond, immersed in the cortex, about 80×40 μm (up to 110×65 μm), tetraspores zonately arranged.

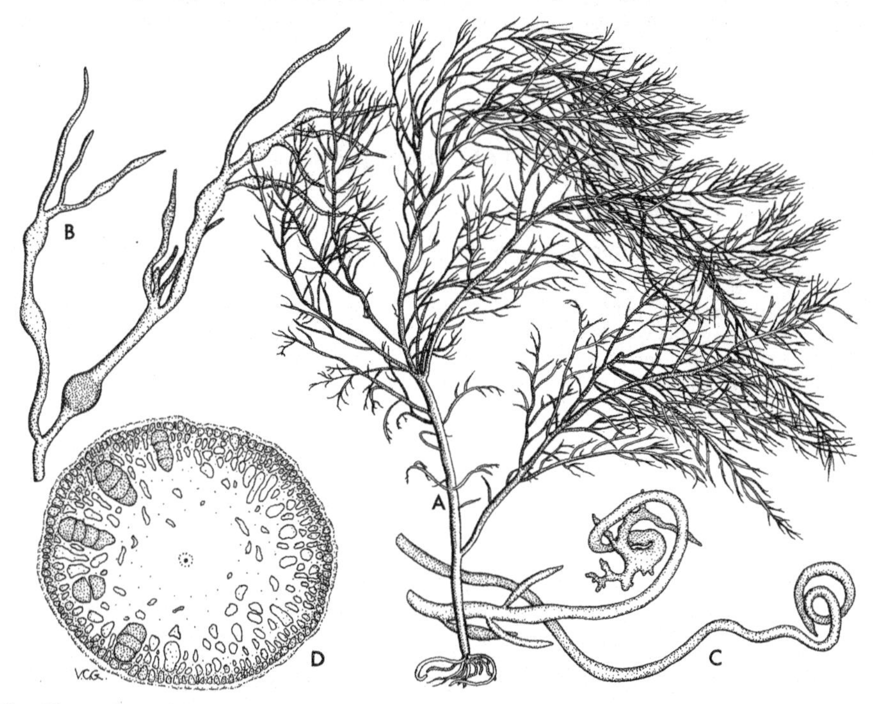

Fig. 70 *Cystoclonium purpureum*
A. Habit (June) ×$\frac{2}{3}$; B. Branch with cystocarps (July) ×4; C. Branches with tendrils (Nov.) ×4; D. T.S. branch with tetrasporangia (July) ×80.

Epilithic in both open and shaded situations from the midlittoral to a depth of at least 24 m.

Generally distributed throughout the British Isles.

Arctic Ocean (50°E) to Atlantic France (Vannes); Baltic; Greenland; Arctic Canada to U.S.A. (New Jersey).

Most plants are probably annual, though some may regenerate from the holdfast. Young plants appear in autumn; they remain small until the following spring and become fertile in the summer, old plants persisting in the drift throughout the winter, spermatangia recorded for August, cystocarps for June–September and tetrasporangia for July–August.

Although specimens are usually much branched and bushy, little branched, straggly plants are sometimes found, and drifting plants lose most of their branchlets. Mature plants sometimes develop spirally twisted branchlets which behave as tendrils.

This species can be difficult to distinguish from *Dumontia incrassata* (O. F. Müller) Lamour. when young. The apical cell of *C. purpureum* divides very obliquely whilst that of *D. incrassata* divides horizontally or nearly so. *C. purpureum* superficially resembles *Polysiphonia elongata* (Huds.) Spreng., which differs in having four primary pericentral cells arranged in the manner typical of the genus, and is also occasionally confused with *Gracilaria verrucosa* (Huds.) Papenf., q.v.

Galls, probably of bacterial origin, are quite frequent in this species, (Chemin, 1927). Kylin applied the name *Choreocolax cystoclonii* to the galls which he thought constituted a separate parasitic organism.

RHODOPHYLLIS Kützing, **nom. cons.**

RHODOPHYLLIS Kützing (1847), p. 23.

Type species: *R. bifida* (Lamouroux) Kützing (1847), p. 23 (=*R. divaricata* (Stackhouse) Papenfuss (1950) p. 190).

Bifida Stackhouse (1809), p. 95, 97.
Inochorion Kützing (1843) p. 443.
Wigghia Harvey (1846), pl. 32.
Stictophyllum Kützing (1847) p. 1.
Leptophyllium Nägeli (1847), p. 236.

Thallus erect, compressed or flattened, usually richly dichotomously or laterally branched or irregularly lobed; apical cells 2 sided, often grouped together giving the impression of a marginal meristem, medulla obviously filamentous, little developed, sometimes appearing as indistinct veins, cortex thin, formed of one or two cell-layers.

Gametangial plants monoecious; spermatangia superficial, carpogonial branches 3 celled, auxiliary cells visible before fertilization, gonimoblast forming either a large fusion cell *(R. divaricata)* or a small-celled sterile tissue before producing filaments whose cells become carposporangia arranged in rows; cystocarps scattered over frond or sometimes restricted to margins or marginal outgrowths, cortex thickening to form a pericarp with or without a pore; tetrasporangia scattered over younger parts of thallus, immersed in cortex, zonate.

One species in the British Isles:

Rhodophyllis divaricata (Stackhouse) Papenfuss (1950), p. 190.

Lectotype: original description (Hudson, 1778) in the absence of material (see Dixon & Irvine, 1977). Hampshire.

Bifida divaricata Stackhouse (1809), p. 97.
Fucus bifidus Hudson (1778), p. 581, nom. illeg., non *F. bifidus* S. G. Gmelin (1768), p. 201.
Rhodophyllis bifida Kützing (1847), p. 23.
Rhodophyllis appendiculata J. Agardh (1852) p. 389.

Thallus consisting of a very small attachment disc giving rise to one or a cluster of erect or somewhat prostrate fronds of exceedingly variable shape with or without a narrow stipe-like portion below, fronds adhering to and sometimes fusing with the same or occasionally other species, usually by rhizoids produced secondarily from the frond surface; blade either dichotomously branched at a wide angle, the divisions narrow (*c.* 1–3 mm) and parallel-sided, or cuneate and more irregularly divided, sometimes broad and fanlike, frequently proliferous from the margins, especially in older plants; up to 70 mm long and 100 mm in overall breadth; young plants thin and delicate, 35–70 μm thick, often conspicuously rose-red in colour, but older plants much darker and more cartilaginous.

Structure uniaxial but axial filament not visible, apical cells often crowded together, dividing obliquely; medulla little developed, a sparse network of filaments 7–15 μm in

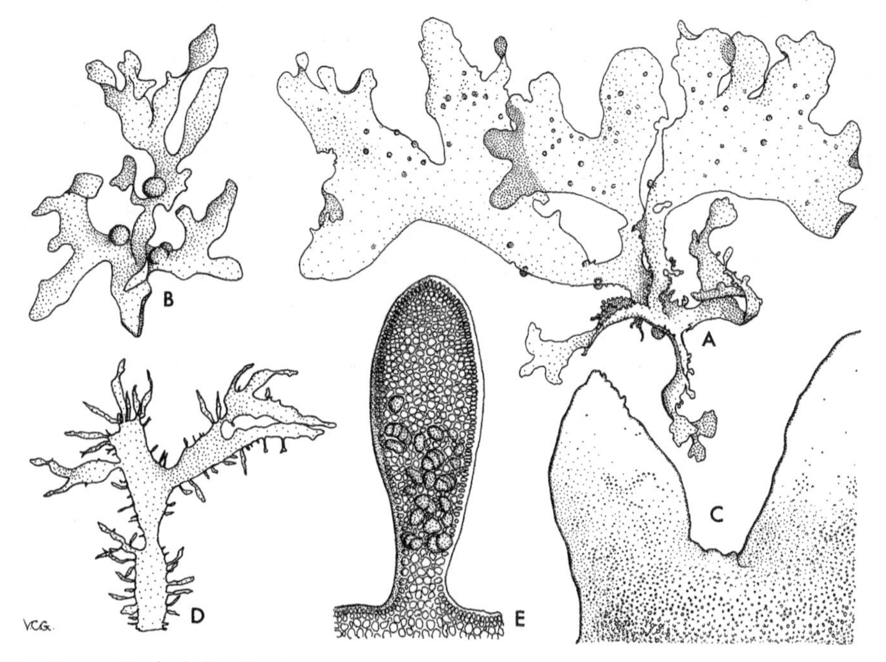

Fig. 71 *Rhodophyllis divaricata*
A. Habit (Apr.) ×2; B. Part of plant with cystocarps (Aug.) ×6; C. Part of apical region of blade with tetrasporangia (July) ×12; D. Part of regenerating plant with proliferations (Jan.) ×2; E. Surface view of proliferation with tetrasporangia (Jan.) ×80.

diameter; cortex of a single layer of isodiametric cells usually about 15–20 μm in diameter but occasionally up to 30 μm or more except at the margins.

Gametangial plants monoecious; spermatangia in superficial sori, each cortical cell producing 4 widely separated mother cells from which 1–2 spermatangia arise; gonimoblast with a large fusion cell and filaments giving rise to rows of carposporangia, the carposporangial mass surrounded by a thick cortical pericarp without a pore and protruding to form almost spherical cystocarps 240–400 μm scattered over the frond or restricted to the margins, carpospores 26–33 × 15–26·5 μm; tetrasporangia occurring in the younger portions of the blade or proliferations, 35–45 × 23–35 μm, with zonately arranged tetraspores noticeably flattened before discharge.

Epilithic and epiphytic in the sublittoral to at least 30 m, occasionally in lower littoral pools.

Generally distributed throughout the British Isles, although records for eastern coasts of England and Scotland are comparatively few.

Norway (Trondelag) to Morocco; Mediterranean.

Most plants annual, some persisting over the winter, new growth forming marginal processes which may bear tetrasporangia. No information available on occurrence of spermatangia; cystocarps recorded from April–November with a peak in July and August; tetrasporangia found between March–December, sometimes occurring in plants less than 10 mm in length.

This species is extremely variable in appearance, almost defying description. Additionally, the shape is obscured by the fronds frequently occurring in tangled clusters. The causes of this variation are not known.

Rhodophyllis appendiculata was distinguished from *R. divaricata* by having the tetrasporangia restricted to marginal proliferations. There appear to be no other distinguishing features. *R. divaricata* frequently shows similar proliferations especially in a regenerating specimen. Tetrasporangia are apparently produced only in young, actively growing regions; they would therefore be restricted to the proliferations in such a case. The type specimen of *R. appendiculata* (Herb. Alg. Agardh. no. 27253) is a regenerating specimen of *R. divaricata*.

The typical form of this species, with the frond divided at a wide angle and with narrow parallel subdivisions, is comparatively easy to recognize. In other cases, however, plants are often mistakenly referred to the Delesseriaceae. This is the only species in the British Isles outside the Delesseriaceae which sometimes displays cortical cells of a size large enough to be confused with those in members of this family. The species can be recognized, however, by the presence of one or both of the following features: (a) zonate tetrasporangia; (b) a network of internal filaments visible after treatment with a stain such as iodine in potassium iodide. In the Delesseriaceae the veins are superficial and the apices and margins appear much more regular in outline because of the occurrence of intercalary cell division.

PLOCAMIACEAE Kützing

PLOCAMIACEAE Kützing (1843), p. 442 [as Plocamieae].

Thallus with erect fronds, bilaterally compressed, richly laterally branched with sympodial organization, repeatedly pinnate with alternating groups of 2–5 branchlets; apical cell distinct, axial filament sometimes visible throughout; structure pseudoparenchymatous with large cells inwards and small cells outwards; carpogonial branch 3 celled, supporting cell functioning as auxiliary cell, gonimoblast developing outwards, most cells becoming carposporangia, a few inner cells enlarging and remaining sterile, gonimoblast composed of several gonimolobes, enveloping filaments absent, cystocarps either scattered along margin of frond, or borne in special short fertile branches and then shortly stalked, without a special pore, tetrasporangia immersed in cortex of special fertile branches, zonate.

Plocamium was for a long time placed in the Rhodymeniales. It was removed to the Gigartinales by Bliding (1928) because its structure is uniaxial, the auxiliary cell is the supporting cell of the carpogonial branch (not a daughter cell) and the tetrasporangia are zonate.

PLOCAMIUM Lamouroux **nom. cons.**

PLOCAMIUM Lamouroux (1813), p. 137 (reprint p. 49).

Type species: *P. vulgare* Lamouroux (1813), p. 137 (reprint p. 50) (=*P. cartilagineum* (Linnaeus) Dixon (1967), p. 58).

Neridea Stackhouse (1809), p. 58, 86.
Thamnophora C. Agardh (1822), p. 225.
Thamnocarpus Kützing (1843), p. 450.

Thallus bilaterally compressed, sympodially branched, branches repeatedly pinnate with alternating groups of 2–5 branchlets; structure uniaxial, apical cell distinct, axial filaments sometimes visible, medulla and cortex pseudoparenchymatous throughout, compact, inner cells large, outer cells smaller.

Gametangial plants dioecious; spermatangia in superficial sori; carpogonial branches 3 celled, supporting cell acting as auxiliary cell, gonimoblast developing outwards, most cells becoming carposporangia, innermost cells sterile, gonimoblast fairly loose, composed of several lobes, cystocarps occurring either on margins or in special branches and then shortly stalked, without a pore; tetrasporangia immersed in cortex of special small branches, zonate.

The record of *Plocamium biserratum* Dickie from Swanage, September 1892 (Batters, 1892) requires reinvestigation. The specimens certainly show the serrated margins typical of the species, which is nowadays usually considered to be conspecific with *P. corallorhiza* J. Ag.

One species in the British Isles:

Plocamium cartilagineum (Linnaeus) Dixon (1967), p. 58.

Lectotype: L (910.184.14). Type locality doubtful (see Dixon, 1967).

Fucus cartilagineus Linnaeus (1753), p. 1161.
Fucus plocamium S. G. Gmelin (1768), p. 159.
Fucus coccineus Hudson (1778), p. 586.
Plocamium coccineum (Hudson) Lyngbye (1819), p. 39.
Plocamium vulgare Lamouroux (1813), p. 137 (reprint p. 50).

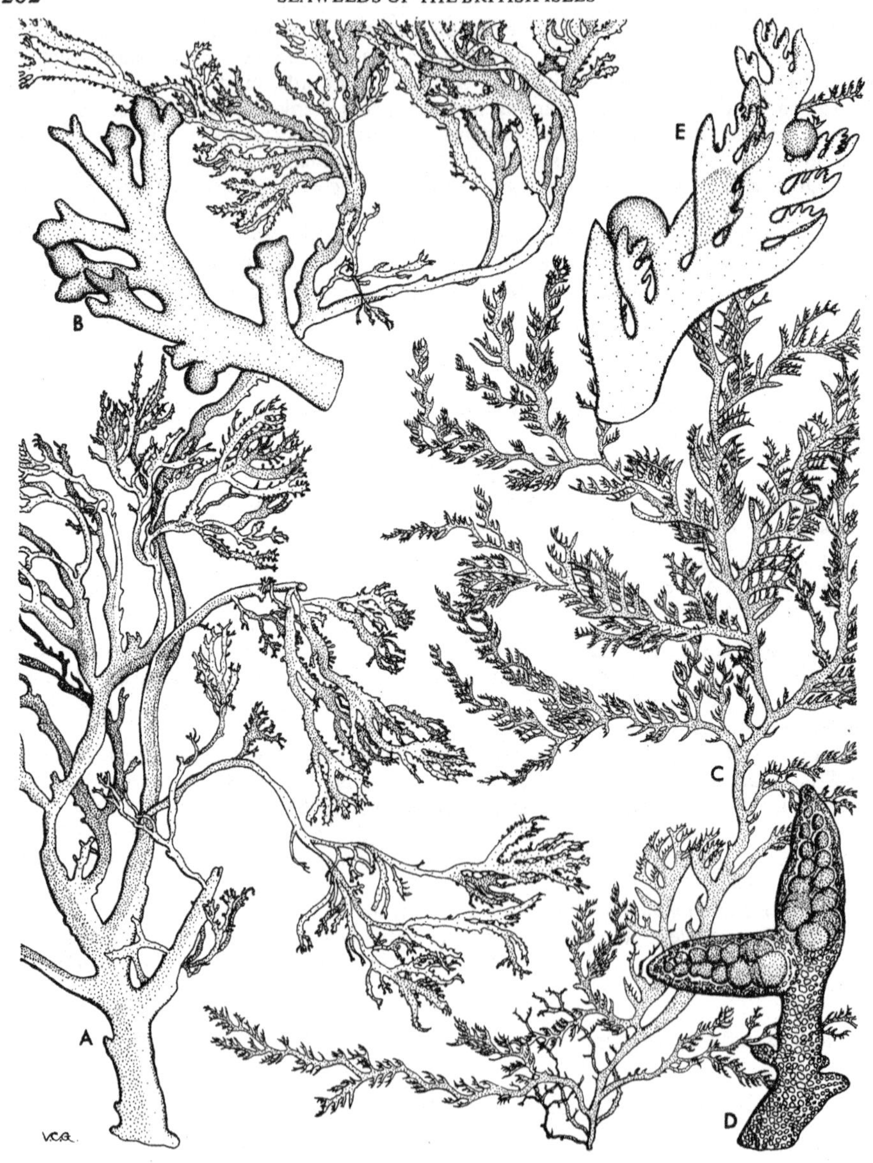

Fig. 72 *Sphaerococcus coronopifolius*
A. Habit (Sep.) ×1; B. Part of plant with cystocarps (Sep.) ×12.
Plocamium cartilagineum
C. Habit (Sep.) ×1; D. Part of plant with cystocarps (Sep.) ×12; E. Surface view of branchlet with tetrasporangial sori (Aug.) ×80.

Thallus consisting of a small attachment disc from which arises an erect or more or less prostrate flattened frond up to 150 mm in length, sometimes bushy and entangled, especially when prostrate, cartilaginous, somewhat translucent, rose red to brownish red, often with a blue sheen, bleaching to a vivid orange pink colour or paler, especially when drifting; repeatedly alternately sympodially branched from the base in a very characteristic manner and furnished with 2–5 small curved branches on the adaxial side of each alternation; lower branches up to 2 mm broad, upper branches narrower, apices markedly acute.

Structure uniaxial but axial filament not visible; medulla composed of thin-walled isodiametric or somewhat longitudinally elongated cells 55–130 µm in diameter surrounded by 1–2 layers of cortical cells 7–15 µm in diameter in surface view.

Gametangial plants dioecious; spermatangia in sori covering the surface of the youngest branchlets, about 4·5×3 µm; gonimoblast developing outwards, lobed, most cells becoming carposporangia, cortex forming a thick pericarp, cystocarps up to 1 mm in diameter, borne singly on the margins of the frond, scattered throughout the plant, without a pore, carposporangia 33–40×22–29 µm; tetrasporangia developing in succession within branchlets which become swollen and distorted, 33–83×22–66 µm, tetraspores zonately arranged, markedly flattened before discharge.

Epilithic and epiphytic on the stipes of *Laminaria hyperborea* (Gunn.) Fosl. and other algae in the sublittoral to at least 30 m.

Generally distributed throughout the British Isles.

Faeroes; Norway (Nordland) to Sénégal; Mediterranean; Canary Isles. Widespread, recorded for many temperate and warmer seas.

Plants potentially perennial, but many fronds lost annually, persisting in the drift; reproductive structures have been reported at different times of year for different locations, but at present there is insufficient evidence to indicate a pattern of seasonal occurrence throughout the British Isles.

There is much variation in the general appearance of plants: they may be compact, closely branched and comparatively broad, or narrower, straggling and with the branches widely separated. The distinctive branching pattern, with alternating secund groups of branches of limited growth, makes the species easy to identify, however.

Specimens in which the branchlets are reflexed rather than incurved have been distinguished as *P. coccineum* var. *uncinatum* (C. Ag.) J. Ag. Such specimens are not common in the British Isles but occur more frequently in the Mediterranean.

Sphaerococcus coronopifolius Stackh. and *Callophyllis cristata* (C. Ag.) Kütz. may bear a superficial resemblance, but they never show the characteristic branching pattern.

Mature plants are sometimes heavily invested with epiphytic coralline algae and Bryozoa.

SPHAEROCOCCACEAE Dumortier

SPHAEROCOCCACEAE Dumortier (1829), p. 76 emend. Searles (1968), p. 3.

Thallus with erect fronds, terete or compressed, dichotomous or laterally branched; structure uniaxial, axial filament indistinct and medulla pseudoparenchymatous, cortex compact with large cells inwards and small cells outwards; gametangial plants monoecious, carpogonial branches 3 celled, supporting cell acting as auxiliary cell, gonimoblast developing outwards, usually with sterile tissue and a large fusion cell, only last 1–3 cells of filaments forming carposporangia, cystocarps either scattered or in special fertile branches, protruding outwards, without a pore, tetrasporangia reputedly zonate.

SPHAEROCOCCUS Stackhouse

SPHAEROCOCCUS Stackhouse (1797), p. xxiv

Type species: *S. coronopifolius* Stackhouse (1797), p. xxiv.

Rhynchococcus Kützing (1843), p. 403.

Thallus erect, bilaterally flattened, dichotomous or laterally branched; structure uniaxial, axial filament distinct, surrounded by internal rhizoids, forming a medulla, cortex compact, with cells in radial rows, becoming smaller outwards.

Gametangial plants monoecious; spermatangia superficial, in depressions in cortex; carpogonial branches 3 celled, supporting cell acting as auxiliary cell, gonimoblast with a large fusion cell and a few sterile cells, enveloping filaments absent, carposporangia terminal, (occasionally two), cystocarps in short lateral branches, strongly emergent on one side, without a special pore; tetrasporangia reputedly scattered over thallus surface, zonate, not known in British material, reports from elsewhere doubtful.

One species in the British Isles:

Sphaerococcus coronopifolius Stackhouse (1797), p. xxiv.

Holotype: BM-SL. Cornwall.

Fucus coronopifolius Goodenough & Woodward (1797), p. 185 nom. illeg. non *F. coronopifolius* Zoega (1772), appendix p. 19.

Thallus consisting of a comparatively large attachment disc from which arises an erect flattened frond up to 150 mm in length; plants coarse below with a compressed stipe-like portion up to 4 mm in width, more slender above, cartilaginous, almost opaque, dark red, paler in the more delicate parts, much brighter when drifting; repeatedly irregularly alternately branched, sometimes partly secund, branches more or less distichous, often curving and fringed with small branchlets, apices acute.

Structure uniaxial, apical cell visible only in young branches, cells of axial filament producing a whorl of 4 laterals which develop unequally to give the distichous frond composed of medullary cells up to 100 μm in diameter and cortical filaments about 5 cells in length, the ultimate cells 6–9 μm in diameter in surface view.

Gametangial plants monoecious, spermatangia formed from a small group of elongated mother cells in depressions in the thallus surface; gonimoblast developing outwards from a large fusion cell and pushing out the overlying cortex, cystocarps about 350–400 μm in diameter, terminal cells becoming carposporangia, 21–51 × 15–30 μm, enveloping filaments and pore absent; cystocarps developing near apices of young branches and so appearing beaked; tetrasporangia unknown in the British Isles. (Fig. 72, p. 202.)

Epilithic in the sublittoral to 15 m.

Attached plants widely distributed on southern and western shores, northwards to Isle of Man and eastwards to Isle of Wight; Clare. Drift specimens have been found throughout the British Isles.

British Isles to the Canary Isles; Mediterranean, Black Sea.

The frequent occurrence of plants in the drift suggests that individuals do not persist after fruiting, though their age (one or more years) at maturity is not known. Spermatangia recorded for August and cystocarps for April–December with a peak in September–October.

The fringed appearance of the margins varies with the number of small branchlets present, but the overall aspect of the fronds remains fairly constant.

As pointed out by Newton (1931), a fishbone patterning of the thallus can be detected with a hand lens; this feature helps to distinguish the species from others with a similar aspect such as *Plocamium cartilagineum* (L.) Dixon and *Callophyllis cristata* (C. Ag.) Kütz.

Tetrasporangia possibly occur in an unrecognized encrusting phase.

GRACILARIACEAE (Nägeli) Kylin

GRACILARIACEAE (Nägeli) Kylin (1930) p. 54.
Gracilarieae Nägeli (1847), p. 240.

Thallus with erect fronds, terete, compressed or flattened, irregularly dichotomous or lobed or divided; apical organization obscure, medulla pseudoparenchymatous, without an obvious axial filament, cortex compact, inner large-celled, outer small-celled; carpogonial branch 2 or 3 celled, extending to exterior, borne laterally on a cell of outer cortex, daughter cell of supporting cell reputedly functioning as auxiliary cell, after fertilization carpogonium fusing with neighbouring cells to form a fusion cell sometimes accompanied by small sterile cells, gonimoblast developing outwards, enveloping filaments absent, outer layers of gonimoblast tissue becoming carposporangia usually arranged in distinct rows, cystocarps either scattered or along margins, protruding hemispherically, pericarp thick, with a pronounced pore; tetrasporangia in cortex either scattered or in sori, cruciate.

This is one of the few families in which the thallus structure is always pseudoparenchymatous throughout, the medullary cells being comparatively large and isodiametric. Eight genera were included by Kylin (1956); only one of these, *Gracilaria*, occurs in Britain. *Gracilaria* is of economic importance elsewhere but not in Britain.

GRACILARIA Greville **nom. cons.**

GRACILARIA Greville (1830), p. liv, 121.

Type species: *G. confervoides* (Linnaeus) Greville (1830), p. liv, 121 (=*G. verrucosa* (Hudson) Papenfuss (1950), p. 195).

Ceramianthemum Donati (1758), p. 24, 27.
Ceramion Adanson (1763), p. 13.
Plocaria C. G. Nees (1820), p. 42.

Thallus with erect fronds, terete, compressed or flattened, dichotomously or laterally branched; apical organization obscure, medulla pseudoparenchymatous, cells large inwards, smaller outwards; cortex of 2 or 3 layers of irregularly arranged small cells.

Gametangial plants dioecious; spermatangia usually in pits formed by invagination of cortex (apparently in superficial sori in *G. bursa-pastoris*), carpogonial branches 2 celled, supporting cell bearing one or more sterile cells or branches, one of these cells functioning as auxiliary cell, gonimoblasts developing outwards from a large lobed fusion cell, outermost cells forming radial rows of carposporangia, carposporangial mass protruding as a large hemispherical cystocarp, pericarp formed from much thickened cortex, with a prominent pore and often attached to gonimoblast by isolated filaments (trabeculae); tetrasporangia in cortex, scattered, cruciate.

As Taylor (1960) commented, the species in this genus are notoriously variable. The distinction between *G. verrucosa* and *G. bursa-pastoris* is not at all clear at present; the differences described by Sjöstedt (1926) need to be re-evaluated by means of population studies of British material, especially of *G. bursa-pastoris*. The characters he used are indicated by * in the key. Taylor added, 'This name (*compressa* = *bursa-pastoris*) has been applied to *verrucosa*-like specimens in which there seems to be a distinct flattening of the axis at the forks and perhaps a little elsewhere. It will be difficult to sharply distinguish between it, *G. verrucosa* and *G. foliifera* var. *angustissima*'.

The species *Cordylecladia erecta* (Grev.) J. Ag. has sometimes been placed in the genus *Gracilaria* (Kylin, 1956, Jones, 1962). It is more correctly placed in the Rhodymeniales, however (Feldmann, 1967). Species in which trabeculae are absent were segregated into the genus *Gracilariopsis* by Dawson (1949). The apical development of *Gracilaria* appears to be difficult to elucidate; Sjöstedt described it as multiaxial but Kylin (1930) followed Killian (1914) in believing that a 3 sided apical cell is present in the very early stages but that this is rapidly obscured.

Key to Species

1 Frond flattened, broadening above up to 10 mm between dichotomies . *G. foliifera*
  Frond terete or compressed, remaining narrow, not more than 5 mm broad . 2

2 Colour conspicuously reddish; compressed, brittle, (*cortex 1–2 layered, containing hair-producing cells 18–22 μm in surface view, hairs present in young and old plants; spermatangia in superficial sori) . . *G. bursa-pastoris*
  Colour dark purplish; not or scarcely compressed, very elastic, (*cortex 3 layered, containing hair-producing cells 10 μm in surface view, hairs deciduous, present only in young plants; spermatangia in pits) . . *G. verrucosa*

Gracilaria bursa-pastoris (S. G. Gmelin) Silva (1952), p. 265.

Lectotype: original illustration (Gmelin (1768), pl. VIII, fig.3), in the absence of specimens. Mediterranean.

Fucus bursa-pastoris S.G. Gmelin (1768), p. 121.
Sphaerococcus compressus C. Agardh (1822), p. 308.
Gracilaria compressa (C. Agardh) Greville (1830), p. 125.

Plants closely similar to those of *G. verrucosa* but somewhat stouter and more or less compressed throughout, noticeably redder in colour and very succulent and brittle.

Fig. 73 *Gracilaria bursa-pastoris*
A. Habit (Aug.) ×1; B. Branch with cystocarps (Aug.) ×4.
Gracilaria verrucosa
C. Habit (Aug.) ×1; D. Branch with cystocarps (July) ×4; E. T.S. with tetrasporangia (Apr.) ×80; F. T.S. with spermatangia (June) ×300.

Medulla consisting of large, thin-walled isodiametric cells 200–300 μm in diameter, surrounded by 1–2 layers of much smaller cortical cells 6–9 μm in diameter in surface view, interspersed in all parts of the plant with specialised hair-bearing cells, about 18–22 μm in diameter, hairs said to be more persistent than those of *G. verrucosa*.

Gametangial plants dioecious; spermatangia said not to be in pits, but to occur in small colourless sori on the surface of the plants; carposporophyte development and tetrasporangia as in *G. verrucosa*, cystocarps often said to be larger than those of *G. verrucosa*.

Epilithic in sheltered places in the upper sublittoral, often associated with sand deposition.

Apparently restricted to southern coasts between Cornwall and Sussex.

British Isles to Sénégal; Mediterranean; Cape Verde Isles. U.S.A. (Florida) to Brazil; India; Japan; probably widespread in warmer waters.

Presumably perennial; cystocarps August–October; tetrasporangia July–August.

Very variable in branching, degree of compression and thallus width.

Discrimination between *G. bursa-pastoris* and *G. verrucosa* is difficult especially in the case of dried specimens. Reports of *G. bursa-pastoris* should be accepted with caution.

'Under the impression that it was really the *Fucus lichenoides* which is eaten in the east, Mrs Griffiths . . . made a pickle and a preserve from fresh specimens, and in both cases it proved excellent.' (Greville, 1830).

This species is used commercially in Japan, India and Cuba, mixed with other species, both for food and as an agaroid raw material (Levring *et al.*, 1969).

Gracilaria foliifera (Forsskål) Børgesen (1932), p. 7.

Lectotype: C (see Børgesen, 1932 fig. 1). Arabia (Mocha).

Fucus foliifer Forsskål (1775), p. 191.
Fucus multipartitus Clemente (1807), p. 311.
Gracilaria multipartita (Clemente) Harvey (1846), pl. XV.

Thallus consisting of a small attachment disc giving rise to one or more erect fronds up to 250 mm long; translucent, cartilaginous and very brittle, dull purple or reddish-brown, sometimes bleached; stipe compressed, expanding gradually into a cuneate laciniate blade up to 1 mm or more thick and up to 10 mm broad between dichotomies; branching up to 6 times di-tri-chotomous at a very narrow angle (10–50°) and in the plane of the blade, frequently proliferous from the margins.

Medullary cells large and thin-walled, up to 200 μm in diameter, passing abruptly into a cortical layer 2–3 cells deep, the outermost being 6–9 μm in diameter in surface view.

Gametangial plants dioecious; spermatangia *c.* 3 μm in diameter, in pits *c.* 50 μm in diameter; cystocarps scattered over the blade and projecting strongly from the face and margins, to 2 mm in diameter; trabeculae usually present; carposporangia 35–45 × 20–35 μm in diameter, carpospores escaping through a prominent pore; tetrasporangia developing in the cortex of the younger parts of the blade, 30–45 × 20–35 μm, with cruciately arranged tetraspores.

Epilithic in situations varying widely in degree of exposure to wave action; recorded from the upper sublittoral to a depth of 15 m, tolerating sand and mud cover.

Cornwall, south Devon and Dorset. In view of the diverse habitats in which the species occurs in a particular geographical location, it is curious that its distribution in the British Isles is so restricted.

British Isles to Sénégal; Mediterranean; Canada (Nova Scotia) to Uruguay; Red Sea, Indian Ocean; probably widely distributed in warmer waters.

Probably perennial, though there is some evidence of dying back to the stipe and regeneration; spermatangia recorded for April and September, cystocarps for April–November and tetrasporangia for April and September.

Although there is some variation in the degree of branching and frond width, the appearance is quite characteristic in the British Isles.

A similar habit is shown by some forms of *Palmaria palmata* (L.) O. Kuntze. These differ in having a much thinner frond (80–120 μm) and in never bearing cystocarps, whilst the cortical cells are larger and more regular in surface view.

This species is commercially used as an agar or agaroid raw material in U.S.A. etc. (see Levring *et al.*, 1969).

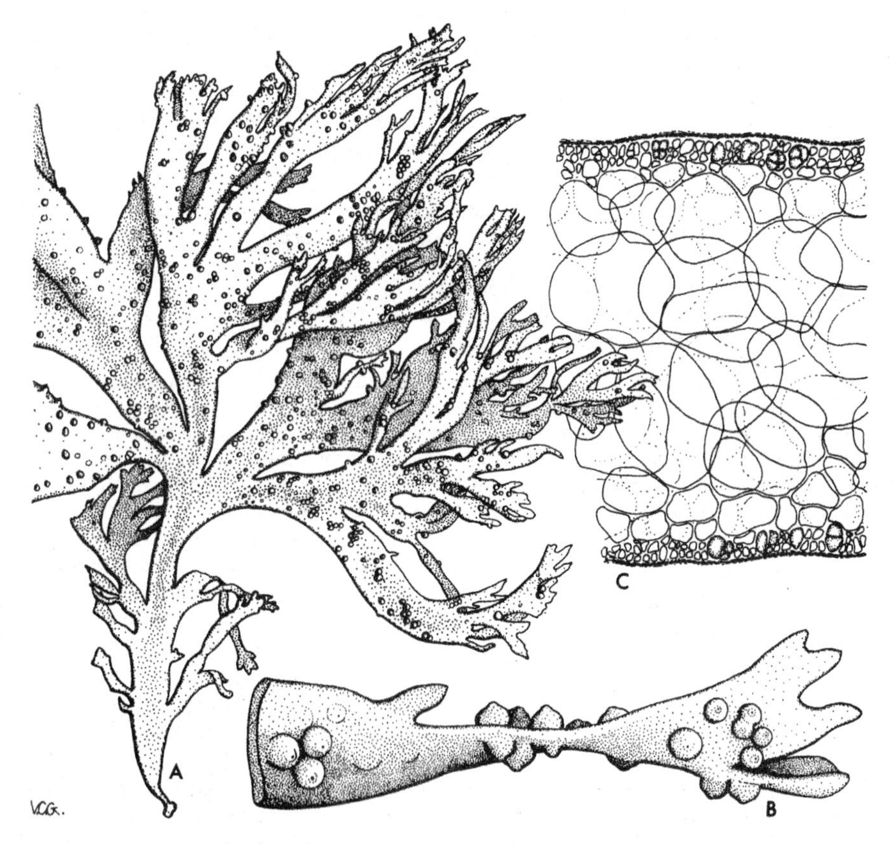

Fig. 74 *Gracilaria foliifera*
A. Habit of plant with cystocarps (Sep.) ×1; B. Branch of same ×4; C. T.S. branch with tetrasporangia (Sep.) ×80.

Gracilaria verrucosa (Hudson) Papenfuss (1950), p. 195.

Lectotype: Hudson's description, in the absence of material. England.

Fucus verrucosus Hudson (1762), p. 470.
Fucus confervoides Linnaeus (1763), p. 1629, non *F. confervoides* Hudson (1762), p. 474.
Gracilaria confervoides Greville (1830), p. 123, nom. illeg.

Thallus consisting of a discoid holdfast which produces one or more erect fronds; habit variable, branched and bushy or long and straggling; dark brownish purple to green, frequently bleached, translucent; up to 600 mm in length and 1–3 mm in diameter, terete throughout, tapering towards the apices, repeatedly subdichotomously or irregularly branched, sometimes with many secondary proliferations.

Medullary cells large and thin-walled to 250 (450) μm in transverse diameter, surrounded by 3 layers of cortical cells *c.* 9 μm in surface view, interspersed with specialized cells 10 μm in surface view, which produce deciduous hairs.

Gametangial plants dioecious; spermatangia *c.* 3 μm, occurring in pits to 50 μm in diameter in the surface of the younger parts of the frond; cystocarps hemispherical, about 0·5–1 mm in diameter, furnished with a conspicuous pore through which the successively ripening carpospores, 30–40 μm in diameter, escape; trabeculae usually present, tetrasporangia scattered, developing from the primary cortical cells, 30–33 × 22–30 μm, tetraspores cruciately arranged. (Fig. 73, p.207.)

Epilithic in the lower littoral and sublittoral in areas protected from wave action and recorded to a depth of 15 m; sometimes loose-lying, tolerant of sand cover and some variation in salinity.

Generally distributed throughout the British Isles.

Norway (Trondelag) to South Africa; Baltic; Mediterranean; Canada (Prince Edward Island) to Brazil; Indian and Pacific Oceans.

Perennial, maximum growth occurring in May and June, but retarded by high light intensity. In North Carolina it was found that growth occurred almost entirely between May and November, when the mean temperature was above 10°C. Reproductive structures occur throughout the year with maxima as follows: spermatangia in August, cystocarps in early September and tetrasporangia in July (Jones, 1958, 1959; Causey *et al.*, 1946).

There is considerable variation in the degree of branching and hence in the general appearance of plants, no doubt due to environmental factors; for example, damage resulting from sand movement or animal grazing induces regeneration and proliferation. Decay, particularly of cystocarpic plants, occurs after spore-shedding has ceased.

Spermatangial pits are formed from a single cortical cell. This gives rise to a system of branched filaments which line the pit and whose cells function as spermatangial mother cells (Yamamoto, 1973).

This species is sometimes confused with *Cordylecladia erecta* (Grev.) J. Ag., *Cystoclonium purpureum* (Huds.) Batt. and *Gigartina acicularis* (Wulf.) Lamour. *C. erecta* is bright red in colour and consists of several fronds arising from an expanded basal disc; the medullary cells are longitudinally elongated and there are also differences in the tetrasporangial sori and cystocarps. *C. purpureum* has a branched holdfast and usually more slender ultimate branches, each with a distinct apical cell; the tetrasporangia are zonate and the cystocarps embedded. *G. acicularis* is a more rigid plant with acuminate apices; the cortical cells are very small and the medulla is obviously filamentous.

Plants are occasionally found infected with the parasite *Holmsella pachyderma* (Reinsch) Sturch, q.v. Galls occur frequently and have been thought to be of bacterial origin (Chemin, 1931).

This is the most important species of *Gracilaria* used as raw material for agar (see Levring *et al.*, 1969); it is not commercially exploitable in Britain, however.

PHYLLOPHORACEAE Nägeli

PHYLLOPHORACEAE Nägeli (1847), p. 248.

Thallus erect, encrusting or parasitic, erect fronds terete, compressed or flattened; branching dichotomous, sometimes proliferous, structure multiaxial, medulla compact and pseudoparenchymatous, cortex compact with small cells in radial rows; supporting cell functioning as auxiliary cell, gonimoblast initially developing outwards, often with a network of sterile filaments amongst which lie carposporangia in irregular rows, enveloping filaments absent, cystocarps scattered over thallus, either embedded or protruding to a greater or lesser extent, pore present or absent; tetrasporangia in rows in superficial sori, cruciate.

In addition to five species of the type genus *Phyllophora,* which show a considerable diversity in apparent life history patterns, there are in the British Isles species of the genus *Gymnogongrus,* in which much interest has been aroused by the posthumous publication of Schotter's fascinating, but incomplete studies (1968), *Ceratocolax,* a supposed parasite, *Stenogramme,* unique in having a false midrib of cystocarps, and *Schottera* (=*Petroglossum* pro parte), finally unequivocally recorded for Britain (Guiry & Hollenberg, 1975) after some years of uncertainty. To these we append the imperfectly known encrusting form *Erythrodermis allenii,* distinguished from other crusts by having cruciate tetrasporangia in superficial rows, a feature known to occur only in this family. The sporangia are small and of a size which falls within the range usual in this family. Batters, in the original description, comments that 'the tetraspores are borne in true nemathecia, not unlike those of *Phyllophora membranifolia*' (=*P. pseudoceranoides*). *Ahnfeltia plicata* is also included although recently (Farnham & Fletcher, 1974) discovered to have an encrusting tetrasporangial phase. The tetrasporangia are not in rows and contain zonately arranged tetraspores, suggesting that the species may in fact be misplaced here. *Phyllophora* and *Ahnfeltia* are of economic importance elsewhere, but not in the British Isles.

Species of *Rhodymenia,* especially *R. pseudopalmata* (Lamour.) Silva, are frequently confused with members of this family. Plants found growing epiphytically, e.g. on *Laminaria* stipes, almost without exception belong to the genus *Rhodymenia.* In the case of epilithic plants, no field characters of general application have been found, however. In difficult cases such as sterile specimens the identification can be confirmed by microscopic examination of the surface cells of the blade. In the Phyllophoraceae the cells are closely compacted in an even layer of fairly uniform size in any one sample, and range in diameter from 4–7 μm. In *Rhodymenia* spp. the cells are more widely spaced in an irregular layer; they appear more rounded and vary in size in a single sample between 4–12 μm, usually in the higher end of this range.

AHNFELTIA Fries

AHNFELTIA Fries (1836), p. 310.

Type species: *Ahnfeltia plicata* (Hudson) Fries (1836), p. 310.

Sterrocolax Schmitz (1893), p. 394.

Thallus of gametangial phase with erect fronds, terete, irregularly dichotomously branched, of a horny consistency; structure multiaxial, cortex compact, composed of very small cells arranged in radial rows, giving way gradually to a medulla of somewhat broader cells in which rows are only slightly less prominent.

Carpogonial branches and auxiliary cells in shallow outgrowths, apparently abortive; terminal cells of outgrowth filaments developing into monosporangia.

Thallus of tetrasporangial phase crustose, at least in British species, and closely similar to plants described as *Porphyrodiscus simulans* Batt.; tetrasporangia in superficial sori, zonate.

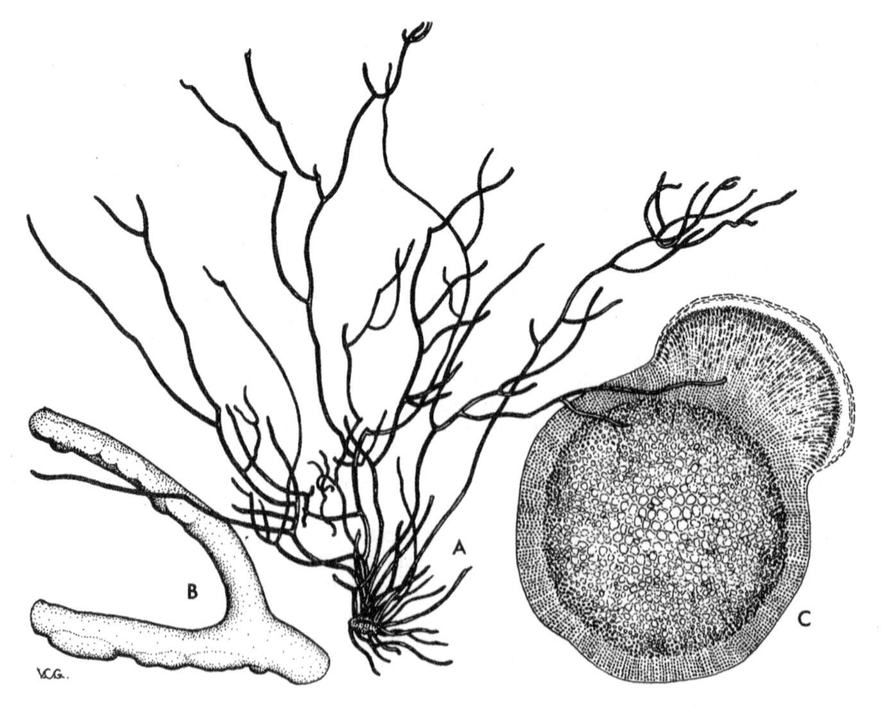

Fig. 75 *Ahnfeltia plicata*
A. Habit (Feb.) ×1; B. Branch with monosporangial outgrowth (Feb). ×8; C. T.S. same ×80.

One species in the British Isles:

Ahnfeltia plicata (Hudson) Fries (1836), p. 310.

Lectotype: BM. An unlocalized, undated Hudson specimen accepted as of provisional lectotype status. England.

Fucus plicatus Hudson (1762), p. 470.
Sterrocolax decipiens Schmitz (1893), p. 394.

Gametangial thallus consisting of a discoid holdfast, often quite extensive, to 10 mm, composed of cubical cells in vertical rows; producing erect fronds terete, solid and of horn-like consistency, almost black (bleached when buried in sand), growing in somewhat tangled, wiry clumps up to 150 mm in length and about 0·5 mm in diameter; branching variable, dichotomous, secund or completely irregular, divaricate, apices blunt.

Structure multiaxial, consisting internally of pseudoparenchymatous medullary cells up to 45 μm in diameter surrounded by small cortical cells arranged in compact radial rows, 3–4 μm in diameter at the surface.

Dioecious; spermatangia occurring in small pale individuals which do not bear outgrowths, apparently functionless; carpogonial branches in yellowish outgrowths borne terminally or laterally on the end-cells of the outgrowth filaments and apparently producing a secondary series of filaments the terminal cells of which produce monosporangia.

Tetrasporangial plants crustose, similar to the basal discs, tetrasporangia 18–25×5–9 μm, occurring in mucilaginous superficial sori, with zonately arranged tetraspores, sometimes apparently on the same crust as the erect fronds.

Erect fronds epilithic in the lower littoral in pools and emergent, sublittoral to 12 m; in shelter or with some exposure to wave action; tolerant of sand cover. For habitat of crustose phases see *Porphyrodiscus simulans* Batt.

Plants with erect fronds generally distributed throughout the British Isles.

Northern Russia (170°E) to Portugal; Baltic; Arctic Canada to U.S.A. (New Jersey); widely distributed in the Pacific and Indian Oceans. For the distribution of crustose phases see *Porphyrodiscus simulans*.

Erect plants perennial; older fronds show growth zones like annual rings, the boundary cells having thickened radial walls; the conditions giving rise to this effect are not known; gametangial outgrowths recorded between December and March (Gregory, 1934); encrusting phases probably perennial.

Although to some extent varying in habit and branching, this species is comparatively easy to distinguish by its tough, wiry consistency. It is usually almost black in colour, but becomes bleached after burial in sand.

For a comparison of the encrusting basal regions, which sometimes bear sori of tetrasporangia, with *Porphyrodiscus simulans* see Farnham & Fletcher (1974). Schmitz applied the name *Sterrocolax decipiens* to the gametangial outgrowths which he thought constituted a separate parasitic organism.

Gymnogongrus griffithsiae (Turn.) Martius, q.v., is occasionally confused with this species.

Galls, probably bacterial, are common in older plants.

This species is an important raw material for Russian agar.

CERATOCOLAX Rosenvinge

CERATOCOLAX Rosenvinge (1898), p. 34.

Type species: *C. hartzii* Rosenvinge (1898), p. 34.

Thallus inconspicuous, base endophytic (? parasitic), consisting of filaments ramifying between surface and medullary cells of host; erect fronds with short branches, pinkish, consistency softer than that of host; structure multiaxial, medulla pseudoparenchymatous, surrounded by a cortical layer of several cells.

Gametangial plants monoecious; spermatangia developing in pits in cortex; carpogonial branches at interface between cortex and medulla, 3 celled, supporting cell functioning as auxiliary cell, gonimoblast producing rows of cruciate carpotetrasporangia; tetrasporangial plants unknown.

One species in the British Isles:

Ceratocolax hartzii Rosenvinge (1898), p. 34.

Lectotype: C. Greenland (between Julianehaab and Huidenaes).

Thallus up to 5 mm in diameter, composed of a cluster of short, often densely and irregularly divided, terete branches up to 1 mm in diameter arising from a filamentous basal portion embedded in the tissues of *Phyllophora truncata,* the filaments ramifying between the surface and medullary cells of the host, forming pit connections with them; pinkish, not gelatinous but of a softer consistency than the *Phyllophora*; consisting of a medulla of rounded cells up to 65 μm and a cortex of rows of radially elongated cells 4–5 μm in surface view.

Gametangial plants monoecious; spermatangia in pits in the surface layer, spermatia released through pores in the outer wall, 3 μm in diameter; gonimoblast forming an outgrowth after fertilisation, containing rows of carpotetrasporangia about 9–26×8–16 μm, carpotetraspores cruciately arranged.

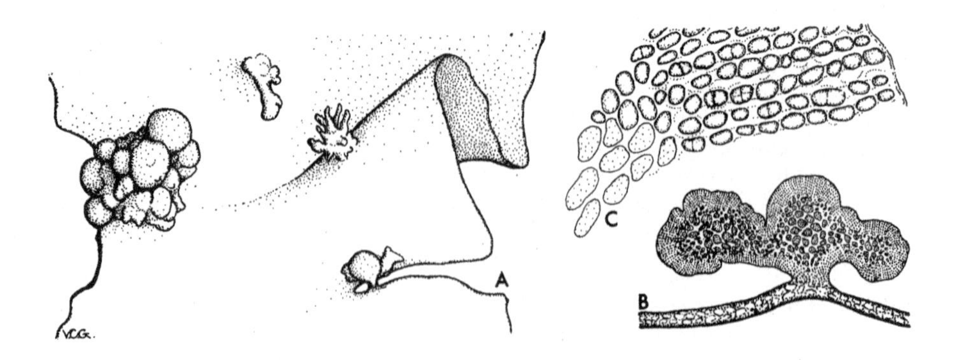

Fig. 76 *Ceratocolax hartzii*
A. Four plants on *Phyllophora truncata* (Newfoundland, Feb.) ×8; B. V.S. of carpotetrasporangial outgrowth (Apr.) ×16; C. Same with carpotetrasporangia ×300.

In the sublittoral to 10 m.

Clare; Shetland.

Circumboreal in the Arctic Ocean, southwards to the British Isles, Denmark and Baltic. Western Atlantic southwards to U.S.A. (Rhode Island). Distributed throughout distributional range of its host except England, Wales, mainland Scotland and Black Sea.

Little information is available for the British Isles; elsewhere the species has been recorded throughout the year and is apparently perennial; spermatangia recorded for March–April and August; carpogonia for March–August; and carpotetrasporangia (divided) for March–July.

The erect fronds are sometimes in the form of an irregular cushion but more often consist of branched finger-like processes.

The first report of the occurrence of *C. hartzii* in the British Isles, on *Phyllophora crispa* (Huds.) Dixon (Boalch, 1964), was found to be erroneous (Newroth & Taylor, 1968). Further records increasing the known distribution of the species in the British Isles are to be expected.

ERYTHRODERMIS Batters

ERYTHRODERMIS Batters (1900), p. 378.

Type species: *E. allenii* Batters (1900), p. 378.

Thallus encrusting, orbicular or indefinite in outline, adhering closely to substrate; composed of one or very few layers of filaments radiating in a fanlike manner from several points, cells polygonal.

Gametangia unknown; tetrasporangia in rows in short vertical filaments grouped in superficial sori, cruciate.

One species in the British Isles:

Erythrodermis allenii Batters (1900), p. 378.

Lectotype: BM (Slide 8542). Devon (Queen's Ground, Plymouth Sound).

Thallus a crust of 6–7mm in diameter and about 15 μm thick; consisting of 1–3 layers of prostrate filaments composed of cells 6–9 μm in diameter and 6–12 μm long.

Gametangia and carposporophytes unknown; tetrasporangial sori slightly elevated, composed of simple or branched erect filaments each consisting of a row of 4–6 tetrasporangia, 6–9 μm in diameter, with cruciately arranged tetraspores.

Originally discovered on broken earthenware, but later found also on stones and coralline algae; sublittoral, 7–11 m.

Known only from S. Devon. Reported occurrence in Cork not confirmed.

Data on seasonal growth and form variation insufficient for comment. Tetrasporangia occur in February and March.

Richardson & Dixon (1970) compared the tetrasporangial phase of *Thuretellopsis peggiana* Kylin (Cryptonemiales, Dumontiaceae) with *E. allenii*, noting, however, that plants of the former bear tetrasporangia singly rather than in rows. This latter comment is critical in view of our inclusion of *E. allenii* in the Phyllophoraceae on the basis of the occurrence of tetrasporangia in rows. Newly collected material of *E. allenii* has not been available for study but the type material has been compared with Richardson & Dixon's drawings and there appear to be differences with regard not only to the size, shape, division and disposition of the tetrasporangia but also in the arrangement of the cells in a surface view of the crust. Their material corresponds more closely with the type material of *Rhododiscus pulcherrimus* Crouan frat. (Cryptonemiales, Peyssonneliaceae), differing only in the thickness of the crust.

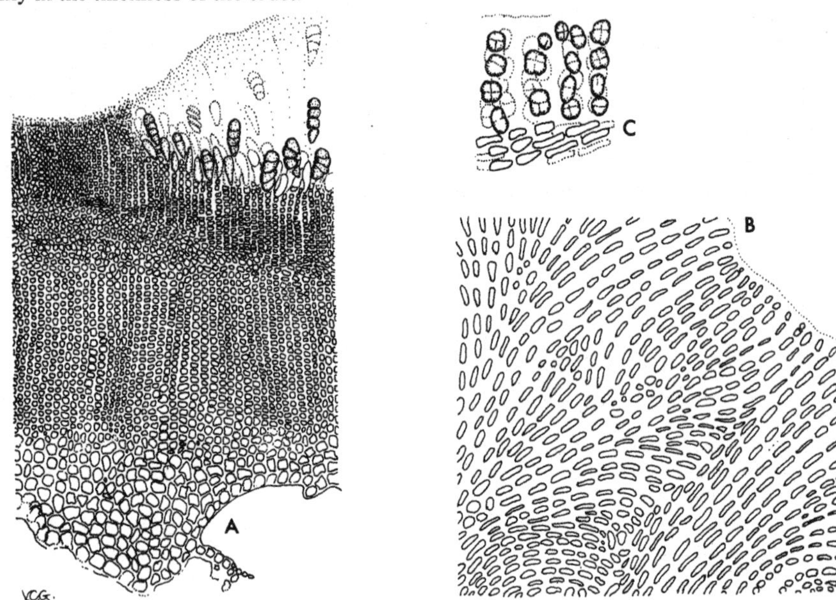

Fig. 77 *Porphyrodiscus simulans* (Type)
A. V.S. of crust with tetrasporangia (Feb.) ×300.
Erythrodermis allenii (Type)
B. Surface view of crust margin (Mar.) ×300; C. V.S. of crust with tetrasporangia (Mar.) ×300.

GYMNOGONGRUS Martius

GYMNOGONGRUS Martius (1828), p. 27.

Type species: *G. griffithsiae* (Turner) Martius (1828), p. 27 (see Dixon, 1964).

Tylocarpus Kützing (1843), p. 411.
Oncotylus Kützing (1843), p. 411.
Pachycarpus Kützing (1843), p. 412.

Thallus with erect fronds, terete, compressed or flattened, branching usually repeatedly dichotomous, sometimes irregular, cartilaginous or nearly horny; structure multiaxial, medulla pseudoparenchymatous throughout, compact with large cells, cortex small celled, mostly in radial rows.

Gametangial plants dioecious; spermatangia in superficial sori, carpogonial branches 2 or 4 celled, supporting cell acting as auxiliary cell, gonimoblasts either developing into immersed cystocarps containing carposporangia interspersed with rhizoids, and protruding outwards on one or both sides, with a pericarp formed by a local thickening of cortex, without a pore; or producing rows of cruciate carpotetrasporangia within wart-like superficial outgrowths; tetrasporangial plants unknown.

Two species are at present recognized in the British Isles, *G. crenulatus* (Turn.) J. Ag. and *G. griffithsiae* (Turn.) Martius. Specimens of *Gymnogongrus* from the British Isles bearing the name *G. patens* (Good. & Woodw.) J. Ag. have been found to belong to *G. crenulatus*. *G. patens* cannot, however, be regarded as a synonym of *G. crenulatus* as no authentic material of *Fucus patens* Good. & Woodw. (1797, p. 173) can be traced. The identity of the species for which the name was used is therefore a matter for conjecture and there is no evidence that J. Agardh was correct in assigning it to the genus *Gymnogongrus*.

Key to Species

1 Frond terete or occasionally somewhat flattened towards apices, up to 0·5 mm in diameter; carpotetrasporangial outgrowths large in proportion to branches, sometimes completely encircling them, up to 3 mm in extent . . . . *G. griffithsiae*

Frond flattened except at extreme base, up to 4 mm broad; carpotetrasporangial outgrowths small in proportion to branches, scattered over both surfaces, 1–2 mm in extent . . . . *G. crenulatus*

Gymnogongrus crenulatus (Turner) J. Agardh (1851) p. 320.

Lectotype: BM-K (see Turner, 1802a, pl. 8). Portugal (Oporto).

Fucus crenulatus Turner (1802a), p. 130.
Fucus norvegicus sensu Turner (1802), p. 222 non Gunnerus (1772), p. 122.
Chondrus norvegicus Lamouroux (1813), p. 127 (reprint p. 39), pro parte, non *Fucus norvegicus* Gunnerus.
Gymnogongrus norvegicus J. Agardh (1851), p. 320, pro parte, non *Fucus norvegicus* Gunnerus.
Fucus devoniensis Greville (1821), p. 396.
Gymnogongrus devoniensis (Greville) Schotter (1968), p. 54.
Chondrus celticus Kützing (1843), p. 399.
Actinococcus peltaeformis Schmitz (1893), p. 387.

Thallus consisting of an expanded attachment disc, up to 10 mm in diameter, from which arise one to several erect fronds which are stipitate below; stipes with a short terete portion, sometimes branched, expanding gradually into a flattened blade which is repeatedly (to 7 times) dichotomously branched in one plane, but often twisted on its axis; apices usually broadly rounded; blade parallel-sided and about 4 mm broad between the dichotomies, not normally proliferous, up to 100 mm long, brownish-red.

Structure multiaxial; medulla composed of colourless or occasionally pigmented cells about 25 μm in transverse diameter and 75 μm long; cortex of usually three layers of isodiametric cells about 6 μm in diameter in surface view.

Gametangial plants dioecious; spermatangia grouped in small sori produced by the secondary growth of a group of cortical cells, carpogonial branches developing in the cortex, trichogynes protruding and elevating the cuticle; gonimoblast filaments at first growing inwards, then usually forming a protruding outgrowth composed of parallel filaments; outgrowth when mature irregularly shaped, 1–2 mm in diameter, containing rows of carpotetrasporangia about 10×20 μm with spores cruciately arranged; gonimoblast filaments occasionally producing immersed cystocarps up to 600 μm in diameter, containing normal carposporangia, carpospores *c.* 10 μm; tetrasporangial plants unknown.

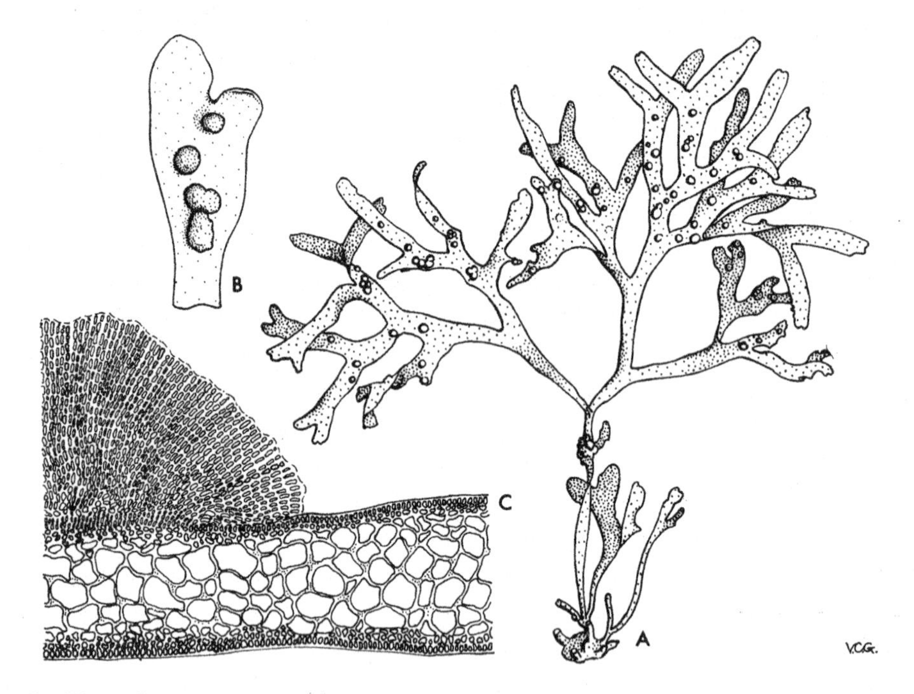

Fig. 78 *Gymnogongrus crenulatus*
A. Habit of carpotetrasporangial plant (Apr.) ×1; B. Branch of same with carpotetrasporangial outgrowths ×4; C. T.S. of same with undivided carpotetrasporangia ×80.

Epilithic, lower littoral in pools and emergent, and sublittoral to 13 m, tolerant of sand cover.

Southern and western shores of the British Isles, extending northwards to Anglesey and eastwards to Kent; in Ireland northwards to Mayo, Down, and eastwards to Waterford; reports for eastern England and Scotland not confirmed.

British Isles to Mauretania; Mediterranean; Canada (New Brunswick) to U.S.A. (N. Massachusetts).

Perennial, internal cystocarps recorded for February and August, carpotetrasporangial outgrowths perennial. In northern France carpospores shed March–July and carpotetraspores in April (Chemin, 1927a, 1929).

Although not usually proliferously branched, the pattern of the dichotomies often becomes irregular with some variation in the width of the blades.

Schotter (1968) treats plants with internal cystocarps as a separate species, *G. devoniensis* (Grev.) Schotter. Such plants are rare and have been found only in Cornwall, Devon, Galway, Mayo, Down and France.

Plants are sometimes confused with *Chondrus crispus* Stackh. *G. crenulatus* is brownish red in colour, pinkish when young, and never iridescent. The blade divisions are usually parallel-sided and the axils rounded whilst the internal pseudoparenchymatous construction distinguishes specimens in difficult cases.

Schmitz applied the name *Actinococcus peltaeformis* to the outgrowths which he thought constituted a separate parasitic organism.

Plants are typically encrusted with species of Bryozoa, Foraminifera and calcareous algae.

Gymnogongrus griffithsiae (Turner) Martius (1828), p. 27.

Lectotype: BM-K (see Turner, 1808, pl. 37). Devon (probably near Chit Rock, Sidmouth).

Fucus griffithsiae Turner (1808), p. 80 [as *Griffithsii*].
Actinococcus aggregatus Schmitz (1893), p. 385.

Base a disc sometimes extending to 20 or 30 mm, erect fronds tufted, to 50 mm long, terete, up to 0·5 mm in diameter and little tapering; apices usually flattened and slightly expanded; dark purplish brown to black (bleached after burial in sand), stiff and cartilaginous but somewhat mucilaginous; branching often corymbose, repeatedly and usually regularly dichotomous.

Structure multiaxial; medullary cells up to 25 μm in transverse diameter, longitudinally elongated, surrounded by radial rows of very small, compact radially elongate cortical cells, terminating in cells *c.* 2 μm in diameter in surface view.

Spermatangia unknown; gonimoblast filaments at first growing inwards and later outwards to form carpotetrasporangial outgrowths usually developing near a dichotomy and sometimes entirely surrounding the branch for 2–3 mm; carpotetrasporangia in rows from most cells of the outgrowth filaments, only dividing just before spore-release; 10–11×7–8 μm, spores cruciately arranged. Plants with tetrasporangia or with immersed cystocarps and normal carposporangia unknown in this species.

Epilithic, littoral in pools and emergent, upper sublittoral; tolerant of sand cover.

Southern and western shores of the British Isles, extending eastwards to Sussex and northwards to at least Argyll; in Ireland eastwards to Dublin and northwards to Mayo; Orkney report awaiting confirmation.

British Isles to Mauretania and Canary Isles; Mediterranean. U.S.A. (Massachusetts) to Uruguay.

Perennial; carpotetrasporangial outgrowths present throughout the year; spores shed June–December (Gregory, 1934).

Schmitz applied the name *Actinococcus aggregatus* to the outgrowths which he thought constituted a separate parasitic organism.

Plants may resemble *Ahnfeltia plicata* (Huds.) Fries, q.v., but are usually less wiry,

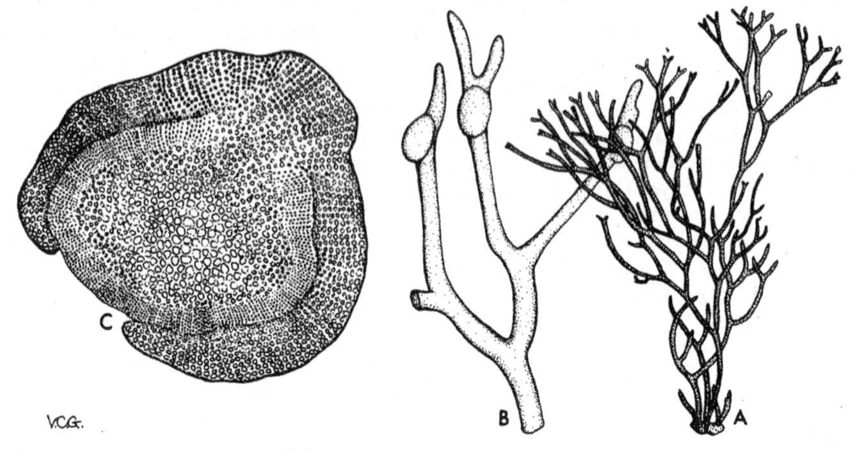

Fig. 79 *Gymnogongrus griffithsiae*
A. Habit (Aug.) ×1; B. Branch with carpotetrasporangial outgrowths (Feb.) ×8; C. T.S. of same with undivided carpotetrasporangia ×80.

more regularly dichotomously branched and frequently flattened towards the apices. The reproductive outgrowths produce carpotetraspores, not monospores, and are not restricted to the period December–March. In transverse section, the cortical cells appear thick-walled and radially elongated, (*c.* 4×2 μm) not thin-walled and quadrate (*c.* 2 μm) as in *A. plicata*. The outgrowth cells are subquadrate (*c.* 8×6 μm) whereas in *A. plicata* they become very elongated radially (*c.* 6×2 μm) immediately before monospore production.

According to Schotter (1968), *Gymnogongrus pusillus* (Mont.) Feldm. & Maz. is a closely similar species in which a gonimoblast producing immersed carposporangia occurs. This species has not yet been found outside the Mediterranean.

PHYLLOPHORA Greville **nom. cons.**

PHYLLOPHORA Greville (1830), p. lvi, 135.

Type species: *P. rubens* Greville (1830), p. lvi, 135, (=*P. crispa* (Hudson) Dixon (1964), p. 63), non *Fucus rubens* Linnaeus.

Agarum Link (1809), p. 7, non *Agarum* Bory, (1826), p. 193.
Membranifolia Stackhouse (1809), p. 55, 75.
Prolifera Stackhouse (1809), p. 56, 77.
Epiphylla Stackhouse (1816), p. x.
Acanthotylus Kützing (1843), p. 413.
Coccotylus Kützing (1843), p. 412.
Phyllotylus Kützing (1843), p. 412.
Colacolepis Schmitz (1893), p. 417.

Thallus with erect fronds arising from a discoid base, cartilaginous; stipitate, stipe terete, blade always flattened; usually dichotomous, sometimes irregularly branched, sometimes with leaflike proliferations; structure multiaxial, with a compact pseudoparenchymatous medulla composed of large cells, bounded by 2–4 layers of smaller cells forming a compact cortex.

Gametangial plants monoecious or dioecious; spermatangial sori at apices or on specialized bladelets or papillae, spermatangia produced in small pits derived from division of outer cortical cells; carpogonial branches in cortex at apices or in specialized outgrowths, consisting of 3–4 cells with supporting cell acting as auxiliary cell; carposporophytes developing inwards, carposporangia produced in clusters from gonimoblast filaments, without enveloping filaments; cystocarps protruding externally, pericarp without a pore; tetrasporangia in radial rows in elevated sori, cruciate.

The variation in form and colour does not appear to be entirely related to ecological factors, and would repay further study. Several species can tolerate low salinity but in such areas (e.g. Loch of Stenness, Orkney) their appearance is atypical, showing extremely narrow blades.

Species of *Phyllophora* may be confused with other members of the Phyllophoraceae such as *Gymnogongrus crenulatus* (Turn.) J. Ag., *Schottera nicaeensis* (Lamour. ex Duby) Guiry & Hollenb. and *Stenogramme interrupta* (C. Ag.) Mont. ex Harv., and also with *Rhodymenia* spp., especially *R. pseudopalmata* (Lamour.) Silva. It is not easy to give a clear-cut summary of the differences, but a comparison of the nature, position and season of occurrence of the reproductive organs gives a guide to the identification of specimens. In *Rhodymenia* spp. the cortical cells are slightly larger (6–12 μm) and more widely separated in surface view and the medullary cells are considerably larger.

Key to Species

1 Terete stipe short, rarely up to 10 mm long; blade more or less parallel-sided . . . 2
Terete stipe considerably more than 10 mm long; blade more or less fan-shaped . . . 3

2 Mature thallus more than 35 mm tall; branching dichotomous and from face of blade; margins undulate; cystocarps, spermatangia and tetrasporangia in outgrowths on blade surface . . . *P. crispa*
Mature thallus not more than 35 mm tall; branching by one or two dichotomies only; margins plane; cystocarps and spermatangia in marginal outgrowths, tetrasporangia unknown . . . *P. traillii*

3 Blade branching wide-angled, outline broadly fan-shaped; cystocarps and spermatangia in marginal outgrowths, tetrasporangia in sori at base of blades . . . *P. pseudoceranoides*
Blade branching narrow-angled, outline with an angle of less than 90° at junction with stipe; cystocarps unknown; (carpo)tetrasporangia either in sori central in the blades or in spherical outgrowths at or near apices of blade lobes . . . 4

4 Basal disc of mature thallus more than 10mm in diameter; transition from blade to stipe often abrupt; stipe often bent or varying in diameter; blade sparingly branched; tetrasporangia in flat sori *c.* 3 mm in diameter in centre of blades . . . *P. sicula*
Basal disc of mature thallus less than 10 mm in diameter; transition from blade to stipe gradual; stipe of uniform diameter; blade often proliferating laterally or apically; carpotetrasporangia in spherical outgrowths 1–2 mm in diameter at or near apices . . . *P. truncata*

Phyllophora crispa (Hudson) Dixon (1964), p. 56.

Lectotype : BM-SL (see Dixon, 1964, fig. 2). England.

Fucus crispus Hudson (1762), p. 472.
Fucus epiphyllus O. F. Müller (1777), pl. 708.
Phyllophora epiphylla (O. F. Müller) Batters (1902), p. 65.
Fucus nervosus De Candolle (1805), p. 29.
Phyllophora nervosa (De Candolle) Greville (1830), p. lvi.
Phyllophora rubens Greville (1830), p. lvi, 135, pro parte, non *Fucus rubens* Linnaeus (1753), p. 1162.
Colacolepis incrustans Schmitz (1893), p. 417.

Thallus consisting of a small discoid holdfast and erect fronds with short terete stipes rarely up to 10 mm long and blades up to 100–150 mm long and about 10 mm broad, cartilaginous, bright red to pink, with parallel sides, undulating margins and rounded apices; branching dichotomous and proliferous from the surface of the blade, the new blade sometimes connected to the old one by a short stipe, and often with midribs, formed by the secondary growth of the cortical layer, appearing as an extension of the stipe in the midline of the blade.

Structure multiaxial, medulla composed of large thick-walled cells, up to 65 μm in diameter surrounded by a cortical layer of 3 cells, the outermost measuring 2–6·5 μm in diameter in surface view.

Gametangial plants dioecious; spermatangia lining the wall of pits in the surface layers of special outgrowths arising from the surface of the blade, spermatia 4×2 μm; cystocarps 1–2 mm in diameter, shortly pedicellate, without a pore and with a distinctive rugose surface; carposporangia 7–9·5 μm in diameter; tetrasporangia in rows in outgrowths on the surface of the stipe of peltate bladelets which arise from the surface of the blade, 6–9 μm in diameter, with cruciately arranged tetraspores.

Epilithic, rarely epiphytic on *Laminaria* stipes, in shady places in the lower littoral, and sublittoral to 30 m.

Generally distributed throughout the British Isles, but not common on the east coast of England.

Iceland; Norway (Trondelag) to Portugal; Mediterranean; Black Sea.

Fronds perennial, sometimes showing 5 or 6 successive proliferations, regeneration occurring after erosion or animal grazing; spermatia released in September–October; cystocarps recorded for September–March, the carpospores probably usually released in January; tetrasporangia recorded for August–March, the tetraspores usually released in January.

There is considerable variation in the appearance of plants resulting from the number, size, and disposition of the proliferations which are marginal, terminal and from the blade surface. Rosenvinge (1931) illustrates much incised narrow fronds from areas in the Kattegat with reduced salinity.

Schmitz applied the name *Colacolepis incrustans* to the cystocarpic outgrowths which he thought constituted a separate parasitic organism.

Plants of *Schottera nicaeensis* (Lamour. ex Duby) Guiry & Hollenb. (q.v.) have been misidentified as young *P. crispa*.

P. crispa forms the basis of the agar industry in the Black Sea (Levring *et al.*, 1969).

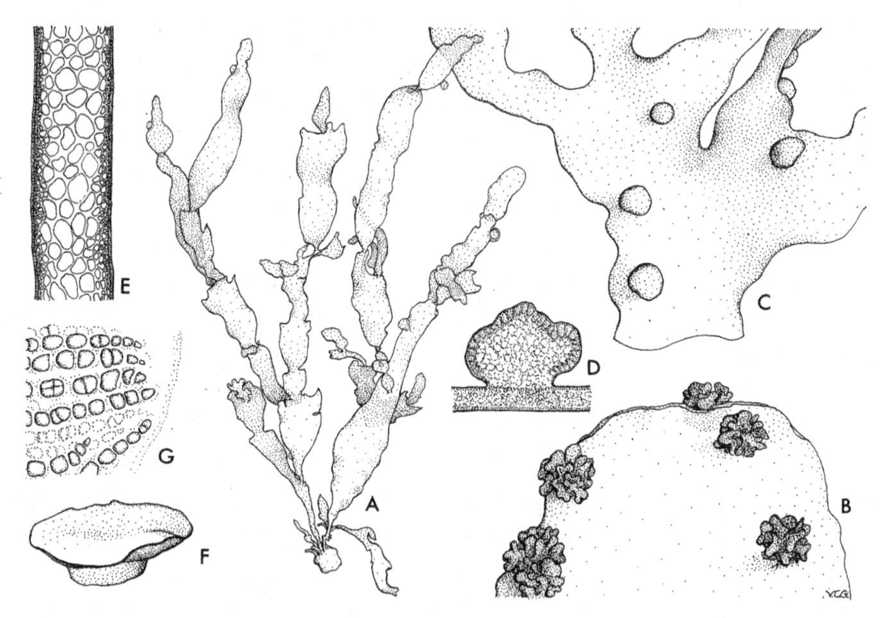

Fig. 80 *Phyllophora crispa*
A. Habit (Oct.) ×1; B. Part of frond with carposporangial outgrowths (Nov.) ×8; C. Part of frond with spermatangial outgrowths (Nov.) ×4; D. V.S. spermatangial outgrowth (Sep.) ×16; E. T.S. blade (Nov.) ×80; F. Tetrasporangial outgrowth (Jan.) ×12; G. T.S. stipe of same with tetrasporangia ×300.

Phyllophora pseudoceranoides (S. G. Gmelin) Newroth & A. R. A. Taylor (1971), p. 95.

Lectotype: BM-SL (see Newroth & Taylor, 1971). England.

Fucus pseudoceranoides S. G. Gmelin (1768), p. 119.
Fucus membranifolius Goodenough & Woodward (1797), p. 120, nom. illeg., non *F. membranifolius* Withering (1796), p. 106.
Phyllophora membranifolia Endlicher (1843), p. 38.

Thallus consisting of a small discoid holdfast producing erect fronds up to 100 mm long, composed of a terete stipe, which is usually comparatively long, and a fan-shaped blade; purple or reddish-brown; branching irregularly dichotomous and adventitious from the stipes; blade often considerably divided at the apices, apices usually rounded.

Structure multiaxial, medulla composed of large thick-walled cells up to 65 μm in transverse diameter decreasing in size gradually outwards, surrounded by a cortical layer 3 cells deep, with the outermost measuring 2–6 μm in diameter in surface view.

Gametangial plants usually dioecious; spermatangia produced in pits in the cortex of special bladelets forming a fringe around the upper parts of the blades, spermatangia 3–4 μm; cystocarps 1–3 mm, pedicellate (rarely sessile), urn-shaped, occurring on the stipe or on the lower margins of the blade, carpospores *c.* 10 μm when released; tetrasporangia occurring in rows in a wedge-shaped sorus on one or both sides of the blade, 10–13 × 8–11 μm; tetraspores cruciately arranged.

Epilithic, in pools in the lower littoral, sublittoral to 12 m.

Generally distributed throughout the British Isles.

Iceland; Norway (Nordland) to Atlantic France; Baltic; Canada (Newfoundland) to U.S.A. (Delaware).

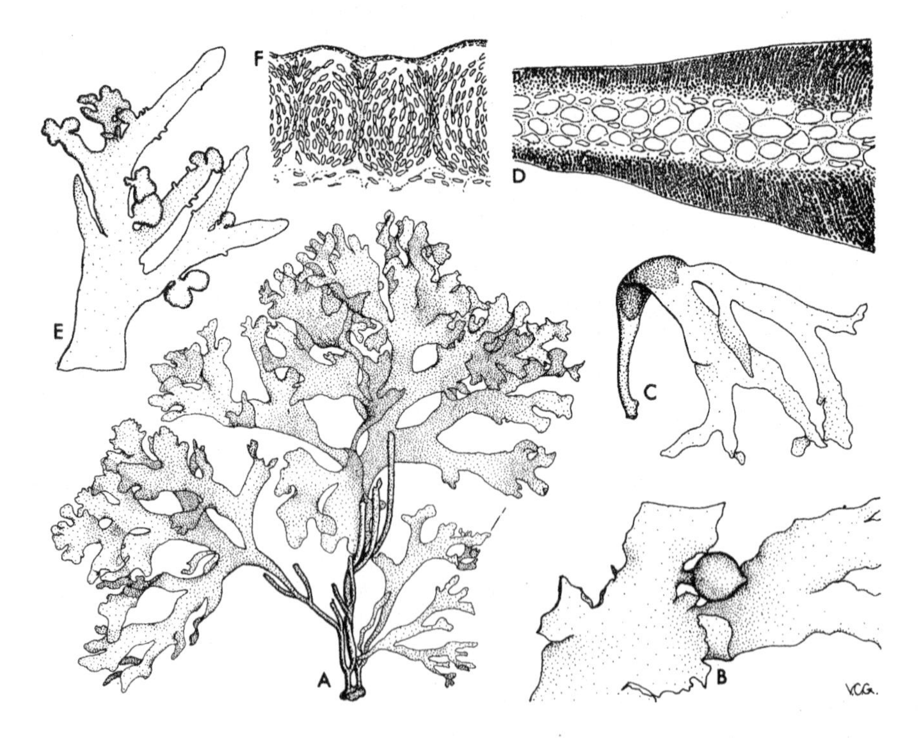

Fig. 81 *Phyllophora pseudoceranoides*
A. Habit (Apr.) ×1; B. Part of plant with carposporangial outgrowths (Feb.) ×4; C. Part of plant with tetrasporangial sorus (Feb.) ×4; D. V.S. of same with undivided tetrasporangia ×80; E. Part of plant with spermatangial outgrowths (Oct.) ×4; F. T.S. of same with spermatangia ×300.

Perennial, the persistent stipes and sometimes the blade margins regenerating new fronds; spermatangia recorded for June–October, spermatia usually released in June–July; cystocarps recorded throughout year, maturing over a period of many months, the carpospores being released between November–May; tetrasporangial sori also present throughout the year, but tetraspore release apparently restricted to November–January.

Fronds in spring and early summer have thin, pale blades; these become thick, dark, somewhat eroded and covered with epiphytes as the season advances. Occasionally plants are found with a very short stipe, especially when growing on vertical rock faces or when partially emergent. For a description of the successive development of new blades over 3–4 years, see Rosenvinge, 1931.

Phyllophora sicula (Kützing) Guiry & L. Irvine (1976), p. 284.

Lectotype: L (941.183.58). Sicily

Phyllotylus siculus Kützing (1847), p. 5.
Fucus membranifolius var. *roseus* Turner (1809), p. 6, nom. illeg.
Chondrus brodiaei var. *simplex* Greville (1830), p. 133, nom. superfl.
Phyllophora palmettoides J. Agardh (1849), p. 144.

Thallus consisting of an expanded attachment disc 10–20 mm in diameter, giving rise to erect fronds up to 50–70 mm long, consisting of terete stipes usually one third of total length of the frond and bright red blades; blades usually with a single dichotomy and obtuse apices, often arising abruptly at or near the ends of previously broken stipes, and frequently twisted on their axes.

Structure multiaxial, medulla consisting of large cells up to 90 μm in diameter with an abrupt transition to the cortical layer of 1–2 cells measuring 2–7 μm in diameter in surface view.

Gametangial plants unknown; tetrasporangia occurring in an irregularly rounded sorus in the centre of the blade, sporangia in rows, 14–18×6–9 μm, tetraspores cruciately arranged.

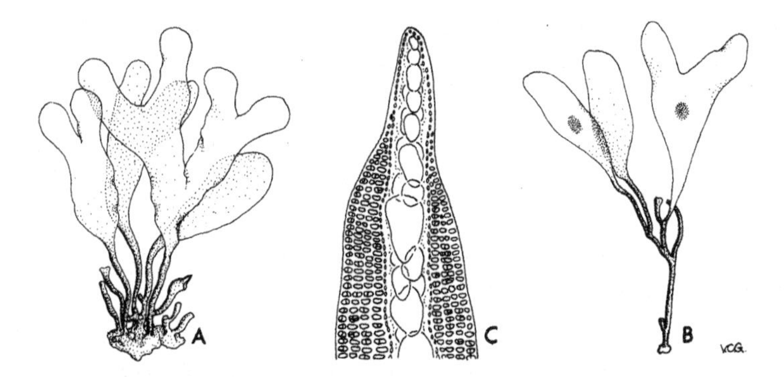

Fig. 82 *Phyllophora sicula*
A. Habit (Feb.)×1; B. Plant with tetrasporangial sorus (Feb.) ×1; C. T.S. blade with undivided tetrasporangia (Jan.) ×80.

Epilithic, lower littoral in pools, rarely emergent, sublittoral to 15 m; tolerant of sand cover.

Restricted to the southwest of the British Isles, extending northwards to Anglesey and eastwards to Dorset; Cork.

British Isles to Portugal; Mediterranean.

Stipe perennial regenerating and producing new blades to replace damaged and eroded ones; tetrasporangial sori recorded for December–April, the tetraspores usually released in February–March.

Apparently with comparatively little morphological variation.

In the absence of gametangial plants, the generic assignment of this species must be regarded as provisional.

The species has been confused with *P. truncata* (q.v.), *Schottera nicaeensis* (Lamouroux ex Duby) Guiry & Hollenb. (q.v.) and *Rhodymenia pseudopalmata* (Lamour.) Silva, (see p. 221).

Phyllophora traillii Holmes ex Batters (1890), p. 334 (reprint p. 114).

Lectotype: BM (see Newroth & Taylor, 1971, fig. 2). Northumberland (Berwick).

Thallus consisting of a small discoid holdfast producing an erect frond up to 30 mm tall, light to dark red; stipe terete, 1–3 mm in length, expanding rapidly into a blade about 1·5–3·5 mm broad, either oblong with nearly parallel sides or tapering slightly towards an obtuse apex; simple or once or twice irregularly dichotomously branched, stipe rarely branched; sometimes with a midrib as an extension of the stipe, formed by the secondary growth of the cortex.

Structure multiaxial, medulla compact, composed of large cells up to 60 μm in diameter, with thick walls, decreasing in size outwards to a cortex of 2–3 layers of cells 2–7 μm in surface view.

Gametangial plants dioecious; spermatangia produced in the walls of pits in the surface layers of special bladelets produced in a fringe around the upper margins of the blade; cystocarps in bladelets fringing the lower margins of the blade, 1–2 mm in diameter, compressed, without a pore; carpospores 9–10 μm in diameter after release; tetrasporangia unknown.

Epilithic, in shady pools and crevices in the lower littoral and sublittoral to 15 m.

Generally distributed throughout the British Isles, comparatively infrequently recorded because of its small size.

Faeroes; British Isles to Atlantic France; Denmark, Sweden, ? Baltic; Canada (New Brunswick) to U.S.A. (Connecticut).

Presumably perennial but little is known about seasonal growth; spermatangia recorded between July and February, spermatia observed to be released in August; cystocarps present throughout the year, maturing over a long period, carpospores apparently released only in December–February.

Comparatively little form variation. Fronds may be simple or dichotomous, with or without marginal proliferations.

The marginal outgrowths containing reproductive bodies characterize this species; in their absence plants can be identified only provisionally because of their small size.

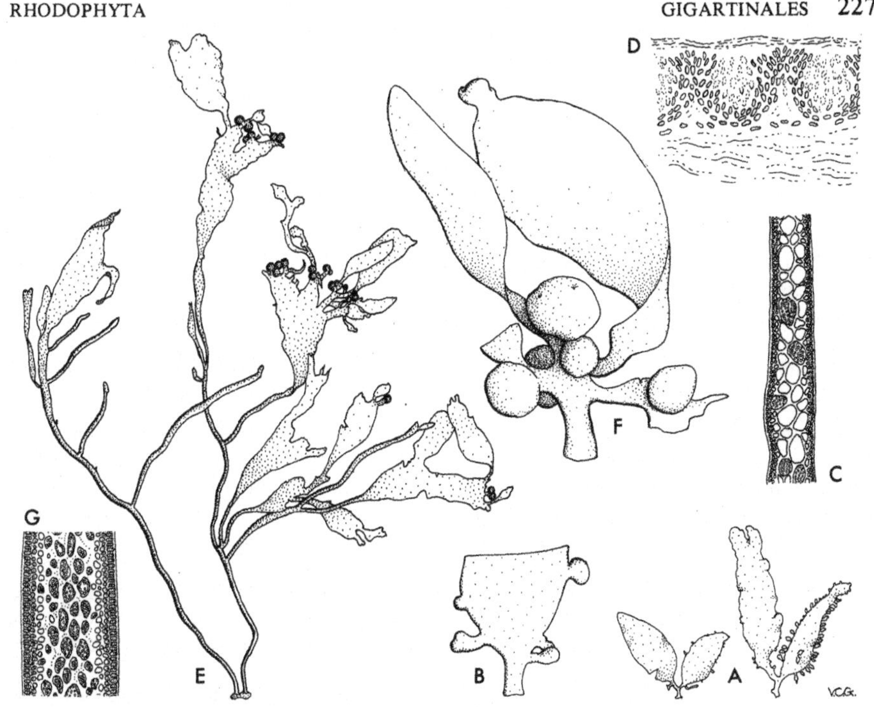

Fig. 83 *Phyllophora traillii*
A. Habit (Aug.) ×1; B. Part of plant with carposporangial outgrowths (Aug.) ×8; C. T.S. blade (Aug.) ×80; D. T.S. outgrowth with spermatangia (undated) ×300.
Phyllophora truncata
E. Habit of carpotetrasporangial plant (Aug.) ×1; F. Part of same with carpotetrasporangial outgrowths ×6; G. T.S. blade ×80.

Phyllophora truncata (Pallas) A.D. Zinova (1970 [1971]), p. 103.

Lectotype: LE. Arctic seas.

Fucus truncatus Pallas (1776), p. 760.
Fucus brodiaei Turner (1809), p. 1.
Phyllophora brodiaei (Turner) Endlicher (1843), p. 38.
Sphaerococcus interruptus Greville (1829), p. 423.
Phyllophora interrupta (Greville) J. Agardh (1862), p. [3].
Chaetophora subcutanea Lyngbye ex Hornemann (1834), pl. 2135, fig. 2.
Actinococcus subcutaneus (Lyngbye ex Hornemann) Rosenvinge (1893), p. 822.

Thallus consisting of a small discoid holdfast producing groups of erect fronds to 150 mm long; bright red when young, darker when mature; stipes terete, branched, up to 90 mm long, expanding gradually into a di- or poly-chotomously divided blade, lobes 3–20 mm broad, usually simple, with rounded or truncated apices; apices producing further stipitate blades.

Structure multiaxial, medulla compact, composed of large cells, up to 65 μm in transverse diameter, decreasing in size outwards to a cortex of 2–3 layers of cells 2–5 μm in diameter in surface view.

Gametangial plants monoecious; spermatangia produced in the walls of pits in the surface layers of the youngest parts of the blade; spermatia 3 μm in diameter; after fertilization auxiliary cell producing spherical outgrowths up to 2 mm in diameter on the margins of primary or secondary blades at or near the apex, containing radial rows of carpotetrasporangia 13–23 × 11–15 μm, spores cruciately arranged.

Epilithic, in lower littoral pools, sublittoral to at least 10 m; often at the lower limit of algal vegetation.

Northern shores of the British Isles, extending southwards to Caernarvon and Northumberland; in Ireland southwards to Kerry and Wexford.

Iceland; North Russia to British Isles and Denmark; Baltic; Black Sea; circumboreal in Arctic Ocean; southwards to U.S.A. (New Jersey).

Perennial, the stipes and blade apices regenerating and producing new blades, conspicuous in spring; spermatangia recorded between May–August, spermatia usually released in June; carpotetrasporangial outgrowths present throughout the year, maturing over a long period, two successive generations sometimes occurring on the same plant; carpotetraspores apparently released between November and March.

There is considerable variation in the appearance of plants resulting from the size and degree of branching of the stipe and blade. The most extreme forms are met with in areas of low salinity, e.g. Loch Stenness, Orkney; in these, the blades are very narrow (1–2 mm) and elongated and the species can only be recognized by its characteristic carpotetrasporangial outgrowths.

For a description of the successive development of new blades over a number of years, see Rosenvinge, 1931.

The carpotetrasporangial outgrowths have been described as a parasitic organism, *Actinococcus subcutaneus* (Lyngb. ex Hornem.) Rosenv.

As it is usually considered that the genus *Phyllophora* is characterized by the restriction of spermatangia and carpogonial branches to special papillae or bladelets, one might question whether *P. truncata* is correctly assigned.

The close similarity between this species and *P. sicula* has led to confusion. *P. truncata* reaches its southern limit within the British Isles whilst *P. sicula* reaches its northern limit, the two overlapping only in Anglesey.

The absence of *P. truncata* from southern counties was not fully appreciated until comparatively recently, earlier reports being attributable to *P. sicula*. *P. sicula* has tetrasporangial sori centrally placed in the blade, although it is rarely fertile, while plants of *P. truncata* usually show apical carpotetrasporangial outgrowths.

Plants are occasionally found infected with the parasite *Ceratocolax hartzii* Rosenvinge, q.v.

PORPHYRODISCUS Batters

PORPHYRODISCUS Batters (1897), p. 439.

Type species: *P. simulans* Batters (1897), p. 439.

Thallus encrusting, closely adhering to substrate by whole undersurface, firm and cartilaginous; consisting of closely compacted vertical rows of isodiametric cells; gametangia unknown, tetrasporangia terminal on short superficial filaments, zonate, grouped into flat, wart-like sori.

Farnham & Fletcher (1974) have shown that *Ahnfeltia plicata* has an encrusting base, closely resembling *P. simulans,* in which zonate tetrasporangia are borne.

One species in the British Isles:

Porphyrodiscus simulans Batters (1897), p. 439.

Lectotype: BM (Slide 9915). Northumberland (Berwick).

Thallus encrusting, up to 300 mm in diameter and to 300 μm thick, dark purple, shining when dry, rather mucilaginous, more so in the region of the sori; cells closely compacted, not obviously arranged in filaments but usually in vertical rows, isodiametric, mainly about 4–6 μm in diameter, often larger adjacent to the substrate.

Gametangial plants unknown, (but see *Ahnfeltia plicata*); tetrasporangia produced in groups from the surface cells, becoming embedded in thick mucilage and forming raised sori up to 300 μm in diameter; spindle-shaped, 20–25 × 5–8 μm, with zonately arranged tetraspores. (Fig. 77, p. 216.)

Epilithic in the lower littoral and sublittoral.
Widely distributed in the British Isles.
British Isles and northern France; Canada (Newfoundland, Nova Scotia).

Data on seasonal growth and form variation insufficient for comment; tetrasporangia recorded for November–April.

This species is apparently indistinguishable from the basal crust of *Ahnfeltia plicata* (Huds.) Fries, q.v. It is also similar to *Hildenbrandia* spp., but the thallus of *P. simulans* is more purplish in colour and more mucilaginous especially in the region of the sori.

SCHOTTERA Guiry & Hollenberg

SCHOTTERA Guiry & Hollenberg (1975), p. 152.

Type species: *S. nicaeensis* (Lamouroux ex Duby) Guiry & Hollenberg (1975), p. 153.

Thallus with erect fronds arising from prostrate terete branches, flattened, ligulate, simple or dichotomous, stipitate, sometimes with few to many lateral and apical proliferations, firm and cartilaginous; structure multiaxial, medulla of large thick-walled cells and cortex mostly 1–2 layers of small cells.

Gametangial plants dioecious, spermatangia forming superficial sori; carpogonial branches 4 celled, trichogyne short, supporting cell acting as auxiliary cell, gonimoblast filaments penetrating medulla, producing narrow sterile cells which in turn produce 2–4 carposporangia in rows, the sterile cells grouped around large thick-walled cells, the whole giving (in section) a very characteristic appearance to the carposporophyte, cystocarps protruding on both sides of frond, pericarp without pore; tetrasporangia cruciate, in radial rows, aggregated in a series of sequentially developing sori which protrude as bands on both surfaces of frond.

One species in the British Isles:

Schottera nicaeensis (Lamouroux ex Duby) Guiry & Hollenberg (1975), p. 153.

Lectotype: STR (see Guiry & Hollenberg, 1975, fig. 4). France (Marseille).

Halymenia nicaeensis Lamouroux ex Duby (1830), p. 942.
Gymnogongrus nicaeensis (Lamouroux ex Duby) Ardissone & Strafforello (1877), p. 186.
Rhodymenia nicaeensis (Lamouroux ex Duby) Montagne (1846), p. 68, non. *R. nicaeensis* sensu Holmes (1883), p. 289.
Petroglossum nicaeense (Lamouroux ex Duby) Schotter (1952), p. 207.

Thallus consisting of prostrate branches giving rise to erect, bright red fronds with a slender terete stipe 300–400 μm in diameter, flattening rapidly into a thin ribbon-like ruffled blade up to 100 mm long and about 60 μm thick, once or twice dichotomously branched, whose margins and extremities frequently bear numerous terete, cirrose proliferations when mature, proliferations elongating and often forming secondary attachments to the substrate; young individuals with simple or once divided blades.

Structure multiaxial, medullary cells to 50 μm, fairly regularly arranged, in 2–6 layers, cortical cells in 1–2 layers, 6–8 μm in diameter in surface view.

Gametangial plants dioecious; spermatangial sori fairly extensive, occurring on plants only 10–20 mm in length; carpogonial branches developing on plants a few millimetres long, gonimoblast filaments developing inwards, carposporangia in terminal rows of 1–4, carpospores 7–8 μm, enveloping filaments absent, cystocarps without a pore, protruding on both sides of the blade at the base; tetrasporangia occurring on plants less than 20 mm long, grouped in oval sori aligned along the axis of the frond, spore liberation beginning at

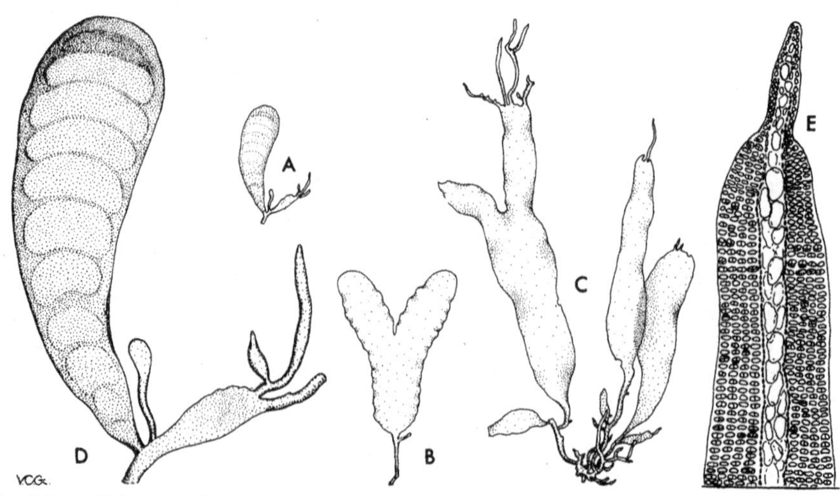

Fig. 84 *Schottera nicaeensis*
A. Habit of plant with tetrasporangial sori (Jan.) ×1; B. Habit (June) ×1; C. Habit (Aug.) ×1; D. Same as A ×4; E. T.S. same with tetrasporangia ×80.

the base of a plant, with empty and colourless sori below and deeply coloured ones containing spores above, tetrasporangia in rows of 2–3, 15–16·5×5–6 μm with cruciately arranged spores.

Epilithic, lower littoral in pools and emergent, sublittoral to 15 m.

Southern and western shores of the British Isles, extending eastwards to Dorset and northwards to the Isle of Man; in Ireland eastwards to Waterford and northwards to Galway, Antrim.

British Isles to Portugal; Mediterranean.

Basal prostrate branches perennial, fronds apparently annual; young fronds appearing from October onwards and becoming fertile when small; fronds continuing to grow after spore and spermatial release, in June and July producing terminal and marginal proliferations which become reflexed, attaching secondarily to the substrate. Tetrasporangial fronds producing a series of sori at the apex and by January appearing as banded plants 10–20 mm long. Gametangial plants unknown outside the Mediterranean, male plants producing spermatangia in winter when 10–20 mm long, female plants showing mature carposporophytes at this stage, their development having begun in fronds only a few millimetres long.

The appearance of plants at different times of the year varies greatly. The contrast between the small banded tetrasporangial plants in winter and the larger sterile dichotomous fronds with numerous terminal and marginal proliferations in summer is particularly striking.

Specimens of *S. nicaeensis* have been misidentified as *Phyllophora crispa* (Huds.) Dixon, *Phyllophora sicula* (Kütz.) Guiry & L. Irvine and *Rhodymenia pseudopalmata* (Lamour.) Silva. Plants of *P. crispa* bear bladelike proliferations from the margins and blade surfaces and the blades are thicker (*c.* 150 μm). In *P. sicula* the base is an expanded disc and prostrate branches are absent. In *R. pseudopalmata* the stipe is broader and compressed, the blade is thicker (*c.* 120 μm) and the cortical cells are larger (6–12 μm) and more widely separated in surface view.

STENOGRAMME Harvey

STENOGRAMME Harvey [as *Stenogramma*] (1840), p. 408.

Type species: *S. californica* Harvey (1840), p. 408 (=*S. interrupta* (C. Agardh) Montagne ex Harvey (1848), pl. clvii).

Thallus with a small attachment disc giving rise to erect fronds, blades stipitate, flattened, repeatedly divided, sometimes proliferating from the margin; structure multiaxial, consisting internally of one or two layers of large medullary cells with one or two layers of small cortical cells on each side.

Gametangial plants dioecious; spermatangia in slightly elevated sori; carpogonial branches 3–4 celled in an interrupted midrib-like ridge, gonimoblasts developing in medullary tissue of this ridge, producing irregular carposporangial masses, cystocarps protruding on both sides of the frond in midline; pericarp formed by local thickening of cortex, each with an obscure pore; tetrasporangia cruciate, formed in rows in sori scattered over thallus surface.

One species in the British Isles:

Stenogramme interrupta (C. Agardh) Montagne ex Harvey (1848), pl. clvii.

Lectotype: LD (Herb. Alg. Agardh 24295). Spain (Cadiz).

Delesseria interrupta C. Agardh (1822), p. 179.
Stenogramme californica Harvey (1840), p. 408.

Thallus consisting of a small conical attachment disc up to 4 mm in diameter from which arise 1–4 rose-red to brownish red erect fronds consisting of a slender terete stipe up to 3 mm long which expands into a fan-shaped blade up to 100 (200) mm long; blade becoming dichotomously or palmately divided, the lobes equal or unequal, and often parallel-sided with rounded or blunt apices, occasionally narrowing towards the apices and then more acute; secondary blades arising from the margins and apices of old worn blades.

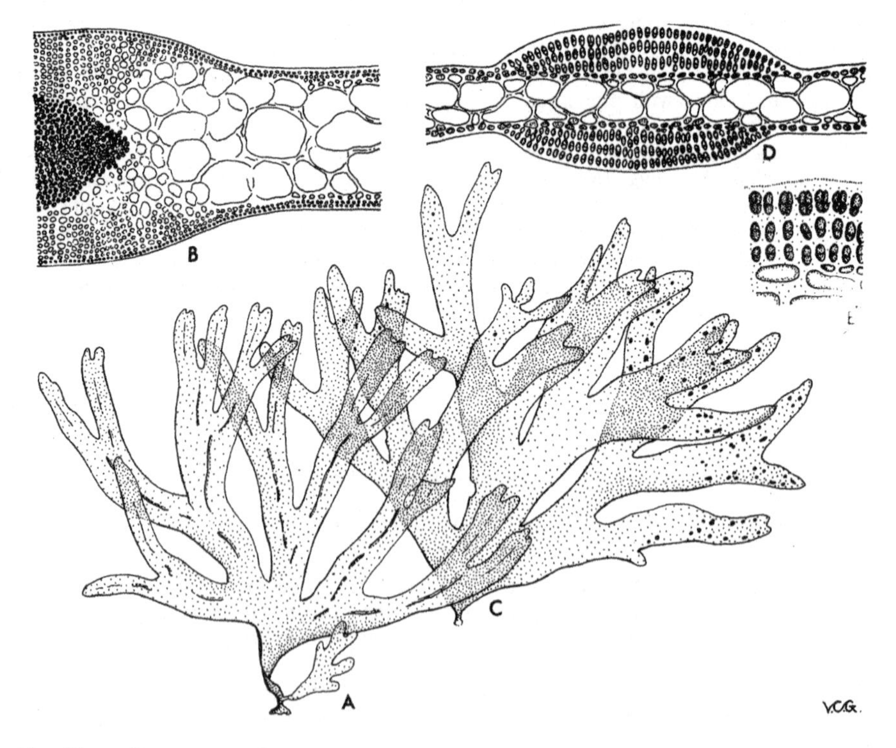

Fig. 85 *Stenogramme interrupta*
A, Habit of plant with cystocarps (Sep.) ×1; B. T.S. of same with cystocarp ×80; C. Habit of tetrasporangial plant (Sep.) ×1; D. T.S. same with tetrasporangia ×80; E. Part of same with tetrasporangia ×180.

Structure multiaxial, medulla consisting of 1–3 layers of large, thick-walled cells up to 140×70 μm in transverse section, surrounded by 1–2 layers of cortical cells 4·5–9 μm in surface view.

Gametangial plants dioecious; spermatangia in sori in younger parts of the blade, extending almost across its width, slightly elevated above the surrounding surface; gonimoblast filaments developing inwards, adjacent carposporophytes forming masses of carposporangia, each 12–15 μm in diameter, the maturing cystocarps forming a conspicuous interrupted ridge, this compound structure opening to the exterior by pores at intervals; tetrasporangial sori occurring as deeply-coloured, oval patches on both surfaces of the blade, about 2×1 mm, tending to be orientated in rows parallel to the long axis, sometimes confluent; tetrasporangia 17–20×9–10 μm, in rows, with cruciately arranged spores.

Epilithic, sublittoral to 13 m, in sheltered areas on small stones among gravel and mud.

Cornwall, Devon, Somerset, Pembroke, Cork, Down, Galway, Kerry.

British Isles to Morocco; Cape Verde Isles; widely distributed throughout the Pacific Ocean.

Perennial; according to Gayral (1966) plants remain sterile during the first year and become fertile during the summer of the second year. Tetrasporangia and cystocarps have been recorded throughout the year in Britain but occur most abundantly in July–September.

There is considerable variation in general shape, degree of branching and extent of proliferation.

The interrupted midrib is often considered to be the distinguishing feature of this species; however, it is absent except in plants with developing cystocarps. Specimens lacking a midrib often bear a strong resemblance to *Rhodymenia pseudopalmata* (Lamour.) Silva (see p. 221) or *Phyllophora sicula* (Kütz.) Guiry & L. Irvine; the former species bears reproductive bodies at or near the frond apices, however, whilst in the latter only tetrasporangia are known and these are borne in single sori centrally placed in a blade. Sterile specimens of *S. interrupta* can be distinguished by the larger size of the medullary cells which are up to 140 μm in transverse diameter, the maximum being less than 100 μm in the other species. Plants of *Phyllophora crispa* sometimes show midribs but these are continuous and are not associated with carposporophyte development.

GIGARTINACEAE Kützing

GIGARTINACEAE Kützing (1843), p. 389 [as Gigartineae].

Thallus erect or encrusting, erect fronds terete, compressed or flattened, dichotomous or laterally branched, irregularly lobed, rarely undivided; multiaxial, medulla and inner cortex rather loose with a reticulate structure, outer cortex compact with small cells often in radial rows; supporting cell functioning as auxiliary cell, gonimoblast initially developing inwards, carposporangia lying amongst a network of narrow sterile filaments, enveloping filaments present or absent, cystocarps embedded either in thallus or in special outgrowths, tetrasporangia embedded in groups within thallus, cruciate, arranged in more or less regular rows.

Chondrus, Gigartina and *Petrocelis* are the representatives of this family in Britain. *Chondrus* is unique in having the tetrasporangial mass completely immersed in the medulla. In some species of *Gigartina* tetrasporangia occur in erect fronds similar to the gametangial ones and are embedded in the inner cortex/outer medullary region. In other species tetrasporangia occur as intercalary cells of the 'cortical' region of a *Petrocelis*-like encrusting plant. Cystocarps occur in the normal frond in *Chondrus* but in *Gigartina* they are in special outgrowths (either papillae or special branches). Cystocarps have been recorded for *Petrocelis hennedyi,* suggesting this may be an autonomous species and so *Petrocelis* has been provisionally retained as a separate genus. Species of *Chondrus* and *Gigartina* are of economic importance in Britain.

CHONDRUS Stackhouse

CHONDRUS Stackhouse (1797), p. xv.

Type species: *C. crispus* Stackhouse (1797), p. xxiv.

Thallus with a discoid holdfast and erect fronds in tufts, consisting of stipe and flattened blade, branching more or less dichotomous; structure multiaxial, medulla more or less distinctly filamentous, cortex of radial filaments.

Gametangial plants dioecious; spermatangia in superficial sori on young branches; carpogonial branches developing in groups in inner cortex, 3-celled on a large supporting (auxiliary) cell, gonimoblast filaments growing amongst medullary filaments and producing carposporangia terminally, enveloping filaments absent; tetrasporangia developing in medulla, in sori obvious externally, cruciate.

One species in the British Isles:

Chondrus crispus Stackhouse (1797), p. xxiv, xv (see Papenfuss, 1950).

Holotype: LINN 1274.68. Atlantic Ocean.

Fucus crispus Linnaeus (1767), p. 134 nom. illeg. non *F. crispus* Hudson (1762), p. 472.

Thallus with a discoid holdfast and erect fronds arising in tufts, consisting of a compressed unbranched stipe expanding gradually into a fanlike blade up to 220 mm long, undulate only in younger parts, membranous to cartilaginous; repeatedly (to 5 times) dichotomous with rounded axils, 2–15 mm broad between dichotomies, occasionally proliferous from the margins, usually expanding but occasionally tapering towards rounded apices; dark reddish or purplish brown, bleaching to a greenish yellow, frequently iridescent at the tips.

Structure multiaxial; medullary filaments thick-walled, 7–16 μm in diameter, to 80 μm in length, cortical cells radially elongated, 4–8 × 3–4 μm.

Gametangial plants dioecious; spermatangia colourless, produced in white or pink superficial sori towards the apices of the youngest parts of the blade which are narrow and slightly flattened, spermatia liberated through distinct pores in the thickened cuticle, 7·5–10×4–5 μm; gonimoblast filaments ramifying among medullary filaments and producing carposporangia terminally, enveloping filaments absent, cystocarps protruding strongly as usually concavo-convex swellings about 2 mm in diameter, without a pore, carposporangia 20–30×14–25 μm; tetrasporangial sori only slightly protruding, oval or linear, 1–5×0·5–1 mm, numerous, particularly in the younger regions, and often confluent; tetrasporangia terminal or intercalary, immersed in the medulla, 17–40×11–30 μm, with cruciately arranged tetraspores.

Almost exclusively epilithic, from midlittoral to 24 m, in shelter and areas exposed to some wave action; tolerant of fluctuating salinity and therefore occurring in estuaries.

Generally distributed throughout the British Isles.

Iceland; North Russia to southern Spain and possibly Morocco and Cape Verde Isles; western Baltic; Canada (Labrador) to U.S.A. (New Jersey); Japan.

Fronds perennial, persisting for up to 6 years in quiet waters, (Harvey & McLachlan, 1973), often regenerating from the holdfast; growth arrested in winter. There is considerable discrepancy in the records of fruiting periods in the literature; Marshall *et al.* (1949) claim that littoral plants are fertile in summer and sublittoral ones in winter; Darbishire (1902) and Lewis (1936), however, record tetrasporangia and cystocarps on littoral plants between November and March whilst Rosenvinge's largely sublittoral Danish plants were fertile in summer.

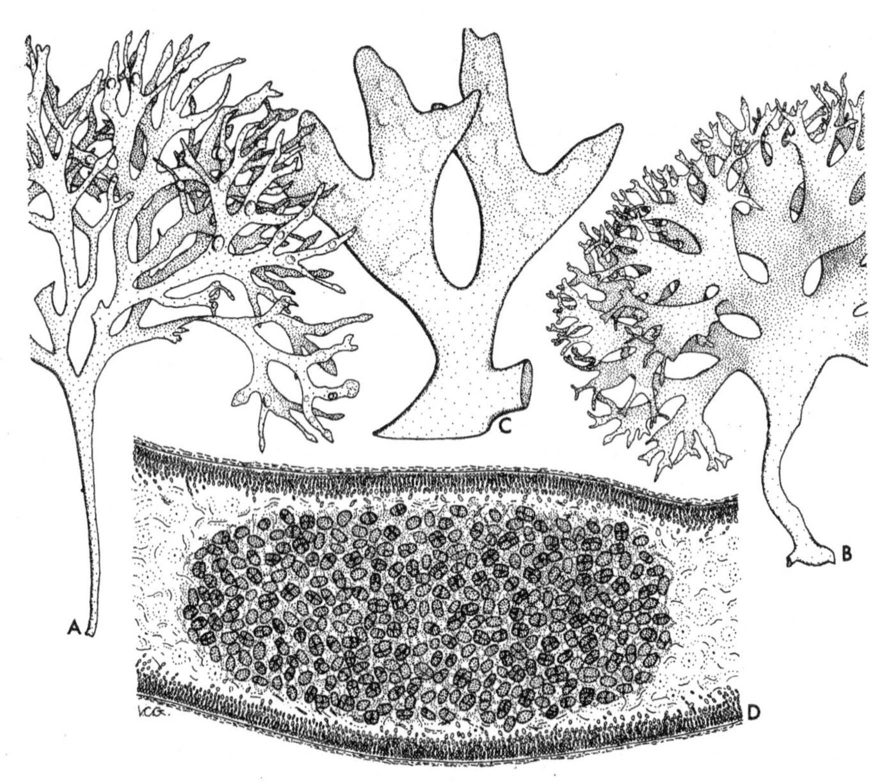

Fig. 86 *Chondrus crispus*
A, B. Two plants showing habit differences (Sep., Nov.) ×1; C. Branch with tetrasporangial sori (Nov.) ×4; D. T.S. same ×80.

Plants of this species are highly variable in appearance. Near low water mark, they are tall, with a long stipe and thick blade but those in pools higher on the shore have a proportionately shorter stipe, are broader and noticeably thinner. On exposed shores the blade lobes are linear; they are broad and overlapping in water of low salinity. See Thomas (1938) for a detailed account of the form range in this species, and the infra-specific names which have been applied. For a comprehensive survey of all aspects of the biology of this species, see Harvey & McLachlan, 1973.

The type of *Fucus norvegicus* Gunn. (1772) belongs to the species currently known as *Chondrus crispus* Stackh. (see Taylor & Chen in Harvey & McLachlan, 1973). The names *Chondrus norvegicus* (Gunn.) Lamour. (1813) and *Gymnogongrus norvegicus* (Gunn.) J. Ag. (1851) have, however, been mistakenly applied to a species of *Gymnogongrus*. These names are therefore candidates for inclusion in the list of rejected names under Article 69 of the International Code of Botanical Nomenclature as modified at the XIIth International Botanical Congress, 1975.

The species closely resembles *Gigartina stellata* (Stackh.) Batt., q.v. *Phyllophora pseudoceranoides* (S. G. Gmel.) Newroth & A. R. A. Taylor and *Gymnogongrus crenulatus* (Turn.) J. Ag. are also similar in habit, but differ internally by having large pseudoparenchymatous medullary cells, the reproductive structures being also very different. On the shore *P. pseudoceranoides* can be distinguished by its usually long terete stipe and short, broadly expanded, but often much subdivided, thin blade; *G. crenulatus* rarely has a long stipe, whilst the lobes are usually parallel-sided and frequently bear verrucose carpotetrasporangial outgrowths.

Fronds are frequently attacked by the fungus *Didymosphaeria danica* (Berlese) Wilson & Knoyle. Galls are also known, produced by nematode worms (Barton, 1901) and bacteria (Schmitz, 1892; Chemin, 1927).

Plants are harvested on both sides of the North Atlantic and processed with *Gigartina stellata* as Irish Moss or Carragheen; for details of widespread and varied applications, see Levring *et al.*, 1969.

GIGARTINA Stackhouse

GIGARTINA Stackhouse (1809), p. 55.

Type species: *Gigartina pistillata* (S. G. Gmelin) Stackhouse (1809), p. 74.

Mammillaria Stackhouse (1809), p. 55.
Chondrodictyon Kützing (1843), p. 396.
Mastocarpus Kützing (1843), p. 398.
Chondracanthus Kützing (1843), p. 399.
Chondroclonium Kützing (1845), p. 302.
Sarcothalia Kützing (1849), p. 739.

Thallus consisting of a discoid base with an erect or occasionally prostrate frond usually divided into a stipe and blade, terete, compressed or flattened; branching lateral or dichotomous; structure multiaxial, with a distinctly filamentous medulla surrounded by a cortex of radial filaments composed of successively smaller cells.

Gametangial plants dioecious; spermatangia in superficial sori, terminal on short, branched filaments; carpogonial branches in cortex of special outgrowths from frond surfaces or margins; carpogonial branch 3-celled, supporting cell functioning as auxiliary cell; enveloping filaments present or absent, gonimoblast filaments producing both nutritive cells and carposporangia in medulla, elevating cortex and protruding externally as more or less globose cystocarps; tetrasporangia developing in rows in more or less deeply immersed sori obvious externally, cruciate; unknown in some species, e.g. *G. stellata*, probably occurring in *Petrocelis*-like encrusting plants.

G. pistillata, *G. acicularis* and *G. teedii* are very restricted in their occurrence in Britain, being apparently at the northern limit of their distribution; our knowledge of them is therefore comparatively limited. *G. pistillata* is related to *G. stellata* but is a reasonably distinct species. *G. teedii* and *G. acicularis* are closely related to each other and specimens are sometimes difficult to ascribe owing to the existence of variants with an apparently intermediate morphology. See Gayral (1958) and Hoek & Donze (1966).

Specimens of *Gigartina* spp. are difficult to preserve and examine microscopically. The cell wall mucilages absorb water, becoming inflated and distorted; the wall thickness shown in illustrations is frequently exaggerated because of this. A more natural appearance can be obtained by preserving the material in alcohol (at least 75 per cent) and examining it in a non-aqueous medium.

Key to Species

1	Frond flattened	2
	Frond compressed or nearly cylindrical	4
2	Frond channelled, U-shaped in T.S.; all but the youngest plants with papillae arising from blade surfaces; branching more or less dichotomous	*G. stellata*
	Frond not channelled, flat in T.S.; papillae marginal when present, branching more or less pinnate	3
3	Branching sparse; fronds arching and reattaching on contact	. *G. acicularis*
	Branching dense; fronds erect, not reattaching	*G. teedii*
4	Fronds erect, tufted; branching primarily dichotomous . .	. *G. pistillata*
	Fronds prostrate, entangled; branching irregular	. *G. acicularis*

Gigartina acicularis (Roth) Lamouroux (1813), p. 136 (reprint p. 44).

Lectotype: Wulfen's description, in the absence of material. Adriatic.

Fucus acicularis Wulfen (1803), p. 63 nom. illeg., non *F. acicularis* Esper (1800), p. 172 [as '132'].
Ceramium aciculare Roth (1806), p. 114.
? *Gigartina falcata* J. Agardh (1851), p. 266.

Thallus consisting of a small attachment disc giving rise to fronds which may be erect or prostrate, erect fronds often arching over, forming secondary small attachment discs upon contact with any substrate; fronds slightly to much compressed, up to 120 mm long and 1–2 mm broad, often with somewhat broader, very pointed apices; cartilaginous, dark reddish-purple (drying black) with the apices often paler, plants sometimes bleached; branching very variable, coarser plants often pinnate with widely separated patent branches, occasionally developing secund adventitious branches later, more slender plants often irregularly branched and entangled.

Structure multiaxial; medullary filaments compact, thick-walled, surrounded by a thick layer of radial cortical filaments with very small cells, about 6–7 μm in surface view.

Spermatangia unknown; cystocarps rare in Britain, paired or solitary, lateral on the younger branches near the apex which is often reflexed, spherical, *c.* 1 mm in diameter, with no obvious pore, carposporangia *c.* 15 μm in diameter; tetrasporangial sori rare in Britain, embedded in the inner cortex and outer medulla producing deeply coloured slight swellings on the younger branches, tetrasporangia 22–40×15–18 μm, with cruciately arranged tetraspores.

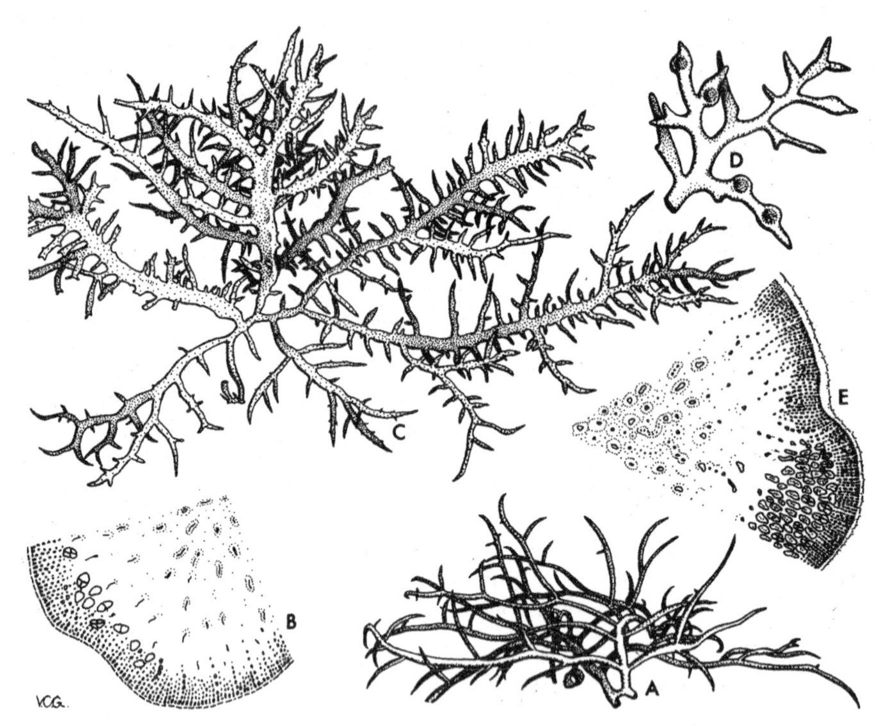

Fig. 87 *Gigartina acicularis*
A. Habit (Sep.) ×1; B. T.S. branch with tetrasporangia (Nov.) ×80.
Gigartina teedii
C. Habit (Sep.) ×1; D. Branch with cystocarps (Aug.) ×4; E. T.S. branch with tetrasporangia (Sep.) ×80.

Epilithic, lower littoral and upper sublittoral in shelter and areas with some exposure to wave action; tolerant of sand cover; sometimes forming a dense turf.

Southern and western shores of the British Isles, eastwards to the Isle of Wight and northwards to Pembroke; in Ireland eastwards to Cork and northwards to Galway.

British Isles to the Cameroons; Azores, Canary Isles; Mediterranean; U.S.A. (North Carolina) to Uruguay.

Little is known of the seasonal growth of this species in the British Isles. Cystocarps have only been recorded in January (Devon) and tetrasporangia only in November (Galway). Gayral (1966) records cystocarps from November to February on the French Channel coast and in August to March in Morocco (1958); Ardré (1969) records tetrasporangia in October in Portugal. It seems reasonable to suppose that plants propagate vegetatively since they readily form secondary attachment discs.

There is a great deal of morphological variation the reasons for which are not fully understood. Large pinnately branched fronds tend to occur on sloping rock surfaces whilst the entangled turf-like plants are found covering flat sandy surfaces. The former, well illustrated by Harvey (1847), have been referred by Ardré (1969) to *G. falcata* J. Ag., though she expresses some doubt that this is a distinct species.

G. acicularis often grows in association with *Gelidium* spp. and *Calliblepharis jubata* (Good. & Woodw.) Kütz., both of which it can closely resemble. It can be distinguished, however, by its multiaxial apices. It is also occasionally confused with *Gracilaria verrucosa* (Huds.) Papenf., q.v. *G. acicularis* is of commercial importance in Portugal and West Africa (Levring *et al.*, 1969).

Gigartina pistillata (S. G. Gmelin) Stackhouse (1809), p. 74.

Lectotype: original illustration in the absence of material (see S. G. Gmelin, 1768, pl. xviii, fig. 1). Type locality: doubtful; sent by Sandifort from Netherlands (The Hague).

Fucus pistillatus S. G. Gmelin (1768), p. 159.
Fucus gigartinus Linnaeus (1759), p. 1344.

Thallus consisting of a group of erect fronds arising from an attachment disc up to 10 mm in diameter; fronds almost terete below to somewhat compressed above, up to 150 (220) mm long and to 2–3 mm in diameter; dark reddish or brownish purple, sometimes paler above, cartilaginous; branching repeatedly dichotomous (to 4 times), a few secondary lateral branches sometimes appearing in old plants and after damage; branches usually in one plane, but axes twisted.

Structure multiaxial; medulla extensive, composed of thick-walled filaments, surrounded by a thin layer of radial cortical filaments, cell lumens 3–4 μm in surface view, with much mucilage between them.

Spermatangia not recorded for British Isles, gametangial plants dioecious elsewhere; special simple or forked patent reproductive branchlets, usually in a row on the margins of the compressed distal branches, bear spherical cystocarps to 3 mm in diameter, with an obvious pore; carposporangia about 16–20 μm in diameter. Tetrasporangial sori immersed in the inner cortex and outer medulla along the margins of the distal compressed branches, visible externally as deeply-coloured longitudinally elongated swellings 1–2 mm in extent and sometimes confluent; tetrasporangia $22–40 \times 11–25$ μm, with cruciately arranged tetraspores.

Epilithic, upper sublittoral, in pools and occasionally emergent; in areas of shelter and some exposure to wave action; tolerant of sand cover.

Southern and western shores of the British Isles, extending eastwards to Devon and northwards to Pembroke; Cork.

British Isles to South Africa.

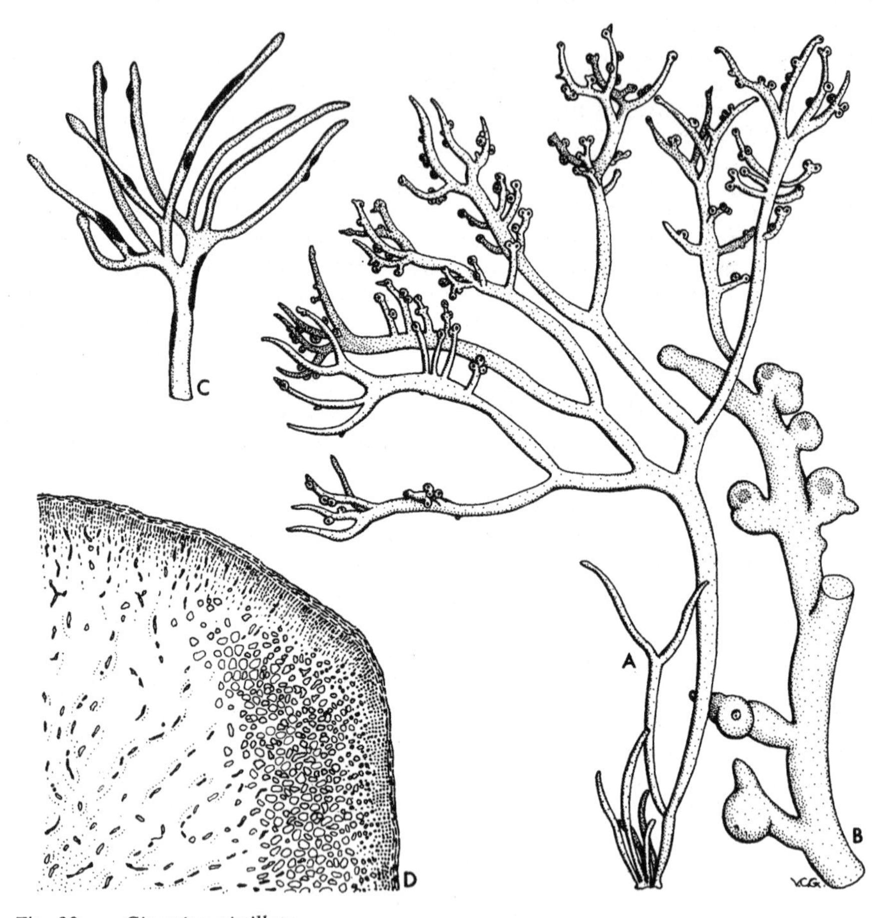

Fig. 88 *Gigartina pistillata*
A. Habit of plant with cystocarps (Apr.) ×1; B. Branch of same ×4; C. Branch of tetrasporangial plant (Apr.) ×4; D. T.S. same with undivided tetrasporangia ×80.

Fronds annual, according to Gayral (1966). Cystocarps recorded from March to December, most commonly in September and October; probably developing during the summer and persisting after spore release. Tetrasporangial records few, April, September and December.

Plants show comparatively little variation in form.

Plants with some branches bearing tetrasporangia and others bearing cystocarps have been recorded for South Africa (Isaac & Simons, 1954) and Spain (Price, unpublished). The species strongly resembles *Furcellaria lumbricalis* (L.) Lamour. and *Polyides rotundus* (Huds.) Grev., especially when the characteristic pinnate branchlets bearing spherical cystocarps are absent; the branches are rather broader, however, and compressed in the upper parts, and the angle of branching tends to be wider (*c.* 60°). The tetrasporangia are much smaller than in the other two species and are embedded in small marginal sori. Sterile plants can be distinguished by the small-celled cortical filaments which disintegrate in fresh water and also when remoistened after drying. This species is used as food and agar raw material (Levring *et al.*, 1969) in some parts of the world.

Gigartina stellata (Stackhouse in Withering) Batters (1902), p. 64.

Lectotype: Stackhouse's description in Withering (1796), (specimen missing).

Fucus stellatus Stackhouse in Withering (1796), p. 99.
Fucus mamillosus Goodenough & Woodward (1797), p. 174.
Gigartina mamillosa (Goodenough & Woodward) J. Agardh (1842), p. 104.
? *Fucus coronopifolius* Zoega (1772), appendix p. 19.
? *Gigartina coronopifolia* (Zoega) Silva (1952), p. 264.

Thallus consisting of an attachment disc sometimes to 50 mm in extent, giving rise to one or more erect fronds to 170 mm, with a narrow stipe-like portion below expanding gradually into a blade, the divisions of which vary from 1–8 mm broad; frond channelled (concavo-

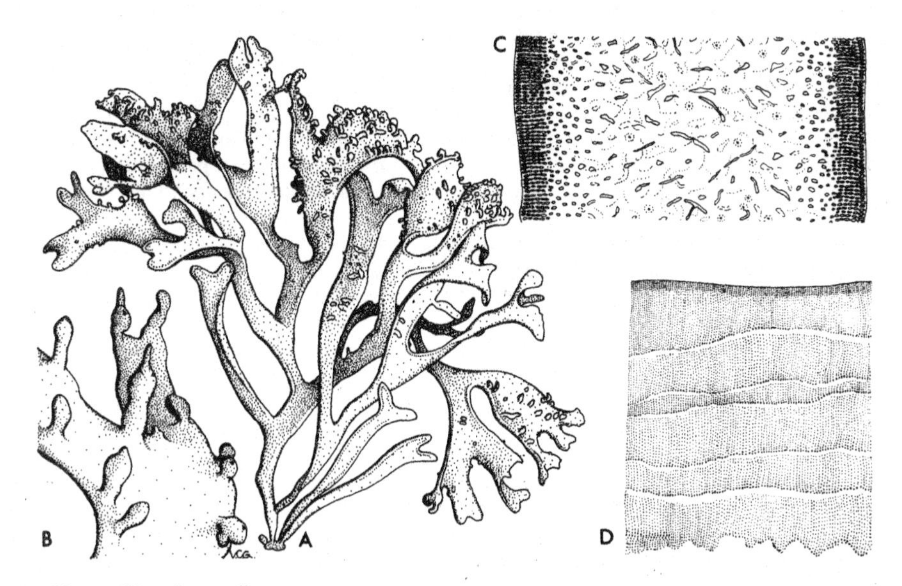

Fig. 89 *Gigartina stellata*
A. Habit of plant with cystocarps (Aug.) ×1; B. Branch of same ×4; C. T.S. blade (Aug.) ×80; D. V.S. basal disc on rock (Apr.) ×80.

convex) throughout with a slightly thickened margin, frequently further twisted and contorted; cartilaginous; reddish brown or purple to almost black or greenish, sometimes bleached; branching repeatedly dichotomous, sometimes rather irregular; mature plants beset with reproductive papillae on one or sometimes both blade surfaces.

Structure multiaxial; medulla consisting of very thick-walled filaments surrounded by radial cortical filaments of very small cells 2–5 μm in surface view.

Gametangial plants dioecious (? sometimes monoecious); spermatangia superficial, occurring on linear papillae, developing terminally on short, branched filaments produced by cortical cells; carpogonial branches occurring in the cortex of special rounded papillae, enveloping filaments very few or absent, cortex elevating and cystocarps protruding externally, up to 1 mm in diameter, furnished with a small lateral pore, carposporangia 12–20 μm; tetrasporangia unknown in the erect fronds, probably occurring in *Petrocelis*-like encrusting plants.

Epilithic in lower littoral and upper sublittoral, flourishing on shores exposed to wave action; occasionally extending into deeper water and higher on the shore in pools, under fucoid cover or where water runs down the shore; tolerant of lowered salinity but not of mud or sand and readily showing damage (bleaching) from insolation.

Generally distributed throughout the British Isles, less frequent in southeastern England because of lack of suitable substrate.

Iceland; Faeroes; North Russia to Rio de Oro; Canada (Newfoundland) to U.S.A. (North Carolina); reports from Pacific difficult to assess owing to the uncertainty of specific limits.

Plants perennial, holdfast persisting for several years, many erect fronds lost during the winter, new fronds forming in spring, conspicuous because of their deep pink colour; these gradually darken and become more cartilaginous, producing the reproductive papillae. Changes occur internally, the cortical filaments increasing in length from 4 to 10 cells, the cell walls thickening and the number of secondary pits increasing. These changes can be correlated with the gel-producing properties of the plant. Male plants have been recorded for February, March, July, September and December. The cystocarps mature in July and August and the maximum period of carpospore discharge appears to be September–December.

Plants are extremely variable in morphology and coloration, the divisions of the frond ranging from linear to broadly fanshaped; for a discussion of these forms and their possible relationships to growth conditions see Marshall *et al.* (1949) where reference is also made to the infraspecific names which have been used.

West (1972) has shown that *Petrocelis franciscana* Setch. & Gardn. represents the tetrasporangial phase of a species of *Gigartina* on the Pacific coast of North America. A similar relationship may exist between *Petrocelis* sp. and *G. stellata,* although Edelstein, Chen & McLachlan (1974) concluded that in eastern Canada the complete life history involves only non-sexual haploid plants.

Although closely similar to *Chondrus crispus* Stackh. and often growing with it, *G. stellata* can be distinguished by its incurved frond with thickened margins and the presence of reproductive papillae in all but the youngest plants i.e. in the spring. At this time, *Chondrus* plants are likely to show either cystocarps or tetrasporangial sori in the fronds.

This species, together with *Chondrus crispus,* is collected as carrageenan raw material on an industrial scale; the gatherers do not distinguish between the two species.

Gigartina teedii (Roth) Lamouroux (1813), p. 137 (reprint p. 49).

Isotype: BM-K. Portugal (Lisbon, *fide* label).

Ceramium teedii Roth (1806), p. 108.

Thallus consisting of small attachment disc and 1–2 more or less erect or arcuate fronds up to 300 mm in length; main axes flattened, up to 3 (5) mm broad, somewhat translucent, reddish purple with green or yellow tinges, frequently bleached above; surface shiny when dry, not becoming black; cartilaginous; primary branching sparse to frequent, irregularly dichotomous, closely beset with secondary pinnate branches, themselves simple or further divided; pinnate branches distichously inserted but projecting at all angles and with long acuminate tips; sometimes showing unequal growth in opposite branches of a pair, the pairs alternating in this respect.

Structure multiaxial, internal anatomy as in *G. acicularis*.

Spermatangia unknown; carpogonial branches occurring near the branch apices lying at the junction of the medulla and cortex, carposporangia terminal, medullary filaments forming a compact layer around the gonimoblast, not interweaving amongst it; cortex thickening and later elevating, cystocarps up to 1 mm in diameter, protruding externally, immediately behind the apex, usually one per branch, carposporangia about 13–20 μm in diameter, spores released through a small pore; tetrasporangial sori embedded, producing small deeply coloured swellings on the branch margins particularly in the younger parts, tetrasporangia 33–40 × 22–30 μm, with cruciately arranged tetraspores. (Fig. 87, p. 238.)

Epilithic in sheltered areas; upper sublittoral to at least 3 m.

Cornwall and South Devon.

British Isles to Angola; Mediterranean; Azores; Cape Verde Isles; Japan.

Little is known of the behaviour of the species in Britain because of its extreme rarity. Cystocarps have been discovered in August and September and tetrasporangia in September in plants dredged from the Yealm estuary, Devon. Gayral (1966) reports cystocarps for the autumn and winter in France and the reproductive period appears longer further south, but nothing is known of the length of the cystocarp maturation period.

There is little form variation in the British Isles.

Gayral (1958) and Donze (1968) describe and illustrate hybrids between *G. teedii* and *G. pistillata* occurring in Morocco and N.W. Spain; these have not so far been encountered in Britain. Hoek & Donze (1966) describe intermediate forms between *G. teedii* and *G. acicularis* on the Côte Basque (S.W. France).

G. teedii is used as food in Japan.

PETROCELIS J. Agardh

PETROCELIS J. Agardh (1852), p. 489.

Type species: *P. cruenta* J. Agardh (1852), p. 490.

Thallus a comparatively thick crust closely adherent to substrate, mucilaginous but firm; consisting of a layer of prostrate filaments giving rise to simple or little branched ascending filaments, which gradually become erect, and short descending filaments below.

Gametangial plants described only for *P. hennedyi*, monoecious; spermatangia lateral near apices of erect filaments, carpogonial branches usually 2 celled, auxiliary cells remote, gonimoblasts scattered among erect filaments, most cells becoming carposporangia; tetrasporangia intercalary in erect filaments, cruciate.

In common with other encrusting algae, this genus is not well understood; further study may indicate that *P. cruenta* and *P. hennedyi* are conspecific.

KEY TO SPECIES

1 Tetrasporangia occurring singly in a filament . . . . *P. cruenta*
  Tetrasporangia occurring in rows of (1–)6–8(–10) in a filament *P. hennedyi*

Petrocelis cruenta J. Agardh (1852), p. 490.

Lectotype: LD (Herb. Alg. Agardh. 27581). France (Brest) (see Dixon & Irvine, 1977).

Cruoria pellita sensu Harvey, (1847), pl. cxvii, non (Lyngbye) Fries (1836), p. 316.

Crusts smooth, to 50 mm or more in extent, and to about 1 mm in thickness, dark brownish or greenish red, drying black and glossy; erect filaments about 4–6 μm in diameter, not much tapering, cells once or twice as long, cell fusions usually present, thallus compact below, loosely held together above by mucilage but filaments firmly united at apices.

Gametangial plants unknown; tetrasporangial primordia about 10×7 μm increasing after division up to about 40×20 μm, elliptical, intercalary, borne singly in erect filaments, tetraspores usually cruciately arranged, sometimes irregular.

Epilithic and on shell, sometimes spreading over encrusting calcareous algae, from midlittoral to upper sublittoral; in semi-exposed to exposed places with some protection from direct wave action.

Generally distributed throughout the British Isles.

Norway (Nordland) to Morocco; western Baltic.

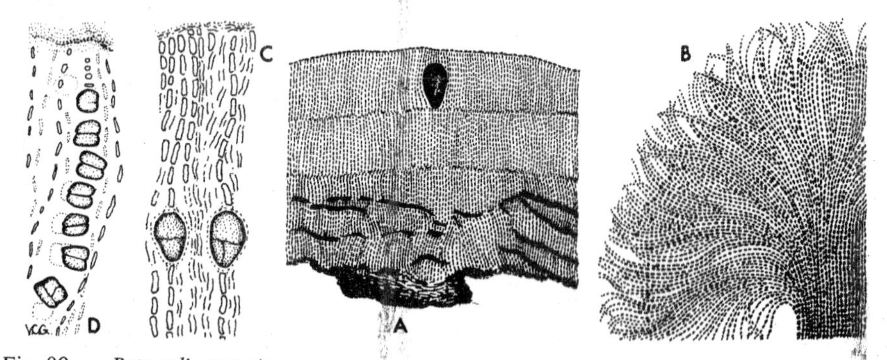

Fig. 90 *Petrocelis cruenta*
A. V.S. crust on rock showing growth zones and endophytic *Codiolum petrocelidis* (Sep.) ×80; B. V.S. crust with intercalary tetrasporangia (Dec.) ×80; C. Part of same ×300.
Petrocelis hennedyi (Type)
D. V.S. crust with tetrasporangia (undated) ×300.

Crust perennial, though little is known about seasonal growth. Fusions between cells of adjacent filaments are common and may form a layer above which filaments are shed. Tetrasporangia initiated in mid-November, dividing in December and persisting until February. Almost all filaments produce sporangia and the upper parts are lost after spore release. Ardré (1969) recorded tetrasporangia between March and October in Portugal.

Attacks by grazing animals, and filament shedding especially after spore release account at least partially for the differences in filament length and other variations in thallus structure.

This species is sometimes confused with *Cruoria pellita* (Lyngb.) Fries, q.v.

The green endophyte *Codiolum petrocelidis* Kuckuck (see Parke & Dixon, 1976) frequently occurs just below the surface of the thallus.

Petrocelis hennedyi (Harvey) Batters (1890), p. 314 (reprint p. 94).

Lectotype: TCD. Isotype: BM. Bute (Cumbrae).

Actinococcus hennedyi Harvey (1857), p. 202.
Petrocelis ruprechtii Hauck (1885), p. 30.

Vegetative plants as in *P. cruenta*.

Reputed gametangial plants monoecious; spermatangia in groups of 2 or 3 on small, usually unicellular, branches near apices of the erect filaments; all gonimoblast cells forming carposporangia 14–17 μm in diameter, in loose spindle-shaped aggregations among the vegetative filaments; tetrasporangia 16–23 × 14–17 μm, intercalary in the erect filaments, in rows of (1–)6–8(–10), without sterile cells between, tetraspores usually cruciately arranged, sometimes irregular.

On the stipe and holdfast of *Laminaria* plants, stones and shells in the lower littoral and upper sublittoral. It is known to a depth of 19 m in Denmark.

Northumberland, Bute, Orkney, Sussex. The species may well prove to be generally distributed in the British Isles.

Norway (Trondelag) to British Isles; Denmark; Baltic. Said to be widely distributed in the northern North Atlantic and Arctic sea.

Little is known of the seasonal aspects of growth. Records indicate that reproduction occurs only between November and March, but more data are required.

The form variation is probably similar to that described for *P. cruenta*.

No vegetative differences have been found to distinguish this species from *P. cruenta* and so only material collected between November and February can be identified. The descriptions of spermatangia, carposporophytes etc. given by Batters (1890) and Rosenvinge (1917) require re-investigation in view of the discovery of a *Petrocelis*-like stage in the life history of a species of *Gigartina* (West, 1972). Rosenvinge also describes single terminal or subterminal sporangia but Denizot (1968) doubts that these belong to the same species.

This species is sometimes confused with *Cruoria pellita* (Lyngb.) Fries, q.v.

The green alga *Codiolum petrocelidis* Kuckuck (see Parke & Dixon, 1976) occurs immersed in the upper part of the thallus.

REFERENCES FOR GIGARTINALES

ADANSON, M. (1763). *Familles des Plantes.* Paris.

AGARDH, C. A. (1817). *Synopsis Algarum Scandinaviae.* Lundae.

—— (1822). *Species Algarum* **1** (2). Lundae.

AGARDH, J. G. (1842). *Algae Maris Mediterranei et Adriatici.* Parisiis.

—— (1849). Algologiska bidrag. *Ofvers. K. Vetensk Akad. Förh. Stockh.* **6**: 79–89.

—— (1851). *Species Genera et Ordines Algarum.* **2**(1). Lundae.

—— (1852). *Species Genera et Ordines Algarum.* **2**(2). Lundae.

—— (1862). *Om Spetsbergens Alger.* Lund.

—— (1876). *Species Genera et Ordines Algarum.* **3**(1). Lundae.

ARDISSONE, F. & STRAFFORELLO, J. (1877). *Enumerazione delle Alghe di Liguria.* Milano.

ARDRÉ, F. (1969). Contribution à l'étude des algues marines du Portugal. *Port. Acta biol.*, ser. B **10**: 137–555 [reprint 1–423].

AUSTIN, A. P. (1960). Life history and reproduction of *Furcellaria fastigiata* (L.) Lamour. *Ann. Bot.* N.S. **24**: 257–274, 296–310.

—— (1960a). Observations on the growth, fruiting and longevity of *Furcellaria fastigiata* (L.) Lam. *Hydrobiologia* **15**: 193–207.

BARTON, E. S. (1901). On certain galls in *Furcellaria* and *Chondrus*. *J. Bot., Lond.* **39**: 49–51.

BATTERS, E. A. L. (1890). A list of the marine algae of Berwick-on-Tweed. *Hist. Berwicksh. Nat. Club* **12**: 221–392 [reprint 1–172].

—— (1892). New or critical British algae. *Grevillea* **21**: 49–53.

—— (1896). New or critical British marine algae. *J. Bot., Lond.* **34**: 384–390.

—— (1897). New or critical British marine algae. *J. Bot., Lond.* **35**: 433–440.

—— (1900). New or critical British marine algae. *J. Bot., Lond.* **38**: 369–379.

—— (1902). A catalogue of the British marine algae. *J. Bot., Lond.* **40** (Suppl.): 1–107.

BLIDING, C. (1928). Studien über die Florideenordnung Rhodymeniales. *Acta. Univ. lund.* Ny Följd, Avd. 2 **24** (3): 1–74.

BOALCH, G. T. (1964). British records of *Ceratocolax*. *Br. phycol. Bull.* **2**: 380–381.

BØRGESEN, F. (1932). Revision of Forsskål's Algae. *Dansk bot. Ark.* **8** (2): 1–14.

—— (1938). *Catenella nipae* used as food in Burma. *J. Bot., Lond.* **76**: 265–271.

BOILLOT, A. (1965). Sur l'alternance de générations hétéromorphes d'une Rhodophycée, *Halarachnion ligulatum* (Woodward) Kützing (Gigartinales, Furcellariacées). *C. r. hebd. Séanc. Acad. Sci., Paris* **261**: 4191–4193.

BORNET, E. (1892). Les Algues de P.-K.-A. Schousboe. *Mém. Soc. natn. Sci. nat. math. Cherbourg* **28**: 165–376.

BORNET, E. & THURET, G. (1876). *Notes Algologiques,* **1**. Paris.

BORY DE ST VINCENT, J. B. (1826). Laminaires. *In: Dictionnaire Classique d'Histoire Naturelle* **9**. Paris.

BREBNER, G. (1896). Algological notes for Plymouth district. *Rep. Br. Ass. Advmt Sci.* **1896**: 485—486.

BUFFHAM, T. H. (1888). On the reproductive organs, especially the antheridia, of some of the Florideae. *J. Quekett microsc. Club,* ser. **23**: 257–266.

CABIOCH, J. (1969). Les fonds de maerl de la baie de Morlaix et leur peuplement végétal. *Cah. Biol. mar.* **10**: 139–161.

CAUSEY, N. B., PRYTHERCH, J. P., MCCASKILL, J. P., HUMM, H. J. & WOLF, F. A. (1946). Influence of environmental factors upon the growth of *Gracilaria confervoides.* *Bull. Duke Univ. Mar. Stn* **3**: 19–24.

CHEMIN, E. (1927). Action des bactéries sur quelques algues rouges. *Bull. Soc. bot. Fr.* **74**: 441–451.

—— (1927a). Sur le développement des spores d'*Actinococcus peltaeformis* Schm. et la signification biologique de cette Algue. *Bull. Soc. bot. Fr.* **74**: 912–920.

—— (1929) Développement des spores issues du cystocarpe de *Gymnogongrus norvegicus* J. Ag. *Bull. Soc. bot. Fr.* **76**: 305–308.

—— (1931). Sur le présence de galles chez quelques Floridées. *Revue algol.* **5**: 315–325.

Clemente y Rubio, S. de R. (1807). *Ensayo sobre las Variedades de la Vid Comun que Vegetan en Andalucia, con un Indice Etimológico y Tres Listas de Plantas en que se Caracterizan Varias Especias Nuevas.* Madrid.

Cotton, A. D. (1912). Marine Algae. *In:* Praeger, R. L., A biological survey of Clare Island in the county of Mayo, Ireland and of the adjoining district. *Proc. R. Ir. Acad.* **31**, sect. 1 (15): 1–178.

Crouan, P. L. & H. M. (1852). *Algues marines du Finistère,* **2.** *Floridées.* Brest.

—— (1858). Note sur quelques algues marines nouvelles de la rade de Brest. *Annls Sci. nat.,* sér. 4, Bot. **9**: 69–75.

—— (1867). *Florule du Finistère.* Paris & Brest.

Darbishire, O. V. (1902). L.M.B.C. Memoirs. 9. *Chondrus. Proc. Trans. Lpool biol. Soc.* **16**: 429–470.

Dawson, E. Y. (1949). Studies on northeast Pacific Gracilariaceae. *Occ. Pap. Allan Hancock Fdn* **7**: 1–54.

Decandolle, A. P. (1805). *Flore Française,* ed. 3, **2**. Paris.

Denizot, M. (1968). *Les algues floridées encroûtantes (à l'exclusion des Corallinacées).* Paris.

Dixon, P. S. (1964). Taxonomic and nomenclatural notes on the Florideae IV. *Bot. Not.* **117**: 56–78.

—— (1967). The typification of *Fucus cartilagineus* L. and *F. corneus* Huds. *Blumea* **15**: 56–62.

Dixon, P. S. & Irvine, L. M. (1977). Miscellaneous notes on algal taxonomy and nomenclature. IV. (in preparation).

Donati, V. (1758). *Essai sur l'Histoire Naturelle de la Mer Adriatique.* La Haye.

Donze, M. (1968). The algal vegetation of the Ria de Arosa, (N.W. Spain). *Blumea* **16**: 159–192.

Drew, K. M. (1958). The typification of *Polyides* and *Furcellaria*. *J. Linn. Soc., Bot.* **55**: 744–752.

Duby, J. E. (1830). *Botanicon Gallicum,* ed. 2, **2**. Paris.

Dumortier, B.-C. (1829). *Analyse des Familles des Plantes avec l'Indication des Principaux Genres qui s'y Rattachent.* Tournay.

Edelstein, T., Chen, L. C.-M. & McLachlan, J. (1974). The reproductive structures of *Gigartina stellata* (Stackh.) Batt. (Gigartinales, Rhodophyceae) in nature and culture. *Phycologia* **13**: 99–107.

Endlicher, S. L. (1843). *Genera Plantarum Secundum Ordines Naturales Disposita.* Suppl. 3. Vindobonae.

Engler, A. (1892). *Syllabus der Vorlesungen über specielle und medicinisch-pharmaceutische Botanik.* Grosse Ausg. Berlin.

Esper, E. J. C. (1800) *Icones Fucorum,* **4**. Nürnberg.

Farlow, W. G. (1875). List of the marine algae of the United States with notes on new or imperfectly known species. *Proc. Am. Acad. Arts Sci.* **10**: 351–380.

Farnham, W. R. & Fletcher, R. L. (1974) Sur la présence à Roscoff du *Porphyrodiscus simulans* Batters, et sa rélation avec l'*Ahnfeltia plicata* (Huds.) Fries. *Trav. Stn biol. Roscoff* N.S. **20**: 9.

Feldmann, G. (1967). Le genre *Cordylecladia* J. Ag. (Rhodophycées, Rhodyméniales) et sa position systématique. *Rev. gén. Bot.* **74**: 357–375.

Feldmann, J. (1954). Recherches sur la strûcture et la développement des Calosiphoniacées (Rhodophycées – Gigartinales). *Rev. gén. Bot.* **61**: 453–499.

Forskål, P. (1775). *Flora Aegyptiaco-Arabica.* Havniae.

Fries, E. M. (1836). *Corpus Florarum Provincialium Sueciae, I Floram Scanicam.* Upsaliae.

Gatty, M. (1872). *British Seaweeds,* **2**. London.

Gayral, P. (1958). *Algues de la Côte Atlantique Marocaine.* Rabat.

—— (1966). *Les Algues des Côtes Françaises. (Manche et Atlantique).* Paris.

GIBSON, R. J. H. (1892). On the structure and development of the cystocarps of *Catenella Opuntia*, Grev. *J. Linn. Soc., Bot.* **29**: 68–76.

GMELIN, S. G. (1768). *Historia Fucorum.* Petropoli.

GOODENOUGH, S. & WOODWARD, T. J. (1797). Observations on the British *Fuci*, with particular descriptions of each species. *Trans. Linn. Soc.* **3**: 84–235.

GREGORY, B. D. (1934). On the life-history of *Gymnogongrus griffithsiae* Mart. and *Ahnfeltia plicata* Fries. *J. Linn. Soc., Bot.* **40**: 531–551.

GREVILLE, R. K. (1821). Description of a new species of *Fucus* found in Devonshire. *Mem. Wernerian nat. Hist. Soc.* **3**: 396–399.

—— (1824). *Flora Edinensis.* Edinburgh.

—— (1829). Descriptiones novarum specierum ex algarum ordine. *Nova Acta Acad. Caesar. Leop. Carol.* **14**: 423–424.

—— (1830). *Algae Britannicae.* Edinburgh.

GUIRY, M. D. & HOLLENBERG, G. J. (1975). *Schottera* gen. nov. and *Schottera nicaeensis* (Lamour. ex Duby) comb. nov. (=*Petroglossum nicaeense* (Lamour. ex Duby) Schotter) in the British Isles. *Br. phycol. J.* **10**: 149–164.

GUIRY, M. D. & IRVINE, L. M. (1976). A first record of the red alga *Phyllophora sicula* (Kützing) comb. nov. (=*P. palmettoides* J. Agardh) for Ireland. *Ir. Nat. J.* **18**: 284.

GUNNERUS, J. E. (1766). *Flora Norvegica,* **1**. Nidrosiae.

—— (1772). *Flora Norvegica,* **2**. Nidrosiae.

HARVEY, M. J. & MCLACHLAN, J. (1973). *Chondrus crispus. Proc. Trans. Nova Scotian Inst. Sci.* **27** (Suppl.): 1–155.

HARVEY, W. H. (1840). *Algae. In:* Hooker, W. J. & Arnott, G. A. W. *The Botany of Captain Beechey's Voyage,* **9**. London.

—— (1846). *Phycologia Britannica,* pl. i–lxxii. London.

—— (1847). *Phycologia Britannica,* pl. lxxiii–cxliv. London.

—— (1848). *Phycologia Britannica,* pl. cxlv–ccxvi. London.

—— (1857). Short descriptions of some new British algae with two plates. *Nat. Hist. Rev.* **4**: 201–204.

HAUCK, F. (1885). *Die Meeresalgen Deutschlands und Oesterreichs. In:* Rabenhorst, L. *Kryptogamen-Flora von Deutschland, Oesterreich und der Schweiz,* ed. 2, **2**. Leipzig.

HOEK, C. VAN DEN & DONZE, M. (1966). The algal vegetation of the rocky Côte Basque (S.W. France). *Bull. Cent. Etud. Rech. scient., Biarritz* **6**: 289–319.

HOLMES, E. M. (1883). *Rhodymenia palmetta,* var. *nicaeensis. J. Bot., Lond.* **21**: 289–290.

HORNEMANN, J. W. (1834). *Flora Danica,* Fasc. **36**. Havniae.

HUDSON, W. (1762). *Flora Anglica.* London.

—— (1778). *Flora Anglica,* ed. 2. London.

ISAAC, W. E. & SIMONS, S. M. (1954). Some observations on *Gigartina pistillata* (Gmel.) Stackh. from Port Alfred with a record of plants bearing both tetraspores and carpospores. *Jl S. Afr. Bot.* **20**: 117–124.

JOHNSTONE, W. G. & CROALL, A. (1859). *The Nature-printed British Seaweeds,* **1**. London.

JONES, W. E. (1958). Experiments on some effects of certain environmental factors on *Gracilaria verrucosa* (Huds.) Papenf. *J. mar. biol. Ass. U.K.* **38**: 153–167.

—— (1959). The growth and fruiting of *Gracilaria verrucosa* (Hudson) Papenfuss. *J. mar. biol. Ass. U.K.* **38**: 47–56.

—— (1962). The identity of *Gracilaria erecta* (Grev.) Grev. *Br. phycol. Bull.* **2**: 140–144.

KILLIAN, C. (1914). Uber die Entwicklung einiger Florideen. *Z. Bot.* **6**: 209–278.

KUCKUCK, P. (1912). Beiträge zur Kenntnis der Meeresalgen. 12. Ueber *Platoma bairdii* (Farl.) Kck. *Wiss. Meeresunters. N.F., Abt. Helgoland* **5**: 187–208.

KÜTZING, F. T. (1843). *Phycologia Generalis.* Leipzig.

—— (1845). *Phycologia Germanica.* Nordhausen.

—— (1847). Diagnosen und Bemerkungen zu neuen oder kritischen Algen. *Bot. Ztg* **5**: 1–5, 22–27.

—— (1849). *Species Algarum.* Lipsiae.

KYLIN, H. (1925). The marine red algae in the vicinity of the Biological Station at Friday Harbor, Wash. *Acta Univ. lund.*, Ny Följd, Avd. 2 **21** (9): 1–87.
—— (1928). Entwicklungsgeschichtliche Florideenstudien. *Acta Univ. lund.*, Ny Följd, Avd. 2 **24** (4): 1–127.
—— (1930). Uber die Entwicklungsgeschichte der Florideen. *Acta Univ. lund.*, Ny Följd, Avd. 2 **26** (6): 1–103.
—— (1932). Die Florideenordnung Gigartinales. *Acta Univ. lund.*, Ny Följd, Avd. 2 **28** (8): 1–88.
—— (1956). *Die Gattungen der Rhodophyceen.* Lund.

LAMOUROUX, J. V. F. (1813). Essai sur les genres de la famille des Thalassiophytes non articulées. *Annls Mus. Hist. nat. Paris* **20**: 21–47, 115–139, 267–293 [reprint 1–84].

LEVRING, T., HOPPE, H. A. & SCHMID, O. J. (1969). *Marine Algae. A Survey of Research and Utilization.* Hamburg.

LEWIS, E. A. (1936). An investigation of the seaweeds within a marked zone of the shore at Aberystwyth, during the year 1933–34. *J. mar. biol. Ass. U.K.* **20**: 615–620.

LIGHTFOOT, J. (1777). *Flora Scotica,* **2**. London.

LINK, H. F. (1809). Nova plantarum genera e classe Lichenum, Algarum, Fungorum. *Neues J. Bot.* **3**: 1–19.

LINNAEUS, C. (1753). *Species Plantarum,* **2**. Holmiae.
—— (1759). *Systema Naturae per Regna Tria Naturae,* ed. 10, **2**. *Regnum Vegetabile.* Holmiae.
—— (1763). *Species Plantarum,* ed. 2, **2**. Holmiae.
—— (1767). *Mantissa Plantarum.* Holmiae.

LYLE, L. (1920). The marine algae of Guernsey. *J. Bot., Lond.* **58** (Suppl. 2): 1–53.

LYNGBYE, H. C. (1819). *Tentamen Hydrophytologiae Danicae.* Hafniae.

MARSHALL, S. M., NEWTON, L. & ORR, A. P. (1949). *A study of Certain British Seaweeds and their Utilisation in the Preparation of Agar.* London.

MARTIUS, C. F. P. VON (1828). *Flora Brasiliensis,* **1**. Stuttgartiae & Tubingae.

MAYHOUB, H. (1974). Cycle du développement de *Calosiphonia vermicularis* (J. Agardh) Schmitz (Rhodophycée, Gigartinale). *C. r. hebd. Séanc. Acad. Sci. Paris,* sér. D **277**: 1137–1140.
—— (1975). Nouvelles observations sur le cycle de développement du *Calosiphonia vermicularis* (J. Ag.) Sch. (Rhodophycée, Gigartinale). *C. r. hebd. Séanc. Acad. Sci., Paris,* sér. D **280**: 2441–2443.

MONTAGNE, J. F. C. (1846). *Algues. In:* Durieu de Maisonneuve, M.C., *Exploration Scientifique de l'Algérie pendant 1840–42, Botanique,* **1**, *Cryptogamie.* Paris.

MUELLER, O. F. (1777). *Flora Danica,* Fasc. **12**. Havniae.

NAGELI, C. (1847). Die neuern Algensysteme. *Neue Denkschr. allg. schweiz. Ges. ges. Naturw.* **9** (unnumbered art. no. 2): 1–275.

NEES VON ESENBECK, C. G. (1820). *Horae Physicae Berolinensis.* Bonnae.

NEWROTH, P. R. & TAYLOR, A. R. A. (1968). The distribution of *Ceratocolax hartzii. Br. phycol. Bull.* **3**: 543–546.
—— (1971). The nomenclature of the north atlantic species of *Phyllophora* Greville. *Phycologia* **10**: 93–97.

NEWTON, L. (1931). *A Handbook of the British Seaweeds.* London.

PALLAS, P. S. (1776). *Reise durch verschiedene Provinzen der Russischen Reichs,* **3**. St Petersburg.

PAPENFUSS, G. F. (1950). Review of the genera of algae described by Stackhouse. *Hydrobiologia* **2**: 181–208.
—— (1966). A review of the present system of classification of the Florideophycideae. *Phycologia* **5**: 247–255.

PARKE, M. (1953). A preliminary check-list of British marine algae. *J. mar. biol. Ass. U.K.* **32**: 497–520.

—— & DIXON, P. S. (1964). A revised check-list of British marine algae. *J. mar. biol. Ass. U.K.* **44**: 499–542.
—— & —— (1968). Check-list of British marine algae – second revision. *J. mar. biol. Ass. U.K.* **48**: 783–832.
—— & —— (1976). Check-list of British marine algae – third revision. *J. mar. biol. Ass. U.K.* **56**: 527–594.
RICHARDSON, N. & DIXON, P. S. (1970). Culture studies on *Thuretellopsis peggiana* Kylin. *J. Phycol.* **6**: 154–159.
ROSENVINGE, L. K. (1893). Grønlands havalger. *Meddr Grønland* **3**: 765–981.
—— (1898). Deuxième mémoire sur les algues du Groenland. *Meddr Grønland* **20**: 1–128.
—— (1917). The marine algae of Denmark. Contributions to their natural history. Part II. Rhodophyceae II (Cryptonemiales). *K. danske Vidensk. Selsk. Skr.* 7 Raekke, Nat. Math. Afd. **7**: 153–284.
—— (1931). The marine algae of Denmark. Contributions to their natural history. Part IV. Rhodophyceae IV (Gigartinales, Rhodymeniales, Nemastomatales). *K. danske Vidensk. Selsk. Skr.* 7 Raekke, Nat. Math. Afd. **7**: 489–628.
ROTH, A. W. (1806). *Catalecta Botanica*, **3**. Lipsiae.
SCHMITZ, F. (1889). Systematische Übersicht der bisher bekannten Gattungen der Florideen. *Flora, Jena* **72**: 435–456.
—— (1892). Knöllchenartige Auswüchse an den Sprossen einiger Florideen. *Bot. Ztg* **50**: 624–630.
—— (1893). Die Gattung *Actinococcus* Kütz. *Flora, Jena* **77**: 367–418.
—— (1894). Kleinere Beiträge zur Kenntniss der Florideen. *Nuova Notarisia* **5**: 608–635.
SCHOTTER, G. (1952). Note sur le *Gymnogongrus nicaeensis* (Duby) Ardissone & Strafforello. *Bull. Soc. Hist. nat. Afr. N.* **43**: 203–210.
—— (1968). Recherches sur les Phyllophoracées. Notes posthumes publiées par Jean Feldmann et Marie France Magne. *Bull. Inst. océanogr. Monaco* **67**: 1–99.
SEARLES, R. B. (1968). Morphological studies of red algae of the order Gigartinales. *Univ. Calif. Publs Bot.* **43**: 1–86.
SILVA, P. C. (1952). A review of nomenclatural conservation in the algae from the point of view of the type method. *Univ. Calif. Publs Bot.* **25**: 241–324.
SJÖSTEDT, L. G. (1926). Floridean studies. *Acta Univ. lund.*, Ny Följd, Avd. 2 **22** (4): 1–95.
SOUTH, G. R., HOOPER, R. G. & IRVINE, L. M. (1972). The life history of *Turnerella pennyi* Schmitz. *Br. phycol. J.* **7**: 221–233.
STACKHOUSE, J. (1797). *Nereis Britannica*. Fasc. 2. Bathoniae & Londini.
—— (1809). Tentamen Marino-Cryptogamicum. *Mém. Soc. Nat. Moscou* **2**: 50–97.
—— (1816). *Nereis Britannica*, ed. 2. Oxonii.
TAYLOR, W. R. (1957). *Marine Algae of the North-Eastern Coast of North America*. Revised edition. Ann Arbor.
—— (1960). *Marine Algae of the Eastern Tropical and Subtropical Coasts of the Americas*. Ann Arbor.
THOMAS, M. (1938). Der Formenkreis von *Chondrus crispus* und seine ökologische Bedingtheit. *Hedwigia* **77**: 137–210.
TRAILL, G. W. (1890). The marine algae of the Orkney Islands. *Trans. Proc. bot. Soc. Edinb.*, **18**: 302–342.
TURNER, D. (1802). *A Synopsis of the British* Fuci. Yarmouth.
—— (1802a). Descriptions of four new species of *Fucus*. *Trans. Linn. Soc.* **6**: 125–136.
—— (1808). *Fuci*, **1**. London.
—— (1809). *Fuci*, **2**. London.
WEST, J. A. (1972). The life history of *Petrocelis franciscana*. *Br. phycol. J.* **7**: 299–308.
WITHERING, W. (1776). *A Botanical Arrangement of All the Vegetables Naturally Growing in Great Britain*, **2**. Birmingham & London.
—— (1796). *An Arrangement of British Plants*, ed. 3, **4**. Birmingham & London.

WOODWARD, T. J. (1797). Observations upon the generic character of *Ulva* with descriptions of some new species. *Trans. Linn. Soc.* **3**: 46–58.

WULFEN, F. X. (1803). *Cryptogama aquatica.* Lipsiae.

YAMAMOTO, H. (1973). The development of the male reproductive organ of *Gracilaria verrucosa* (Huds.) Papenfuss. *Bull. Jap. Soc. Phycol.* **21**: 130–132.

ZINOVA, A. D. (1970 [1971]). Novitates de algis marinis e sinu Czaunskensi (Mare Vostoczno-Sibirskoje dictum). *Nov. Sist. Nizsh. Rast.* **7**: 102–107.

ZOEGA, J. (1772). *Tilhang om de Islandske Urter.* *In:* Olafsson, E. & Povelsen, B. *Reise igjennem Island.* Sorøe.

INDEX TO GENERA

Synonyms in *italics*

CPSIA information can be obtained at www.ICGtesting.com
Printed in the USA
BVOW06s1121011115

424877BV00005B/11/P

9 781907 807084